ENGINEERING LIBRARY

D1310372

Wireless and Mobile All-IP Networks

Yi-Bing Lin and Ai-Chun Pang

WILEY
Wiley Publishing, Inc.

Wireless and Mobile All-IP Networks

Published by
Wiley Publishing, Inc.
10475 Crosspoint Boulevard
Indianapolis, IN 46256

Copyright © 2005 by Wiley Publishing, Inc., Indianapolis, Indiana

Published simultaneously in Canada

ISBN: 0-471-74922-2

Manufactured in the United States of America

10 9 8 7 6 5 4 3 2 1

No part of this publication may be reproduced, stored in a retrieval system or transmitted in any form or by any means, electronic, mechanical, photocopying, recording, scanning or otherwise, except as permitted under Sections 107 or 108 of the 1976 United States Copyright Act, without either the prior written permission of the Publisher, or authorization through payment of the appropriate per-copy fee to the Copyright Clearance Center, 222 Rosewood Drive, Danvers, MA 01923, (978) 750-8400, fax (978) 646-8600. Requests to the Publisher for permission should be addressed to the Legal Department, Wiley Publishing, Inc., 10475 Crosspoint Blvd., Indianapolis, IN 46256, (317) 572-3447, fax (317) 572-4355, e-mail: brandreview@wiley.com.

Limit of Liability/Disclaimer of Warranty: The publisher and the author make no representations or warranties with respect to the accuracy or completeness of the contents of this work and specifically disclaim all warranties, including without limitation warranties of fitness for a particular purpose. No warranty may be created or extended by sales or promotional materials. The advice and strategies contained herein may not be suitable for every situation. This work is sold with the understanding that the publisher is not engaged in rendering legal, accounting, or other professional services. If professional assistance is required, the services of a competent professional person should be sought. Neither the publisher nor the author shall be liable for damages arising herefrom. The fact that an organization or Website is referred to in this work as a citation and/or a potential source of further information does not mean that the author or the publisher endorses the information the organization or Website may provide or recommendations it may make. Further, readers should be aware that Internet Websites listed in this work may have changed or disappeared between when this work was written and when it is read.

For general information on our other products and services or to obtain technical support, please contact our Customer Care Department within the U.S. at (800) 762-2974, outside the U.S. at (317) 572-3993 or fax (317) 572-4002.

Wiley also publishes its books in a variety of electronic formats. Some content that appears in print may not be available in electronic books.

Trademarks: Wiley, the Wiley logo, and related trade dress are trademarks or registered trademarks of John Wiley & Sons, Inc. and/or its affiliates, in the United States and other countries, and may not be used without written permission. ExtremeTech and the ExtremeTech logo are trademarks of Ziff Davis Publishing Holdings, Inc. Used under license. All rights reserved. All other trademarks are the property of their respective owners. Wiley Publishing, Inc., is not associated with any product or vendor mentioned in this book.

About the Authors

Yi-Bing Lin is chair professor of the Department of Computer Science and Information Engineering (CSIE), National Chiao Tung University (NCTU). He also serves as vice president of the Office of Research and Development, NCTU (2005–2006). Dr. Lin is a senior technical editor of *IEEE Network*, an editor of *IEEE Transactions on Wireless Communications*, an associate editor of *IEEE Transactions on Vehicular Technology*, an editor of *IEEE Wireless Communications Magazine*, an editor of *ACM Wireless Networks*, program chair for the 8th Workshop on Distributed and Parallel Simulation, general chair for the 9th Workshop on Distributed and Parallel Simulation, program chair for the 2nd International Mobile Computing Conference, guest editor for the *ACM MONET* special issue on Personal Communications, guest editor for *IEEE Transactions on Computers* special issue on Mobile Computing, guest editor for IEEE Transactions on Computers special issue on Wireless Internet, and guest editor for *IEEE Communications Magazine* special issue on Active, Programmable, and Mobile Code Networking. Lin is co-author of the book *Wireless and Mobile Network Architecture* (with Imrich Chlamtac; published by John Wiley & Sons). Lin received 1997, 1999, and 2001 Distinguished Research Awards from the National Science Council, ROC; the 2005 Pan, Wen-Yuen Outstanding Research Award; the 2004 K.-T. Li Breakthrough Award; and the 1998 Outstanding Youth Electrical Engineer Award from CIEE, ROC. He also received the NCTU Outstanding Teaching Award in 2002. Lin is an Adjunct Research Fellow of Academia Sinica, and is an adjunct chair professor of Providence University. He is also an IEEE Fellow, an ACM Fellow, an AAAS Fellow, and an IEE Fellow.

Ai-Chun Pang received B.S., M.S., and Ph.D. degrees in Computer Science and Information Engineering from National Chiao Tung University. She joined the Department of Computer Science and Information Engineering, National Taiwan University, Taipei, Taiwan, as an Assistant Professor in 2002 and an Associate Professor in 2006. Since August 2004, Dr. Pang has also joined in the Graduate Institute of Networking and Multimedia, National Taiwan University, Taipei, Taiwan. Her research interests include design and analysis of personal communications services networks, mobile computing, voice over IP, and performance modeling.

Credits

Executive Editor
Carol Long

Development Editor
Luann Rouff

Production Editor
Pamela Hanley

Copy Editor
Luann Rouff

Editorial Manager
Mary Beth Wakefield

Vice President & Executive Group Publisher
Richard Swadley

Vice President and Publisher
Joseph B. Wikert

Composition
TechBooks

Proofreading
Publication Services

Contents

Acknowledgments

Many mobile service ideas in this book came from Herman C.-H. Rao. Many of Yi-Bing Lin students have contributed to the original drafts of the chapters: Di-Fa Chang, Yuan-Kai Chen, Ming-Feng Chen, Whai-En Chen, Vincent W.-S. Feng, Yieh-Ran Haung, Meng-Ta Hsu, Wei-Che Huang, Phone Lin, Ya-Chin Sung, Sheng-Ming Tsai, Meng Hsun Tsai, Lin-Yi Wu, and Quency Wu. The book chapters were reviewed by Shiang-Ming Huang, Jui-Hung Weng, Che-Hua Yeh, Chien-Chun Huang-Fu, Hsin-Yi Lee, Yung-Chun Lin, and Shih-Feng Hsu.

Acknowledgments from Yi-Bing Lin: My wife, Sherry Wang, has played a major role in my career path and has encouraged me to complete this book. Without Sherry, my parents, and my daughters, Denise and Doris, I would not have accomplished as much as I have.

My Ph.D. thesis advisor, Edward D. Lazowska, has coached me since I was graduate student. Edward always gives me great advice at critical times.

As a career advisor, Imrich Chlamtac has guided me on mobile telecommunications research. I enjoyed working with Imrich in writing the book [Lin 01b].

I would also like to thank Ian F. Akyildiz, Jean-Loup Baer, Chun-Yen Chang, Jin-Fu Chang, Che Fu Den, H. T. Kung, Leoard Kleinrock, William C. Y. Lee, D.-T. Lee, and Lin-Nan Lee for their support throughout my professional career.

Acknowledgments from Ai-Chun Pang: I am grateful to my parents and my husband, Jack Cherng, for their encouragement and firm support.

Also, I would like to thank my colleagues at the Embedded System and Wireless Networking Laboratory/National Taiwan University, especially Professor Tei-Wei Kuo and Professor Cheng-Shi Chih.

We both would like to thank our editor at Wiley, Carol Long - a special thank you.

The writing of this book was supported in part by NSC Excellence project NSC93-2752-E-0090005-PAE, NTP VoIP Project, ITRI-NCTU Joint Research Center (in particular, Paul Lin and Sheng-Lin Chou), Intel, MOE NICE program, and IIS/Academia Sinica.

Yi-Bing Lin
Ai-Chun Pang

Introduction

Mobile telephony has become an integral part of our lives. Today, a society devoid of mobile networks is more or less unthinkable, and people have high expectations of future mobile services. Existing mobile networks host an endless stream of voice and data information. In our previous book, *Wireless and Mobile Network Architecture* [Lin 01b], we elaborated on how voice and data are delivered through the second generation (2G) and the 2.5 generation (2.5G) mobile networks. In this book, we focus on the third generation (3G) and beyond 3G all-IP networks for advanced mobile applications.

In the future, all telephony services will eventually be delivered over IP due to the low costs and the efficiencies for carriers to maintain a single, unified IP-based telecommunications network. This book emphasizes the all-IP aspect of wireless and mobile core networks.

> **Pre-reading:** The reader is encouraged to review Chapters 9 and 18 in [Lin 01b] as pre-reading for this book. Equivalent materials can also be found at
>
> **http://liny.csie.nctu.edu.tw/supplementary**

This book consists of 17 chapters. At the end of each chapter are some review questions and modeling questions. The review questions help readers to refresh their memory about key points of the chapter, and the modeling questions are designed for readers who have queueing and probability modeling experience. The book is outlined as follows:

> **Chapter 1** presents two platforms, NCTU-SMS and iSMS, that integrate IP networks with the short message mechanism of mobile networks.

The NCTU-SMS platform was developed at National Chiao Tung University (NCTU). The iSMS platform was developed by AT&T and FarEasTone. iSMS supports a middleware for creating and hosting wireless data services based on SMS. The iSMS hardware architecture can be easily established with standard Mobile Stations (MSs) and personal computers. We have developed agent-based middleware with API, which results in a lightweight solution to allow quick deployment of added-value data services in iSMS.

Chapter 2 introduces mobility management evolution from General Packet Radio Service (GPRS) to Universal Mobile Telecommunications System (UMTS). In GPRS, some radio management functions are handled in the core network. These functions have been moved to the radio access network in UMTS. This architectural change results in a clean design that allows radio technology and core network technology to develop independently. The GPRS mobility management functionality has been significantly modified to accommodate UMTS. This chapter emphasizes the differences between the GPRS and the UMTS procedures.

Chapter 3 presents a modular, compatible, and flexible software framework to implement GPRS session management (SM) functions at SGSN. These functions can be implemented in the modules and are compatible with the current and upcoming GPRS standards. Based on the proposed software architecture, we describe the behavior of the SGSN in the Packet Data Protocol (PDP) context activation, deactivation, and modification procedures. We also show how to extend the design of the SM software architecture for UMTS.

Chapter 4 elaborates on the GGSN session management functions for IP connection, which includes the Access Point Name (APN) and IP address allocation, tunneling technologies, and QoS management. In GPRS/UMTS, the GGSN serves as a gateway node to support IP connection toward the external Packet Data Networks (PDNs). We describe how the IP address can be dynamically allocated by the GGSN, a DHCP server, or a RADIUS server, and how the GGSN can interwork with the external PDN through either leased lines or virtual private network (VPN). For VPN interworking, tunneling is required to connect the GGSN with the external PDNs. We also show the tunneling alternatives, including IP-in-IP tunneling, GRE tunneling, and L2TP tunneling. Finally, we elaborate on the QoS processing of a user data packet at the GGSN.

Chapter 5 discusses real-time Serving Radio Network Controller (SRNC) switching mechanisms. In 3GPP TR 25.936, SRNC duplication (SD) and core network bi-casting (CNB) were proposed to support real-time multimedia services in the UMTS all-IP network. Both approaches require packet duplication during SRNC relocation, which significantly consumes system resources. This chapter describes a fast SRNC relocation (FSR) approach that provides real-time SRNC switching without packet duplication. In FSR, a packet buffer mechanism is implemented to avoid packet loss at the target RNC. Our performance study indicates that packet loss at the source RNC can be ignored in FSR. Furthermore, the expected number of buffered packets at the target RNC is small, which avoids long packet delay. FSR can be implemented in the GGSN, the SGSN, and the RNC without introducing new message types to the existing 3GPP specifications.

Chapter 6 describes mobility and session management mechanisms for UMTS and cdma2000, and compares the design guidelines for these mobile core network technologies. We first introduce network architectures and protocol stacks for UMTS and cdma2000. Then we elaborate on UMTS and cdma2000 mobility management and session management, focusing on the differences between the design guidelines of these two systems. The IP mobility mechanisms implemented in UMTS and cdma2000 are also demonstrated and compared. Based on the UMTS and cdma2000 mobility management mechanisms, we describe an architecture and inter-system roaming procedures for UMTS-cdma2000 IP-level interworking. In this approach, a mobile user can roam between the UMTS and cdma2000 systems without losing the ongoing communication session.

Chapter 7 introduces the UMTS charging protocol. In UMTS, the GTP' protocol is used to deliver the call charging records from GPRS support nodes to charging gateways. This chapter describes a UMTS charging protocol implemented in NCTU. To ensure that the mobile operator receives the charging information, availability of the charging system is essential. We also elaborate on connection failure detection for GTP' availability.

Chapter 8 presents IP-based SS7 signaling transport. We describe the MTP-based approach and the SCTP-based approach. While the MTP-based approach is used in GSM, GPRS, and UMTS R99, the SCTP-based approach is used in the UMTS all-IP network. We compare these two approaches from three perspectives: message format, connection setup, and data transmission.

Chapter 9 elaborates on security and availability issues for UMTS. We show how UMTS mutual authentication can be effectively conducted by sending multiple authentication vectors from the authentication center to the SGSN, and then performing the authentication procedure locally between the SGSN and the MS. We then describe how potential fraudulent usage can be detected by UMTS registration and call setup procedures. We observe that the movement pattern of a legal user significantly affects the detection time of potential fraudulent usage. We also describe how mobile users may be eavesdropped when they are not having a phone conversation. Then we show how eavesdropping can be avoided. Finally, we address the checkpoint approaches to support backup HLR records in a nonvolatile database. We show how checkpoint can be utilized for HLR failure restoration.

Chapter 10 describes how signaling protocols such as H.323 and Media Gateway Control Protocol (MGCP) are utilized in non-all-IP mobile networks to offer VoIP services. We first introduce GSM-IP, a VoIP service for GSM. A new MGCP package named GSM-IP is introduced, which supports media gateways connected to standard GSM BTSs. Based on the signaling protocol translation mechanism in the MGCP signaling gateway, we describe how to interwork the MGCP elements with HLR, VLR, and MSC in the GSM network. Then we present the message flows for registration, call origination, call delivery, call release, and inter-system handoff procedures for the GSM-IP service. We show the feasibility of integrating GSM and the MGCP-based VoIP system without introducing any new MGCP protocol primitives. Then we describe vGPRS, a VoIP service for GPRS, which allows standard GSM and GPRS MSs to receive VoIP service. In this approach, a new network element called the VoIP Mobile Switching Center is introduced to replace standard GSM MSC. The vGPRS approach is implemented by using standard H.323, GPRS, and GSM protocols. Thus, existing GPRS and H.323 network elements are not modified. We describe the message flows for vGPRS registration, call origination, call release, call termination, and intersystem handoff procedures. We also show that for international roaming, vGPRS can effectively eliminate tromboning (two international trunks in call setup) for an incoming call to a GSM roamer.

Chapter 11 discusses short message multicast mechanisms in the Circuit Switched (CS) domain and a multicast mechanism in the

Packet Switched (PS)-domain MMS. The PS multicast approach is based on the cell broadcast service architecture. We propose a new interface between the cell broadcast center and the SGSN to track the current locations of the multicast members. Then we describe location tracking procedures (including attach, detach, and location update) for multicast members, and the multicast message delivery procedure. The implementation and execution of the multicast table are so efficient that the cost for updating this table can be ignored compared to the standard mobility management procedures.

Chapter 12 introduces the Session Initiation Protocol (SIP). We show how SIP supports user mobility, call setup, and call release. Based on the SIP protocol, we illustrate how the push mechanism and the prepaid mechanism can be implemented in GPRS/UMTS. In GPRS, the push feature is not supported. That is, an MS must activate the PDP context for a specific service before the external data network can push this service to the MS. An example is VoIP call termination (incoming call) to an MS. However, maintaining a PDP context without actually using it will significantly consume network resources. Using the iSMS Internet platform, we describe a SIP-based push mechanism for SIP call termination of GPRS supporting private IP addresses. A major advantage of this approach is that no GPRS/GSM nodes need to be modified. For another SIP application example, we describe a SIP-based prepaid mechanism to handle the prepaid calls in a VoIP system. Integration of this prepaid mechanism with the existing VoIP platform can be easily achieved by reconfiguring the call server.

Chapter 13 introduces mobile number portability. We describe and analyze number portability routing mechanisms and their implementation costs. We first describe the Signaling Relay Function based solution for call-related and non-call-related routing. Then we describe the Intelligent Network based solution for call-related routing. Cost recovery issues for number portability are discussed in this chapter from a technical perspective. Rules for cost recovery also depend on business and regulatory factors, which vary from country to country.

Chapter 14 elaborates on WLAN-based GPRS Support Node (WGSN), a solution for integrating cellular and WLAN networks. WGSN was developed by the NCTU and the Industrial Technology Research Institute. We address how the cellular/mobile mechanisms are re-used for WLAN user authentication and network access without introducing new procedures and without modifying the existing cellular network components. We also describe the WGSN features and show how they are designed and implemented. Then we discuss how IEEE 802.1X authentication can be integrated in WGSN.

Chapter 15 shows a UMTS all-IP approach for the core network architecture, and briefly describes an Open Service Access (OSA) system developed in NCTU. We introduce the core network nodes, and elaborate on application-level registration, circuit-switched call origination, packet-switched call origination, and packet-switched call termination. We show how UMTS can guarantee end-to-end QoS and radio spectrum efficiency. We note that to provide the expected QoS across domains, operators must agree on the deployment of common IP protocols. The common IP protocols impact roaming (i.e., the interfaces between core networks) and the communications between terminals and networks to support end-to-end QoS and achieve maximum interoperability.

Chapter 16 elaborates on the performance of the IP Multimedia Core Network Subsystem (IMS) incoming call setup, and describes the cache schemes with fault tolerance to speed up the incoming call setup process. Our study indicates that by utilizing the I-CSCF cache, the average incoming call setup time can be effectively reduced, and a smaller I-CSCF timeout threshold can be set to support early detection of incomplete call setups. To enhance fault tolerance, the I-CSCF cache is periodically checkpointed into a backup storage. When an I-CSCF failure occurs, the I-CSCF cache content can be restored from this storage. This chapter also investigates an efficient IMS registration procedure that does not explicitly perform tedious authentication steps. As specified by the 3GPP, after a UMTS mobile user has obtained GPRS network access through GPRS authentication, the "same" authentication procedure must be executed again at the IMS level (during IMS registration) before it can receive the IP multimedia services. We describe a one-pass authentication procedure, which only needs to perform GPRS authentication. At the IMS registration, the one-pass procedure performs several simple operations to verify whether a user is legal. We prove that the one-pass procedure correctly authenticates the IMS users. Our study indicates that this new approach can save up to 50% of the network traffic generated by the IMS registration. This approach also saves 50% of the storage for buffering the authentication vectors.

Chapter 17 describes iMobile, a proxy-based platform for developing mobile services for various mobile devices and wireless access technologies. This platform was developed by AT&T and FarEasTone.

iMobile introduces three abstractions on top of an agent-based proxy: dev-let, which interacts with various access devices and protocols; info-let, which accesses multiple information spaces; and app-let, which implements application and service logic. The let engine arbitrates the communications among dev-lets, app-lets, and info-lets. The let engine also maintains user and device profiles for personalized services. The iMobile vision allows a mobile user to access vast amounts of information available on various wired and wireless networks regardless of where the user is and what device or communication protocol is available. This modular architecture allows developers to write device drivers, information access methods, and application logic independently from each other. We also elaborate on a simplified iMobile platform called iMobile ME. The iMobile ME architecture provides a uniform architecture on mobile devices, and allows these devices to both communicate with and access resources from each other. As mobile devices become more powerful, iMobile ME provides an ideal infrastructure to facilitate P2P mobile computing.

By integrating wireless and mobile technologies with the IP core networks, many IP-based multimedia services can be offered to the mobile subscribers. In doing so, functionality of mobile terminals must be greatly enhanced, and the man-machine interface (MMI) may become very complicated. Bjarne Stroustrup, the inventor of the C++ language, said,

> *"I have always wished that my computer would be as easy to use as my telephone. My wish has come true. I no longer know how to use my telephone."*

When introducing new data services, the MMI must be designed such that the simplicity of "telephone characteristics" is maintained as much as possible. This issue is partially addressed in Chapter 17, and is still open for futher study.

Short Message Service and IP Network Integration

Short Message Service (*SMS*) is a mature wireless communication service [Lin 01b, 3GP04l, Nok 97]. Most modern digital cellular phone systems offer SMS, which is considered a profitable added-value service. A natural extension is to integrate SMS with electronic mail services, which provides linkage between mobile networks and IP networks. Furthermore, several Internet applications over SMS can be implemented on similar platforms. This chapter uses the *Global System for Mobile Communications* (*GSMC*) SMS as an example to illustrate SMS-Internet integration.

GSM SMS provides a connectionless transfer of messages with low-capacity and low-time performance. Each message can contain up to 140 octets or 160 characters of the GSM default alphabet [ETS97a]. The short messages are transported on the GSM *Stand-alone Dedicated Control Channel* (*SDDC*). Since a voice session utilizes GSM radio traffic channels, short messages can be received and sent while the mobile users are in conversation.

The GSM SMS network architecture is illustrated in Figure 1.1. In this architecture, when a Mobile Station (MS) sends a short message, this message is delivered to the GSM radio system, that is, a *Base Transceiver Station* (*BTS*) and then a *Base Station Controller* (*BSC*). The radio system then forwards the message to a *Mobile Switching Center* (*MSC*) called SMS

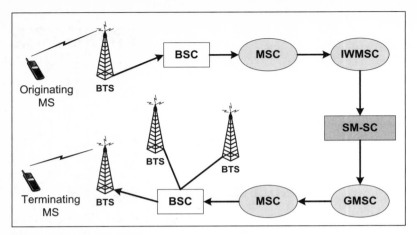

Figure 1.1   GSM Short Message Service Network Architecture

Inter-Working MSC (*IWMSC*). Details for BTS, BSC, and MSC are given in Chapter 2.

The IWMSC passes this message to a *Short Message Service Center* (*SM-SC*). Upon receipt of a short message, the SM-SC may send an acknowledgment back to the originating MS (sender) if the acknowledgment request is specified in the short message. The SM-SC then forwards the message to the destination GSM network through a specific GSM MSC called the SMS *Gateway MSC* (*SMS GMSC*). Following the GSM roaming protocol (see Section 9.2.1), the SMS GMSC locates the serving MSC of the message receiver and forwards the message to that MSC. This MSC broadcasts the message to the BTSs, and the BTSs page the destination MS (receiver). Every short message contains a header in addition to the body. The header includes the originating MS address, the terminating MS address, the serving SM-SC address, a time stamp, and the length of the message body. *Mobile Station ISDN Numbers* (*MSISDNs;* the GSM telephone numbers), are used for addressing.

As a wireless data service, SMS has distinct features, such as handset alert capability and support for ME-specific and SIM-specific data. The MS is considered an "always-on" device that facilitates instant information exchange. No dial-up modem connection is required to access SMS. Furthermore, SMS provides message storage when the recipient is not available. It also allows simultaneous transmission with GSM voice, data, and fax services. On the other hand, SMS has drawbacks, such as narrow bandwidth and long latency of end-to-end transmission. To design an SMS-IP system, the aforementioned SMS strengths and limitations cannot be ignored. In particular, the IP server itself may also be mobile (for example, the iSMS gateway described in this chapter can move when it provides services), and usage of the limited

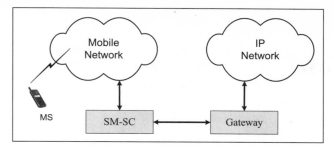

Figure 1.2   SMS-IP Integration with SM-SC

wireless bandwidth must be carefully addressed. To facilitate the development of various SMS applications over the Internet, we need a generic gateway to interwork the GSM network and IP network, and many specific data format converters between IP application contents and the SMS byte codes.

In most commercial implementations, SMS and IP networks are integrated through SM-SC, as illustrated in Figure 1.2. In this figure, a gateway interworks the SM-SC to the IP network, where a specific protocol is essential for communication between the SM-SC and the gateway. Since SM-SC implementation is vendor-specific, the SM-SC-based SMS-IP integration solution depends heavily on the SM-SC vendors. Furthermore, the SMS-IP gateway is maintained and controlled by GSM operators. It is difficult for a third party to deploy new services via SMS without having full cooperation from GSM operators. Also, from the GSM operator's viewpoint, maintaining a reliable, secure, scalable interconnection platform between individual service providers and the SMS-IP gateway will not be an easy task. The SMS-IP approach typically utilizes a centralized gateway, where performance, scalability, and reliability issues must be carefully considered. To address these issues and to further support an environment for quick prototyping and hosting wireless data service, an endpoint SMS-IP integration solution called iSMS was proposed in [Rao01a]. This solution is transparent to an existing SM-SC and GSM network. This chapter first describes a SM-SC-based SMS-IP system called NCTU-SMS. Then we introduce iSMS, a non-SM-SC based SMS-IP system, and elaborate on the designs and implementations of several iSMS applications.

1.1   SMS-IP Integration with SM-SC

In a joint project between the FarEasTone Telecommunications Corporation and National Chiao Tung University (NCTU), we developed a web-based short message system called NCTU-SMS [Hun 04]. NCTU-SMS integrates

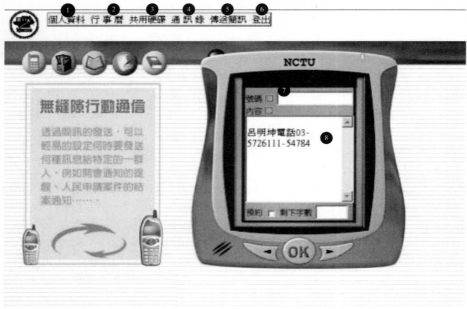

(1) Personal data (2) Calendar (3) Shared disk (4) Phonebook (5) Message delivering options (6) Logout
(7) Destination phone number (8) Message text

Figure 1.3 The Web-based Graphical User Interface for NCTU-SMS

SMS with IP through SM-SC. With a user-friendly *Graphical User Interface (GUI;* Figure 1.3), NCTU-SMS allows users to input short messages through a web site. Several commercial web-based SMS services are now available. Among them, a version of NCTU-SMS was the first system utilized by the Taiwanese government (specifically, E-Land County) for civilian services. This system can, for example, notify a citizen that his/her driver's license should be renewed. At NCTU, students who take the personal communication course receive their test scores and final grades through short messages generated by NCTU-SMS.

1.1.1 NCTU Short Message System

The NCTU-SMS architecture is illustrated in Figure 1.4, and consists of an *Application Client (AC)*, an *Application Server (AS)*, and the SM-SC. A delivery state tree (see Figure 1.5) is used to indicate the status of every short message delivery. The delivery state of a short message is recorded in the Application Client. To send a short message, the following steps are executed:

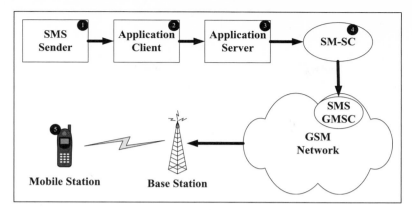

Figure 1.4 The NCTU-SMS Architecture

Step 1. A user (Figure 1.4 (1)) issues a short message through the web-based GUI. Specifically, the user inputs the destination phone number (MSISDN) and types the message text. When the user presses the OK button, the Application Client (Figure 1.4 (2)) generates a message delivery record that records

- the destination MSISDN
- the time T_i when the message is issued,
- the current delivery state,
- the time T_s associated with the delivery state, and
- other parameters.

At this step, the delivery state is **Init** and $T_s = T_i$.

Steps 2 and 3. The Application Client forwards the short message to the Application Server (Figure 1.4 (3)), and the delivery state is changed to **Send-To-AS**. If the communications link between the Application Client and the Application Server is disconnected, the Application Client sets the delivery state to **AS-Failure**. Otherwise, the Application Server receives the message delivery request from the Application Client and performs format checking (for example, the destination phone number format, the message text format, and so on). If the format check fails, an error message is sent back to the Application Client. In this case, the Application Client updates T_s to the current time, the delivery state is set to **AS-Failure**, and the procedure exits. If the format check succeeds, then an acknowledgment message is sent back to the Application Client, and the Application Server forwards the short message to the SM-SC

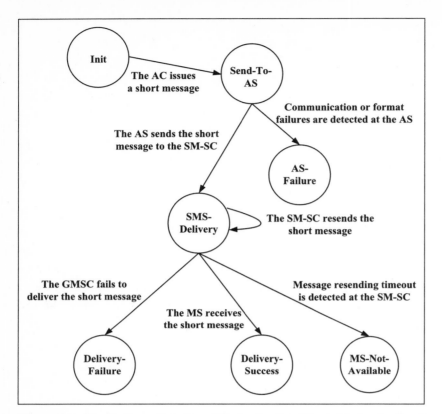

Figure 1.5   The Delivery State Tree (AC: Application Client; AS: Application Server)

(Figure 1.4 (4)). When the Application Client receives the acknowledgment, it updates T_s and sets the delivery state to **SMS-Delivery**.

Step 4. When the SM-SC receives the short message, it delivers the message following the standard SMS procedure illustrated in Figure 1.1. Note that the message may not be actually delivered if errors occur, for example, the mobility database *Home Location Register* (*HLR;* see Chapter 2 and Section 9.2.1) cannot identify the destination MS. In this case, an error message is sent back to the Application Client. The Application Client updates T_s to the current time, the delivery state is set to **Delivery-Failure**, and the procedure exits. In the normal case, the message will be sent to the destination MS, and the next step is executed.

Step 5. When the destination MS receives the short message, an acknowledgment is sent back to the Application Client. The Application Client updates T_s to the current time, the delivery state is set to **Delivery-Success**, and the procedure exits. Sometimes the short message may not be received by the destination MS (for example, the destination MS is power off). In this case, the short message will be periodically resent until either the destination MS receives the message or the SM-SC gives up. In the latter case, an error message is sent back to the Application Client. The Application Client updates T_s, sets the delivery state to **MS-Not-Available**, and exits the procedure.

From the above description, if the delivery is successful, the final delivery state is **Delivery-Success**. If the delivery fails, the final delivery state is either **AS-Failure**, **Delivery-Failure**, or **MS-Not-Available**.

1.1.2 Statistics for SMS Delivery

Several interesting statistics can be derived from the delivery records in the Application Client—for example, the hourly distribution T_i^* of T_i (the time when a short message is issued), where

$$T_i^* = T_i \bmod 24.$$

We are also interested in the distribution of the issued times $T_{i,MNA}^*$ for those failed SMS deliveries with the final state **MS-Not-Available** (MNA). Let $p_i^*(t)$ be the probability that an SMS is issued during the tth hour (where $t = 0, 1, 2, \ldots, 23$) see Fig.1.6. Similarly, let $p_{i,MNA}^*(t)$ be the probability that an **MS-Not-Available** SMS is issued during the tth hour. Figure 1.6 plots the p_i^* and $p_{i,MNA}^*$ curves. The p_i^* curve indicates that most short messages are issued during normal work hours (8:00 A.M.–5:00 P.M.), with a major peak at 9:oo A.M.. It implies that many SMS users tend to send short messages immediately after they arrive at the office. It is also apparent that the number of issued short messages drops during the lunch break. The $p_{i,MNA}^*$ curve follows the same trend as the p_i^* curve. It is interesting to observe that $p_{i,MNA}^* > p_i^*$ during 8:00 P.M.–1:00 A.M.. This phenomenon can be explained by the following human behavior: We first note that in NCTU-SMS, the short messages are dropped after a resent period of 8 hours. Therefore, the **MS-Not-Available** short messages are dropped around the eighth hour after they are issued. Our experience indicates that many people turn off their MSs during 8:00 P.M.–9:00 A.M. Therefore, many short messages issued during 8:00 P.M.–1:00 A.M. will be dropped. At 9:00 A.M., a large number of short messages are issued. We observe a nontrivial phenomenon that $p_{i,MNA}^* > p_i^*$.

From the measured data we collected, Table 1.1 shows the numbers of deliveries, with various final states. For the deliveries with the **AS-Failure** state, they failed due to sender mistakes (for example, inputting the incorrect

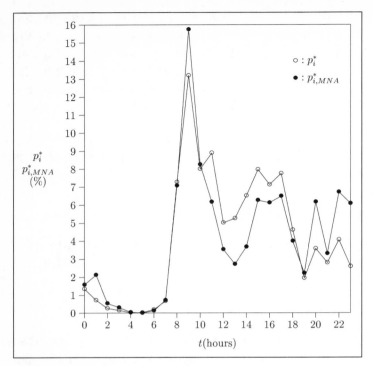

Figure 1.6   The T_i^* and $T_{i,MNA}^*$ Distributions

phone number formats). For the deliveries with the **MS-Not-Available** state, they failed due to unavailability of the receivers. Therefore, to consider the failures resulting from the SMS delivery system, we should only consider the **Delivery-Failure** short messages, and this system failure probability is 3.57%.

The most important statistics are the round-trip transmission delays T_D of short messages, which are defined as

$$T_D = T_s - T_i$$

Table 1.1   Numbers of deliveries, with various final states

STATE	AS-FAILURE	DELIVERY-FAILURE
Number	423	1649
Percentage	0.92 %	3.57 %
STATE	MS-NOT-AVAILABLE	DELIVERY-SUCCESS
Number	4278	39,780
Percentage	9.27 %	86.23 %

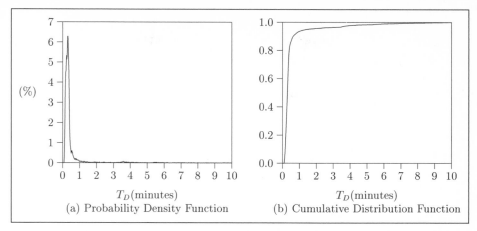

Figure 1.7   The T_D Distribution (**Delivery-Success**; the normalized factor is 1.1)

Figure 1.7 illustrates the T_D probability distribution for successful deliveries, which is normalized in the range [1, 10] (the normalization factor is 1.1). Figures 1.8 and 1.9 illustrate the T_D probability distribution for failed deliveries. That is, consider the **Delivery-Failure** and **MS-Not-Available** short messages as the sample space. Figure 1.8 plots the **Delivery-Failure** part of the T_D distribution, and Figure 1.9 plots the **MS-Not-Available** part. Figure 1.8 indicates that most messages in the **Delivery-Failure** state will be detected within 10 seconds. Figure 1.9 indicates that most messages in the **MS-Not-Available** state are reported within 15 minutes after the 8-hour timeout.

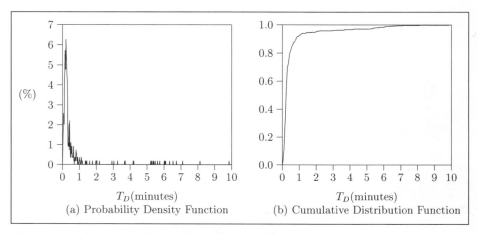

Figure 1.8   The T_D Distribution for **Delivery-Failure**

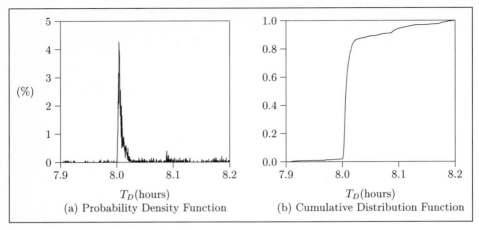

(a) Probability Density Function   (b) Cumulative Distribution Function

Figure 1.9   The T_D Distribution for **MS-Not-Available**

1.2   iSMS System Architecture

Figure 1.10 illustrates the architecture for iSMS, a non-SM-SC based SMS-IP integration solution. In this architecture, an *iSMS gateway* is introduced. No components in the GSM and IP networks are modified. The MS is a commercial handset product that does not require installing any new software.

A major difference between the iSMS architecture and the SM-SC based architecture is that the iSMS gateway connects to an MS instead of the SM-SC. This MS serves as the GSM-compliant modem that provides iSMS wireless access to the GSM network. We refer to this MS as the *MS modem*. In our implementation, the iSMS gateway is a PC running on Windows or UNIX operating systems. The PC-based gateway can be a desktop with high

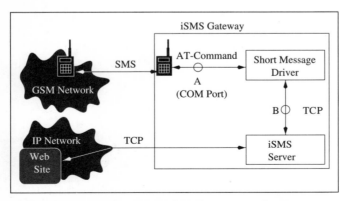

Figure 1.10   iSMS System Architecture (An End point SMS-IP Integration Solution)

reliability and availability. The PC can be replaced by a notebook. In this case, the gateway becomes a mobile server, which may move around just like a GSM MS. The MS modem and the iSMS gateway can be connected by the serial port, the Infrared port (for example, IrDA), the USB interface, or the PCMCIA interface. In iSMS, the data sent from the IP network to the GSM network is automatically packaged into short messages. The short messages can be multicast up to 65,535 receivers. Similarly, an MS can broadcast short messages to several servers connected to the iSMS gateway. The iSMS system is identified by the IP network through the IP address assigned to the gateway, and is addressed by the GSM network through the MSISDN of the MS modem attached to the gateway.

In the iSMS model, a user of an iSMS application is also a GSM customer. There are several alternatives to access iSMS services. For example, a user may make a request by sending a short message. This short message is terminated at the MS modem connected to the iSMS gateway. Upon receipt of the short message, the iSMS gateway executes the corresponding routines and returns the results to the user. The iSMS gateway may also be triggered by pre-defined, customized events. In a home security application, for example, if someone rings the doorbell when the user is not at home, the iSMS gateway may send an alerting short message to the GSM MS of the customer. The iSMS gateway consists of two parts (see Figure 1.10):

- **iSMS servers:** Responsible for service provisioning
- **Short message driver:** Responsible for communication between the GSM network and iSMS servers in the IP networks

The communication protocol between the MS modem and the short message driver (the reference point A in Figure 1.10) is implemented by using the SMS *AT Command Set* [ETS 98c, ETS98b]. The communication functions between the iSMS servers and the short message driver (the reference point B) is implemented through the iSMS communication *Application Programming Interface (API)* based on the TCP socket. These communication mechanisms are elaborated in Section 1.3.

During iSMS system initialization, the short message driver opens the serial port driver (reference point A) for sending/receiving short messages to/from the GSM network (via the MS modem). The short message driver also opens and listens on a pre-defined TCP port (reference point B) for server connection requests. For each connection request, the server will register the MSISDNs of its customers to the short message driver. Messages from registered senders will then be forwarded to the server.

The short message driver performs conversion between the TCP socket API (interface to the iSMS server) and the SMS AT Command Set (interface to the MS modem). The short message driver receives incoming short messages from the serial port and passes these messages to the iSMS servers according to a *registration table*. Depending on the registration status, the short message driver may forward a message to several iSMS servers or drop the message if no server has registered the sender of the message. For outgoing short messages, the driver receives messages from servers, transforms them into a short message format, and then sends them out to the GSM network via the serial port.

An iSMS server may run on the same host as the short message driver or on a remote site. For security reasons, the driver authenticates the servers for communication sessions. A command sent from the MS is pure text in the short message. The short message driver dispatches the message to the appropriate iSMS server based on the *caller Id* (i.e., the MSISDN of the requesting MS) and the mapping in the registration table. The server parses the command in the message body and then invokes corresponding internal functions or external agents to execute the messages. Functions are executed on the same address space as the server, while agents are running on different processes. The caller Id is used to identify the user in the current iSMS version. iSMS security is implemented at the application level. For example, a password may be required as a parameter of every command sent from the MS to the iSMS gateway.

To develop a new service, one implements iSMS servers that communicate with the short message driver by using the functions defined in the iSMS communication API. In the current iSMS version, we have implemented two kinds of servers:

- Email Forwarding Daemon
- Agent Dispatching Server

The iSMS platform allows developers to implement new server types. The relationship among various server types and the short message driver is illustrated in Figure 1.11. In this figure, the Email Forwarding Daemon ((1) in Figure 1.11) relays messages between the MSs and email systems on the IP domain. It supports an interface to Microsoft Exchange Server as well as to standard *Simple Mail Transfer Protocol* (*SMTP*) and Post Office Protocol Version 3 (POP3) [Mye 96].

The daemon converts a short message sent from an MS to email and forwards it to a SMTP server for delivery. The daemon may periodically query a mail server (for example, via a POP3 interface), pick up important

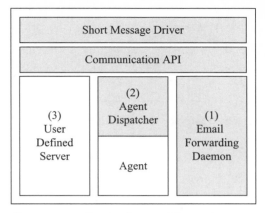

Figure 1.11   iSMS Driver and Server Structure

emails (according to user profiles), and send out SMS notifications to users' mobile stations. The daemon software can be easily generalized to support other mail systems, such as AOL Instant Message.

A server of an agent-dispatching type ((2) in Figure 1.11) consists of an agent dispatcher and several agents, where each of the agents is an iSMS server. The dispatcher invokes the agent corresponding to the SMS message header and passes the message body as the parameter to the agent. Each agent implements one function. When the agent finishes processing of a message, the agent dispatcher collects the results and sends them back to the short message driver.

Depending on the implemented services, each server of an agent-dispatching type may implement its own message parsing rules and maintain a command table with function/agent pairs. The current iSMS version has implemented a general-purpose agent dispatcher platform. In this platform, details of communication between the short message driver and the server are hidden from service developers. The developer only needs to specify the agent dispatching rules and implement the agents that carry out the services.

The iSMS platform allows service developers to implement new server types ((3) in Figure 1.11). In this case, the developer needs to implement the interaction between the short message driver and the servers. We provide a communication API that allows the developer to quickly deploy new servers. This communication API is elaborated in Section 1.3.2.

The system also supports group broadcasting and smart message delivery, including the ring tone, music, and icons. The smart messaging protocol defines the format of an ASCII stream that can be passed via different transport protocols. Smart messaging is considered in iSMS because it has

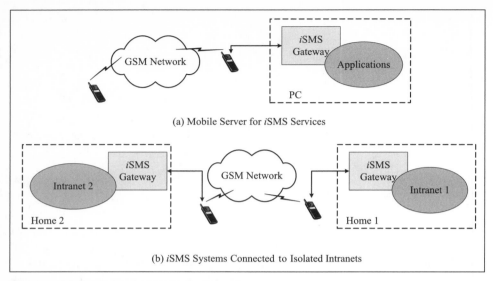

(a) Mobile Server for *i*SMS Services

(b) *i*SMS Systems Connected to Isolated Intranets

Figure 1.12   Variations of iSMS Configuration

been adopted by major GSM handset suppliers for SMS as well as Infrared among PDA and even for Bluetooth [Blu 04].

If all iSMS applications and the iSMS gateway are implemented in a portable notebook that is not connected to any IP network, then the iSMS system becomes a mobile server, as illustrated in Figure 1.12 (a). In this configuration, mobility management of the iSMS server is automatically maintained by the GSM Mobile Application Part (see Chapter 8). In other words, the existing GSM mechanism will transparently track the moving iSMS server, and iSMS does not need to implement any location tracking mechanism. A mobile iSMS server can be used, for example, in a mobile library application in which the library truck moves around a city. The iSMS server connects to the library database in the truck. An iSMS customer can use the MS to check the status of a book from the library database and the location of the book mobile.

Figure 1.12 (b) shows a different iSMS configuration in which the iSMS systems are connected to different isolated Intranets (e.g., homes), which can query each other through a mobile network.

1.3   **iSMS Communication Protocols**

Two communication protocols have been developed in iSMS. A communication protocol between the MS modem and the short message driver is implemented using the SMS AT Command Set [ETS 98c]. iSMS also implements a Communication API that provides functions to support interaction between the short message driver and the iSMS servers through TCP connections.

With the above communication protocols, several types of MS (customer) and Agent (service) interaction are implemented in iSMS:

Query-Response Services: A query (in SMS format) from an MS is sent to the iSMS server, which in turn invokes corresponding agents and sends back results to the original MS. Examples of query-response services include querying stock (see Section 1.4.1), train schedules (see Section 1.4.3), and UPS package status.

Relaying Services: The iSMS server forwards messages and information (such as emails and AOL Instant Messages) to the corresponding mobile users whenever messages are available (see Section 1.4.4). The server maintains the user profiles and actively collects information from different Internet servers.

Notification Services: The iSMS server delivers a notification to the user's MS. Notifications are triggered by events specified by the users. For example, a user may specify information (headline news, stock quotes, and weather information) that he/she wants to receive every day to a scheduling server in iSMS (such as the cron service in UNIX). As another example, an iSMS agent (for example, a stock/auction monitor agent) running on behalf of a user monitors information on the Internet, and delivers alert messages to the user.

Group Communication Service: The iSMS server allows specifying groups, each of which contains more than one MS member. Messages to a group (via SMS or from other services) will be forwarded to all members in the group (see Section 1.4.4).

This section shows how iSMS communication protocols are implemented based on GSM. To accommodate iSMS for non-GSM systems, if the systems support AT commands, then we only need to replace the GSM modem with an appropriate wireless modem (for example, a cdmaOne card phone). If the mobile systems do not support AT commands, the short message driver must be modified. With the communication API structure, the iSMS server and all iSMS service applications need not be changed.

1.3.1   SMS AT Command Set

The short message driver connects to the MS modem by serial ports or a PCMCIA card. In the current iSMS version, the messages are delivered through serial ports. Two serial ports are used to test the driver in the

Table 1.2   AT commands used in iSMS (a partial list)

AT COMMAND	DESCRIPTION
+CNMI	New Message Indications to TE
+CSCA	Service Center Address
+CMGD	Delete Message
+CMGS	Send Message
+CMGL	List Message
+CSMP	Set Text Mode Parameters
+CMT	SMS Message Received
+CPMS	Preferred Message Storage
+CNMA	New Message Acknowledgement to ME/TA
+CMGR	Read Message
+CMGC	Send Command
+CMGW	Write Message to Memory
+CMSS	Send Message from Storage
+CSMP	Set Text Mode Parameters
+CMTI	SMS Message Received Indication
+CBMI	New CBM (Cell-Broadcast Message) Indication
+CDSI	New SMS-STATUS-REPORT Indication

ME: Mobile Equipment
TA: Terminal Adaptor

simulation mode. The NULL port always accepts outgoing short messages successfully. The LOOPBACK port always sends back outgoing messages as incoming short messages. A variable MOBILE_COM_PORT is used in the driver to indicate which port is connected.

The communication protocol between the MS modem and the short message driver is based on the SMS AT Command Set [ETS 98c]. To run this protocol for a specific MS model, one should specify the type of the MS modem. The MS modem setup is achieved by two variables: MOBILE_TYPE and MOBILE_INIT_STRING. Some of the AT commands used in the short message driver are listed in Table 1.2. Every command sent from the short message driver begins with "AT" (for example, "AT+CMGS"). The MS modem's responses are without "AT" (for example, "+CMGS").

When the short message driver receives a message from an iSMS server, the driver divides the message into several segments, with length not exceeding 140 octets. For each receiver, the driver generates a set of SMS packets from the message segments. For example, if the message is divided into four segments and there are three receivers, then the drivers will generate 12 SMS packets. The driver pushes these SMS packets into a FIFO queue, and transmits them sequentially. For every SMS packet, the driver issues the SMS AT command that instructs the MS modem to submit a short message. There are two command modes for the MS: *text mode* and *Packet Data Unit (PDU) mode*. The parameters of AT commands for these two modes are different. In PDU mode, the parameter for sending short message is the entire short message packet. In Table 1.2, the AT commend is +CMGS, with the following format (in PDU mode):

```
AT+CMGS=<length><CR><pdu>
```

where <length> is the length of the actual data unit in octets. The <pdu> is the short message to be delivered. Details of other AT commands can be found in [ETS 98c].

1.3.2 iSMS Communication API

The iSMS API was implemented by using VC++. Through this API, servers and agents for specific applications can be easily developed. For every application, there is an iSMS server that communicates with the short message driver through a TCP connection (TCP port number 1122), and several agents may be created to interact with the iSMS server through command execution. The relationships among the short message driver, iSMS servers, and agents are illustrated in Figures 1.10 and 1.11.

In the iSMS API, a class CsmsdServer implements the following communication functions between a server and the short message driver:

Connect() sets up a communication link from a server to the short message driver. This function takes two arguments: the IP address of the short message driver and the TCP port number of the driver. This function returns the status of connection establishment.

Disconnect() terminates the TCP link between the server and the short message driver.

SetTimeout() sets a timeout period. When a server issues an operation to the short message driver, a timeout period is set. If the socket to the driver is not ready before the specified timer expires, then the operation fails.

Register() specifies the customers of a server with their GSM MSISDNs. The function takes two arguments: an array of phone numbers and the size of the array.

Status() returns the communication status between the server and the driver, as described here:

- SMCMD_READABLE indicates that the server is ready for retrieving a short message from the driver.
- SMCMD_WRITABLE indicates that the server is ready for sending a message to the driver.
- SMCMD_ACK indicates that the message from the server to the driver is successful.
- SMCMD_CLOSED indicates that the connection is closed.

Send() sends data to one or more customers (that is, GSM MSs).

Recv() is invoked by a server for receiving data from a GSM MS.

RecvACK() returns acknowledgment from the driver about the status of message transmission.

Besides the Send function, the CsmsdServer provides two additional functions for facilitating text messages and unstructured binary data delivery:

SendText() is used to send a message with a null-terminated string of ISO-8859-1 characters or traditional Chinese (BIG 5) characters.

SendData() is used to send the unstructured binary data using GSM 8-bit coding.

1.3.3 Implementation of an Echo Server

This section illustrates how to implement an echo server using the class CSmsdServer described in the previous subsection. This demo server simply echoes the incoming short message. The C program is given in Figure 1.13, and is described as follows:

Line 1. A list of MSs in phone_list is saved.

Lines 3–5. The variables (to be elaborated later) are declared.

Lines 6 and 7. The IP address of the short message driver is retrieved, and is stored in the variable lphost.

Lines 8 and 9. The TCP port number is set, and the connection timeout is set to 3 seconds.

Lines 10 and 11 (exception case). If the connection cannot be set up before the timer expires, then exit.

Lines 12–14 (normal case). The connection between the server and the short message driver is established, and the server registers the valid

```
#include <stdlib.h>
#include <smsio.h>
1 char *phone_list[] = { "+886936000001", "0931000001" };
2 int main() {
3 class CSmsdServer server; octet data[1024],dcs,option;
4 char sender[22]; char* da[1]; u_long host;
5 int port, ret, length; LPHOSTENT lphost;
6 lphost = gethostbyname("localhost");
7 if(lphost!=NULL) host=((LPIN_ADDR)lphost->h_addr)->s_addr;
8 port = 1122;
9 server.SetTimeout(3, 0);
10 if (server.Connect(host, port) != INET_SUCCESS)
11   { printf("Failure: connect to smsd\n"); _exit(1); }
12 server.SetTimeout(0, 50);
13 if (server.Register(phone_list, 2) != INET_SUCCESS)
14   { printf("Failure: register valid users\n"); exit(1); }
15 while (1) {
16    Sleep(1000);
17    ret = server.Status();
18    if (ret & SMCMD_CLOSED) break;
19    if (!(ret & SMCMD_READABLE)) continue;
20    ret = server.Recv(sender, data, &length, &dcs, &option);
21    if (ret != INET_SUCCESS) break;
22    printf("Sender: %s\nMessage: %s\n", sender, (char*)data);
23    while (!(server.Status() & SMCMD_WRITABLE)) ;
24    da[0] = sender;
25    server.SendText(da, 1, (const char*)data);
26    while (!(server.Status() & SMCMD_ACK)) ;
27    ret = server.RecvACK();
28    if (ret == SMCMD_NACK_SENDSM) break;
29    printf("\n Sending SMS....successful \n");
   }
30 return 0;
}
```

Figure 1.13 A Simple Echo Server Program

MSs specified in `phone_list`. The registration procedure should be complete in 50 ms, or it is considered a failure.

For successful registration, the server enters a loop (Lines 15–29).

Lines 16–20. In this loop, the server waits to receive an incoming short message for 1,000 seconds (Line 16) and then it reads the short message at Line 20.

Lines 21–25. The server waits until it is allowed to transmit (Line 23). Then it returns the message it just received to the MS (Line 25).

Lines 26–29. If the echo message has been forwarded to the MS, the short message driver acknowledges this operation (Lines 26 and 27).

The whole process repeats until one of the following conditions is met: the connection is closed (Line 18), the server fails to read the text (Line 21), or the server receives a negative acknowledgment (Line 28).

1.4 Examples of Services

This section describes several services implemented in iSMS. Special commands are defined for these services. An online menu can be built in the MS so that a customer can check the menu to figure out the meaning of a special command. Note that we have also built a platform to test the implemented services without actually involving the GSM network (the platform simulates the interaction of the user MS and the iSMS wireless modem).

1.4.1 Accessing the Web from GSM MSs

iSMS supports a service that allows users to surf the web using standard GSM MSs. The web accessing architecture is illustrated in Figure 1.14. In this architecture, a customer (that is, a GSM MS holder) sends out a web query in SMS format to a proxy server connected to the iSMS server. The proxy server and the iSMS server may be running on the same host or on different hosts. Based on the customer's preference, the proxy maps the input query to a proper *Hypertext Transfer Protocol (HTTP)* call and forwards the call to the corresponding web server. The proxy then filters and converts the data returned from the web server to SMS format, and returns the SMS

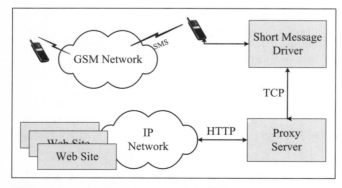

Figure 1.14 iSMS Web Accessing Architecture

message to the original caller. This implementation has several interesting characteristics:

- Standard GSM MSs are used as terminals for surfing the web. No special devices or software are needed.

- The communication medium is SMS. Both queries sent from the MSs (the GSM customers) and data received by the MSs are in SMS format. A single *Uniform Resource Locator (URL)* access is performed by two SMS messages: one from the MS to the proxy containing the URL query and another one from the proxy to the MS including the resulting data. The "instant transmission" property of SMS allows the customer to send an instant web query without long "machine" setup time experienced in the dial-up service to PC.

- Functionality of mapping SMS queries to HTTP calls and converting the data in HTML format to that in SMS format is implemented in the proxy. This application allows customers to tailor the call mapping and data conversion functions. The proxy picks the proper mapping and filter functions based on the customer's profile. The customers may also utilize the default mapping and conversion functions supported by the proxy.

iSMS implements functions for querying Internet information, such as stock quotes, currency exchange rates, and delivery status of FedEx packages:

Stock Quotes

The stock quote query command is of the form

```
QUO { symbol1} { symbol2} ...
```

The QUO command is explained with the following example:

```
QUO http://investor.msn.com/quotes/quotes.asp?Symbol=$1
/bin/quotefilter
```

The first field, QUO, is the keyword for query. The second field, {symbol1}, specifies HTTP call mapping. The third field, {symbol2}, defines the filter/conversion function. If {symbol1} and {symbol2} are not specified, the default call mapping and conversion functions are used.

With the above setup, a customer can enjoy web querying through his/her GSM MS. Note that the customer can turn off and then turn on the MS and the setup is still valid. Suppose that the customer has a GSM MSISDN +886936105401, and the MS modem in the iSMS gateway has the phone number +886931144401. The customer sends out an SMS message to +886-931-144-401 as follows:

```
QUO T
```

This command queries AT&T stock. When the iSMS server (running on the phone number +886-931-144-401) receives the message, it forwards the message to the proxy server, which in turn converts the SMS query into an HTTP call as follows:

```
http://investor.msn.com/quotes/quotes.asp?Symbol=T
```

The proxy sends the HTTP call to the destination web site. When receiving data from the web site, the proxy invokes a filter function (i.e., /bin/quotefilter) that re-formats the data received from the web site. The proxy returns the output of the filter function to the customer's MS 0936-105-401. The SMS data looks like

```
T Last 85 7/8
Change +1 9/16 (+1.85%)
Volume 7.708 M
```

where the first line indicates the last price of AT&T stock, the second line shows the price change since the last closing, and the third line gives the transaction amount.

Currency Exchange Rates

Consider a currency exchange rate query as another example. The currency rate query command issued by the customer is of the form

```
CUR from to
```

The system returns the currency exchange rate between the currency of the country `from` and the currency of the country `to`. For example, the short message

```
CUR USD TWD
```

queries the currency exchange rate between, the American dollar and the Taiwan dollar.

1.4.2 Handset Music Service

Existing MSs can store several music tones in memory. An MS may play various alert tones for incoming calls sent from different caller Ids (either mobile telephones or fixed telephones). Teenage MS holders are particularly interested in having fancy music tones such as the latest pop songs. Nokia, for example, has developed a messaging protocol for this purpose. Based on SMS, the protocol, called *Smart Messaging* [Nok 97], specifies a set of pre-defined message headers for sending music tones, business cards, etc. Each smart message contains a header for the message type. Nokia MSs recognize the header and perform different actions on the message body. Two issues regarding smart messaging deserve attention:

- Each smart message contains a Non-ASCII, binary header, and it is not trivial to input a smart message from regular MSs. It may not be convenient to send a music tone from one MS to another MS.

- Only Nokia MSs understand how to interpret short messages in smart messaging format.

To address these issues, iSMS developed a *Simple Tone Language (STL)* to represent the music notes that are translated into smart message format by the iSMS music agent. More precisely, a GSM customer inputs a short message as a music tone request, containing a music agent name, a receiver of the music tone, and a music tone body in STL. The message is sent to the iSMS server. Upon receipt of the request, the music agent encodes the music notes into a short message and sends it to the receiver. The handset can then store and play the received music tone. The regular expressions [Hop 74] for the STL grammar are described as follows. A STL music tone is expressed as

```
tone = [style] [tempo] [volume] [repeat] (note-
expression)+
```

The components of `tone` are elaborated on as follows:

- The `style` value ranges from `S00` to `S02`. In Nokia's smart messaging specification, `S00` represents *natural style* (rest between notes), `S01` represents *continuous style* (no rest between notes), and `S02` represents *staccato style* (shorter notes and a longer rest period). The default value of `style` is `S00`.

Table 1.3 Lengths of 1/4 note

SYMBOL	LENGTH OF 1/4 NOTE	SYMBOL	LENGTH OF 1/4 NOTE
T00	2.40 sec.	T08	0.95 sec.
T01	2.14 sec.	T09	0.85 sec.
T02	1.90 sec.	T10	0.76 sec.
T03	1.70 sec.	T11	0.67 sec.
T04	1.51 sec.	T12	0.60 sec.
T05	1.35 sec.	T13	0.54 sec.
T06	1.20 sec.	T14	0.48 sec.
T07	1.07 sec.	T15	0.43 sec.

- The `tempo` value ranges from `T00` to `T31`, which represent the lengths of 1/4 note, as given in Table 1.3. The default value is `T08`.

- The `volume` value ranges from `V00` to `V15`. The default value is `V07`. Although the smart messaging specification defines the programmable volume, some handsets, such as Nokia 6150, do not support this feature. In this case, the music volume is adjusted by the handset user, not the message received by the handset.

- The `repeat` value ranges from `R00` to `R15`, where `R00` means repeating infinite times, `R01` means repeating the tone for one time, and so on. The default value is `R01`.

- The `note-expression` component is of the format

    ```
    note-expression = note [scale][duration][duration-
    specifier]
    ```

    where a music note is expressed as

    ```
    note = 0 | 1 | 1# | 2 | 2# | 3 | 4 | 4# | 5 | 5# | 6
    | 6# | 7
    ```

    The `scale` component of the corresponding music note is expressed as

    ```
    scale = "L" ; note-1 is 440 Hz
          |"M" ; note-1 is 880 Hz
          |"N" ; note-1 is 1760 Hz
          |"O" ; note-1 is 3520 Hz
    ```

    where the default value is `M`. The duration of the corresponding music note is expressed as

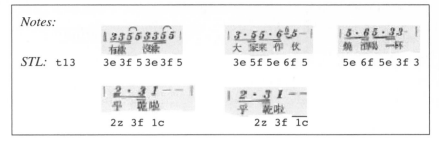

Figure 1.15   The STL Representation of a Taiwanese Song

```
duration = "A" ; full-note
          |"B" ; reserved
          |"C" ; 1/2-note
          |"D" ; 1/4-note
          |"E" ; 1/8-note
          |"F" ; 1/16-note
          |"G" ; 1/32-note
```

where the default value is D. The duration specifier of the corresponding music note is expressed as

```
duration-specifier = "X" ; dotted note
                    |"Y" ; double dotted note
                    |"Z" ; 2/3 length
```

Figure 1.15 illustrates the STL representation of a beer advertisement song in Taiwan.

1.4.3   Train Schedule System

Consider the iSMS train service that provides train schedule query information and ticket reservations. This train schedule server consists of an agent dispatcher and two agents:

- The query agent allows the customers to query particular train schedules.
- The reservation agent allows the customers to reserve train tickets.

At system initialization, the train schedule server registers the customers to the short message driver. This server should allow adding/deleting customers during normal operation. A customer may send a train schedule query (a short message) to the iSMS gateway. Based on the caller Id (the MSISDN of the customer who issues this query), the short message driver forwards the message to the train schedule server. In iSMS, the train schedule query command is of the form

```
tra {FromStation} {ToStation} [options]
```

where {FromStation} and {ToStation} represent the departure station and arrival station, respectively. The field [option] is of the format

```
[option]: {Time1}-{Time2} (S|A)  (F|E)
```

where {Time1}-{Time2} represents the time range (default = CurrentTime-2300), S and A represent the time range for the departure or the arrival time range (the default value is S), and F and E represent Numbered Express (default) and Express, respectively.

From the header of the short message, tra, the agent dispatcher invokes the train schedule query agent using the short message content

```
Taipei - Taichung, 2:00pm - 3:00pm
```

as the parameters. In this query, the customer would like to know the train numbers for the trains from Taipei to Taichung, which depart between 2:00 P.M. and 3:00 P.M. After execution of the request, the agent returns the train numbers to the agent dispatcher. This result is formatted as a short message and is sent back to the original customer. Upon receipt of the result, the customer may issue another short message to reserve tickets for specific train numbers. In this case, the ticket reservation agent will be invoked to handle the request.

1.4.4   Other iSMS Services

In addition to the services described in the previous subsections, other examples of services running on the current version of iSMS include the following:

Online Help Service: An iSMS user types the h command to query the services available in the iSMS system.

Personal Profile: iSMS maintains a personal profile repository for individual registered users. Profiles are organized on a per-user basis (according to the phone numbers). Each entry has the format "keyword=value", where value can be a phone number, an address, a personal note, and so on. A mobile user can add an entry to his/her personal profile by sending a short message to iSMS. For example, a mobile user may send the following short message to iSMS

```
PB Robin +19179075010
```

This message instructs iSMS to add a new entry

```
Robin= +19179075010
```

to his/her profile. Then the user can query the entry with a keyword by sending the following message to iSMS

```
PQ Robin
```

In response, iSMS returns the following message

```
Robin= +19179075010
```

A web interface has been provided, and users may update their profile using regular browsers as well. This repository serves not only as a mobile user's repository on the network, but also as the user profile for other services. For example, email services retrieve the user profiles to locate senders' email addresses by names.

Broadcast Message Service: An iSMS user can broadcast a message to several destinations with the following command:

```
bc {receivers} {message}
```

In this command, {receivers} is a list of MSISDNs of the receivers and {message} is the message to be delivered. In a similar group communication service, the MS can broadcast a message in a group of phone numbers (with command b), create a new group (with command bn), add/delete phone numbers to/from a specific group (with commands ba and bd, respectively), delete a specific group (with command bk), and list all phone numbers in a specific group (with command bl).

Smart Messaging Services: A GSM user uses the following commands to send smart messages. The command

```
rt {receivers} {song}
```

sends a ringtone {song} to the receivers in the list {receivers}. The command

```
lg {receivers} {image name}
```

sends an operator logo {image name} to the receivers. The command

```
ic {receivers} {image name}
```

is used to send a group logo.

Group Communications: iSMS implements a group communication mechanism. A group is identified by a unique name and contains a set of phone numbers. A message sent to a group will be forwarded to all members in the group. iSMS supports the following group communication commands:

- Creating a group with founding members
- Querying members in a named group
- Adding/deleting members to/from an existing group
- Sending messages to a named group

This group mechanism can also be used by other services. For example, a multi-user game based on SMS may communicate with its mobile players using this group communication mechanism.

Mobile Dictionary Service: The command

```
dic {English word}
```

returns the Chinese meaning of the English word {English word}.

Email Service: iSMS implements an agent to SMTP/POP3 servers for relaying email and SMS. A short message from mobile users to iSMS with the format

```
ema {email-address} {message}
```

is transferred to an email and sent to a SMTP server for delivery. Conversely, the agent consults the mail server via the POP3 interface on behalf of a mobile user periodically, and delivers notifications of new messages to a user's MS via SMS. The user reads an email just as if he/she were reading a short message.

Forwarding Service: An iSMS user can execute a command {command} and then forward the results to another user {receiver}. The command is

```
fwd {receiver} {command}
```

These services can be combined to yield more powerful services. For example, a boy may order a song and request to forward the song to play on his girlfriend's MS on her birthday. This action combines the handset music service, forwarding service, and event scheduling service. These services are powerful building blocks when iSMS integrates with applications offered by content providers. Examples of these applications include

mobile banking services, mobile trade services, credit card infoɪ
insurance account information, airline/travel/concert ticket rɪ
news/information, entertainment, and so on.

1.5   Caching for iSMS-Based Wireless Data Access

This section uses a business card application as an example to illustrate
how cache-based wireless data access can be implemented in an SMS-IP
platform such as iSMS. To converge wireless data with the Internet, the
wireless application protocols may integrate a lightweight web browser into
wireless terminals with limited computing and memory capacities. The wireless application protocols implemented in the wireless application gateway
(for example, the iSMS gateway) and the wireless terminal enable a mobile
user to access Internet web applications through a client/server model. A
client application running on the wireless terminal may repeatedly access a
data object received from the application server. To speed up wireless data
access, cache mechanisms have been proposed. For example, the *Wireless
Application Protocol* (*WAP*) [OMA04] user agent caching model tailors the
HTTP caching model to support wireless terminals with limited functions.
In this model, a cache in the wireless terminal is used to buffer frequently
used data objects sent from the WAP gateway. When the data objects in the
application server are modified, the cached objects in the wireless terminals are obsolete. To guarantee that the data presented to the user at the
terminal is the same as that in the application server, a *strongly consistent*
data access protocol [Yin 99] must be exercised. Furthermore, the cache
size in a wireless terminal is typically small, and a cache replacement policy
is required to accommodate appropriate data objects in the cache. We use
the iSMS business card service as an application for wireless data access
study. The business card service is a generalization of the phone book feature in mobile handsets, which is one of the most popular handset features.
Unlike the phone book, the business cards are stored and maintained in a
business card database in the network (that is, iSMS application server), and
the most frequently used business cards are cached in the wireless terminal.
This application offers four major advantages over the phone book feature
in mobile handsets:

- When a user changes the handset, the phone book may not be conve-
  niently transferred to the new handset. This is particularly true when
  the old handset is broken and the phone book is lost. With the iSMS

business card service, the user can access his/her "private" phone book from any handset.

- Besides the private phone book, the business card service can also maintain a "public" phone book database (just like the yellow pages) in the application server.

- After a phone conversation, the business card service allows the call parties to exchange their business cards, which provide much more information than the phone numbers in the phone book feature.

- When the information (for example, phone number) of a user changes, it is not automatically reflected in the phone books of other users. With the business card service, a user can update the business card in the database of the business card application server (typically, through the Internet), and other users always access the correct information.

In the iSMS-based business card application, the format of the business card follows the vCard standard as illustrated in Table 1.4 [Daw 98]. In this format, the FN field is used to specify the vCard object. The N field is a single structured text value, which corresponds, in sequence, to the Family Name, Given Name, Additional Names, Honorific Prefixes, and Honorific Suffixes. A person may have several telephone numbers; for example, work phone number, fax number, and cellular telephone number. In vCard, these numbers are included in the TEL field. We also include a CALENDAR field that allows the user to fill in the schedule of public events he/she wants to share with others. In our iSMS implementation, the size of a vCard can be unlimited in the iSMS application server. However, the vCard is tailored to

Table 1.4 The vCard format in the iSMS-based business card service

FIELD	DESCRIPTION	LENGTH
VERSION	Version of vCard	13 bytes
FN	vCard object name	30 bytes
N	Name information	40 bytes
ORG	Organization information	50 bytes
TITLE	Job title	50 bytes
ADR	Address	120 bytes
TEL	Phone number	130 bytes
EMAIL	Email address	50 bytes
URL	Uniform resource locator	50 bytes
CALENDAR*	Public calendar event	100 bytes

*The CALENDAR field is not defined in the vCard standard.

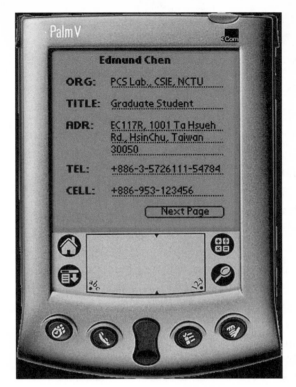

Figure 1.16 An iSMS Business Card Shown in PDA

be of fixed size when it is delivered to the wireless terminal. The length of each field in our implementation is also shown in Table 1.4. The appearance of an iSMS business card in a PDA is illustrated in Figure 1.16. An iSMS business card can be updated, added to, or removed from the database by the application server or by the wireless terminal.

We consider two strongly consistent algorithms for wireless data access: *Poll-Each-Read* (*PER*) and *Call-Back* (*CB*) [Yin 99, Sat 90]. In both algorithms, when a data object (for example, a business card) is updated at the iSMS application server, the valid object is not immediately sent to update the cached copies in the wireless terminal. Instead, the valid object is delivered to a wireless terminal only when a data access operation is actually performed.

In *Poll-Each-Read* [How 88], at a data access request (see (b) and (e) in Figure 1.17), the wireless terminal always asks the iSMS application server to check whether the cached object is valid. If so, the iSMS application server responds affirmatively (Figure 1.17 (f)) and the user accesses the object

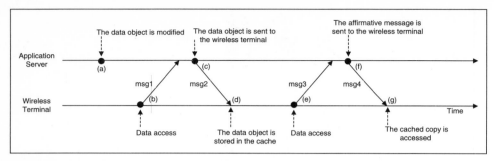

Figure 1.17   Data Access in Poll-Each-Read

in the cache of the wireless terminal (Figure 1.17 (g)). If the data object is updated before the access (Figure 1.17 (a) and (b)), the iSMS application server sends the current data object to the wireless terminal (see Figure 1.17 (c)), and this object is stored in the cache of the wireless terminal (see Figure 1.17 (d)). In this approach, when an object O_i is found in the cache, the wireless terminal still needs to obtain O_i from the iSMS application server if O_i has been invalidated. Thus, a *cache hit* may not be beneficial to PER. Define *effective hit ratio* as the probability that for an access to object O_i, a cache hit occurs in the wireless terminal and the cached object is valid. In PER, the cost for accessing an object with an effective cache hit is a cache affirmative request and acknowledgment exchange between the iSMS application server and the wireless terminal (**msg3** and **msg4** in Figure 1.17). For a cache miss or a cache hit in which the data object is invalidated, the access cost is a request message sent from the wireless terminal (**msg1** in Figure 1.17) and a data object transmission from the iSMS application server to the wireless terminal (**msg2** in Figure 1.17).

In the *Call-Back* approach [How 88, Nel 88], whenever a data object is modified, the iSMS application server sends a message to invalidate the corresponding cached object in the wireless terminals (see (a) in Figure 1.18). The cache storage of the invalidated object is reclaimed to accommodate other data objects, and the wireless terminal sends an acknowledgment to inform the application server that the invalidation is successful (see (b) in Figure 1.18). During the period between points (a) and (d) in Figure 1.18, if other updates to this object occur, no invalidation message needs to be sent to the wireless terminal (because no invalidated copy will be found in the cache). In this approach, all objects stored in the cache are valid, and a cache hit is always an effective hit (if the message transmission delay is ignored). In a data access, if a cache hit occurs, the cached object is used without any communication between the application server and the wireless

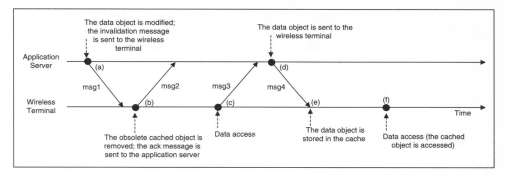

Figure 1.18   Data Access in Call-Back

terminal (Figure 1.18 (f)). For a cache miss, the access cost is a request message (**msg3** in Figure 1.18) sent from the wireless terminal and a data object transmission (**msg4** in Figure 1.18) from the application server to the wireless terminal. It is required to invalidate a cached object at the wireless terminal (if it exists) when the object in the application server is updated (**msg1** and **msg2** in Figure 1.18).

The typical cache size in a wireless terminal is not large. When the cache is full, some cached objects must be removed to accommodate new objects. We consider the *Least Recently Used* (*LRU*) replacement policy. This policy is often utilized to manage cache memory in computer architecture [Sto 93], virtual memory in operating systems [Sil 01], and location tracking in mobile phone networks [Lin 94]. LRU uses the recent past as an approximation of the near future, and replaces the cached object that has not been used for the longest period of time. LRU associates with each cached object the time of its last use. When a cached object must be replaced, LRU chooses the object that has not been used for the longest period of time. Therefore, every object in the cache has a rank. If a cached object has the rank 1, it means that the object is most recently used. If an object O_i has a rank k, it means that $k - 1$ objects are more recently used than O_i is. If $k > K$ where K is the size of the cache, then O_i has been removed from the cache.

The number of data objects in the application server is much larger than the cache size in the wireless terminal. However, our experience in exercising wireless applications [Rao 03] indicates that for an observed period, only a small number N of data objects in the iSMS application server are potentially accessed by a wireless terminal. Although the objects to be accessed vary from time to time, the number N is not significantly larger than the cache size of the wireless terminal. That is, the data access pattern of a wireless terminal exhibits *temporal locality* [Sto 93], which is the tendency for a wireless terminal to access in the near future those data objects referenced in the recent past. Temporal locality may not be observed in wireline Internet access because desktop users typically navigate through several web sites at the same time. Conversely, the restricted user interface of wireless terminals only allows a user to access a small region of data objects for instant information acquisition. Thus, caching can effectively reduce the data access time for the wireless terminals. The cache performance for the iSMS-based wireless terminals was investigated in [Lin03c].

1.6   Concluding Remarks

This chapter first introduced a SM-SC SMS-IP integration system called NCTU-SMS. Then we described iSMS, a non-SM-SC-based platform that integrates IP networks with the short message mechanism of mobile networks. This platform was developed by AT&T and FarEasTone. iSMS supports middleware for creating and hosting wireless data services based on SMS. The iSMS hardware architecture can be easily established with standard MSs and personal computers. We have developed agent-based middleware with an API, which results in a lightweight solution that allows quick deployment of added-value data services. Compared with the solutions that integrate IP networks through SM-SC, iSMS has the following advantages:

Easy installation: The iSMS hardware components are a standard GSM MS and personal computer. The iSMS software can be installed on the PC without special treatment.

Generic platform with personalization: Several applications have been developed for iSMS, including email delivery/forwarding, web access (for example, a stock query or a train schedule query) and handset music service. iSMS users can easily develop new services and tailor the existing services to fit their needs.

Transparency: The iSMS services are transparently run on GSM networks without any maintenance effort by GSM operators.

Immediacy: Unlike the dial-up service to a PC, iSMS offers instant information exchange between the MS and the servers. This property is desirable for a customer to enable IP computing when he/she is moving.

For a GSM operator, iSMS is targeted for small companies that subscribe to the *Closed User Group* (*CUG*) service [Lin 01b]. In each of the small companies, a PC-based iSMS gateway is installed. The iSMS is designed for two types of services:

- For *standard Internet services*, an iSMS gateway serves as an access point for the GSM user to receive existing Internet services (such as stock quotes or Yahoo! surfing).

- *Customized services* (for example, private mails) are tailored for each iSMS gateway. These types of services are exclusive to a closed user group of an iSMS gateway, and cannot be accessed by users outside the gateway (although different CUGs may have the same customized services).

For both types of services, the iSMS gateways do not communicate with each other. Thus, there is no need to manage or administer the distributed iSMS gateways. The iSMS gateway approach has the following advantages:

- **Scalability:** One can easily accommodate new iSMS customers by installing PC-based iSMS gateways at their company sites. Since the radio paths for these iSMS gateways are different, adding new customers does not increase traffic to a specific radio link. It is clear that iSMS is scalable for standard Internet services. To accommodate more users for Internet services, one only needs to add more iSMS gateways. For other approaches such as WAP, all accesses to the Internet should go through a centralized WAP gateway, which may require great maintenance, and the gateway may become a bottleneck. In iSMS, accesses to the Internet are performed through thousands of routes using independent iSMS gateways.

- **Reliability:** When an iSMS gateway at a company fails, one only needs to fix this failed gateway. This failure does not affect other iSMS gateways. To enhance the reliability of iSMS service in a company, one can have duplicated configuration on an iSMS gateway site. In this configuration, an MS (the customer) may register to an operational iSMS gateway and a standby iSMS gateway. This setup is performed at gateway initialization (when the customer subscribes to iSMS). For a standard GSM MS, the phone numbers of both gateways are stored in the phone book of the MS. The MS first tries the operational iSMS gateway for service. If it does not work, then the MS tries the standby iSMS gateway. Message resending can be automatically performed by the GSM divert service. If the GSM MS is equipped with the SIM ToolKit feature, then when the operational gateway fails, the standby gateway will instruct the MS to switch the gateway over the air (by modifying the gateway phone number stored in the MS). When the operational gateway is recovered, the gateway phone number in the MS is switched back by the operational gateway over the air. Gateway restoration is transparent to users.

- **Performance:** Since the customer traffic is distributed to individual iSMS gateways, good performance (for example, short latency time) can be expected. However, the end-to-end short message transmission delay for iSMS is longer than that for NCTU-SMS.

Due to limited SMS bandwidth, most of the iSMS applications are transaction oriented. When high-bandwidth mobile data infrastructures (such as GPRS and UMTS) are available, the data rate of iSMS can be significantly increased, and development of general data services will be essential. The iSMS philosophy has been generalized for iMobile [Rao 01b], a user-friendly environment for mobile Internet. The iMobile platform is described in Chapter 17. Additionally, the reader is referred to Chapter 12 in [Lin 01b] for more details about SMS.

1.7   Questions

1. Describe the procedure of SMS delivery from an MS to the other MS.

2. Describe the NCTU-SMS architecture. What is the NCTU-SMS delivery state tree?

3. Describe the SMS architecture. Does delivery of SMS affects voice call setup? Does it affect voice conversation?

4. Compare the two SMS-IP integration approaches: NCTU-SMS (an approach involving SM-SC) and iSMS (an approach that does not involve SM-SC). Which approach will experience longer transmission delay for end-to-end short message delivery?

5. Describe the three service types defined in iSMS. Can you define a new service type for iSMS?

6. In accordance with Section 1.2, design a new user-defined server and an agent.

7. Describe four services for MS-agent integration in iSMS.

8. Section 1.3.2 defines functions in the iSMS Communication API. Do these functions suffice to support all SMS-IP services? Can you add more functions?

9. iSMS supports the `SendText()` and `SendData()` functions. Give an example to show when `SendData()` is used.

10. Design an iSMS GUI for a PDA that allows adding new iSMS applications without modifying the PDA.

11. What is a strongly consistent data access protocol?

12. Describe the the Poll-Each-Read and the Call-Back approaches. Is one approach always better than the other?

13. Evaluate iSMS in terms of scalability, reliability, and performance.

## CHAPTER 2

Mobility Management for GPRS and UMTS

Universal Mobile Telecommunications System (*UMTS*) is a mobile telecommunications network that evolved from Global System for Mobile Communications (GSM) and *General Packet Radio Service* (*GPRS*). This chapter describes the network architectures of GPRS and UMTS and shows how mobility management evolves from GPRS to UMTS. Readers who are not familiar with GPRS are encouraged to read Chapters 9 and 18 in [Lin 01b]. Equivalent materials can also be found at **http://liny.csie.nctu.edu.tw/supplementary**.

2.1 Network Architectures

The network architecture of the GSM/GPRS/UMTS systems is shown in Figure 2.1. In this figure, the dashed lines represent signaling links, and the solid lines represent data and signaling links. The GSM network consists of the following components:

- *Mobile Stations* (*MSs*; the mobile terminals) that communicate with the network through the *Base Station System* (BSS)

- The BSS consists of the *Base Transceiver Station* (*BTS*) and the *Base Station Controller* (*BSC*). The BTS communicates with the MS through

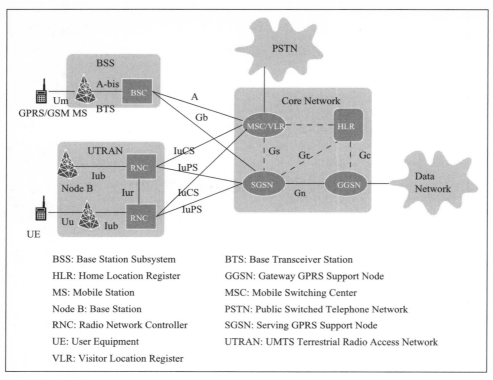

Figure 2.1   GSM/GPRS/UMTS Network Architectures

the radio interface *Um*, based on the Time Division Multiple Access (TDMA) technology. Through the *A-bis* interface, the BTS connects to the BSC. The BSC communicates with exactly one *Mobile Switching Center* (*MSC*) via the *A* interface.

- MSC is a special telephone switch tailored to support mobile applications. The MSC connects the calls from the MSs to the *Public Switched Telephone Network* (*PSTN*).

- The *Home Location Register* (*HLR*) and the *Visitor Location Register* (*VLR*) provide mobility management, elaborated later.

GPRS evolved from GSM, where existing GSM nodes such as BSS, MSC, VLR, and HLR are upgraded. GPRS introduces two new *Core Network* nodes: *Serving GPRS Support Node* (*SGSN*) and *Gateway GPRS Support Node* (*GGSN*).

- The GGSN provides connections and access to the integrated services Internet. It maintains routing information for the GPRS-attached MSs to tunnel *Protocol Data Units* (*PDUs*) to the SGSN through the

Gn interface. The GGSN communicates with the HLR for session management (see Chapter 3) through the *Gc* interface. Note that in most commercial products, the GGSN communicates with the HLR indirectly through the SGSN.

- The SGSN is responsible for the delivery of packets to the MSs within its service area. The SGSN performs security, mobility management, and session management functions by communicating with the HLR through the *Gr* interface. The BSC of the GPRS BSS is connected to the SGSN through the *Gb* interface using the frame relay link.

UMTS evolved from GPRS by replacing the radio access network. The *UMTS Terrestrial Radio Access Network* (*UTRAN*) consists of *Node B*s (the UMTS term for BTS) and *Radio Network Controller*s (*RNCs*) connected by an ATM network. The RNC and the Node B serving an MS are called the *Serving Radio Network Subsystem* (*SRNS*). The *User Equipment* (*UE*; the UMTS term for MS) connects with Node Bs through the radio interface *Uu* based on the WCDMA (Wideband Code Division Multiple Access) technology [Hol 04]. In UMTS, every Node B is connected to an RNC through the *Iub* interface. Every RNC is connected to an SGSN through the *IuPS* interface, and to an MSC through the *IuCS* interface. An RNC may connect to several RNCs through the *Iur* interface. Unlike the RNCs in UMTS, the BSCs in GPRS/GSM do not connect to each other. The IuCS, IuPS, Iub, and Iur interfaces are implemented on the ATM network.

The core network consists of two service domains: the *circuit-switched* (*CS*) service domain (that is, PSTN/ISDN) and the *packet-switched* (*PS*) service domain (that is, the Internet). In the CS domain, an MS is identified by *International Mobile Subscriber Identity* (*IMSI*) and *Temporary Mobile Subscriber Identity* (*TMSI*). TMSI is an alias of IMSI. For security reasons, TMSI is used to avoid sending the IMSI on the radio path. This temporary identity is allocated to an MS by the new VLR at inter-VLR location update, and can be periodically changed by the VLR. In the PS domain, an MS is identified by IMSI and *Packet TMSI* (*P-TMSI*). Based on the service domains, several operational modes are defined for GPRS MSs and UMTS UEs. Three operation modes are defined for GPRS MS:

- *Class A* **MS** allows simultaneous CS and PS connections.
- *Class B* **MS** provides automatic choice of CS or PS connection, but only one at a time.
- *Class C* **MS** supports only PS connection.

Three operation modes are defined for UMTS UE:

- ■ *PS/CS mode* **UE** is equivalent to GPRS Class A MS.
- ■ *PS mode* **UE** is equivalent to GPRS Class C MS.
- ■ *CS mode* **UE** can attach to the CS domain only.

For our purposes, the remainder of this chapter uses the term MS to represent UE.

In terms of the core network evolution from GPRS to UMTS, both the SGSN and the MSC need to be modified. Also, the *Mobility Management* (*MM*) and the *Packet Data Protocol* (*PDP*) contexts of the SGSN and the MS are modified (elaborated in Section 2.4). Other core network nodes such as HLR (specifically, HLR packet domain subscriber data), VLR (specifically, VLR and SGSN association, described later), and GGSN (specifically, PDP contexts) are basically the same.

Figure 2.2 illustrates the mobility management *control planes* between the MS and the SGSN for UMTS and GPRS, respectively. (Note that in the

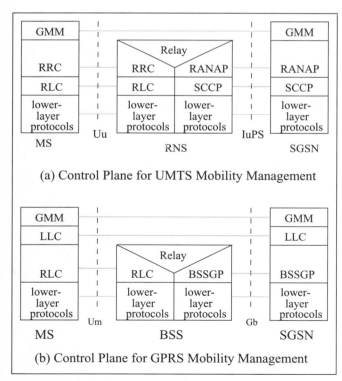

Figure 2.2   Control Planes for UMTS and GPRS

early GPRS version, the control plane was called the *signaling plane*.) The protocol stack for GPRS is described as follows:

- *Radio Link Control (RLC)* provides services for information transfer over the GPRS physical layer. These functions include backward error correction procedures, enabled by the selective retransmission of erroneous blocks.

- *Logical Link Control (LLC)* is a sublayer of OSI layer 2 (see Section 8.1). LLC conveys information between layer 3 entities in the MS and SGSN. It provides services to the GMM. The LLC support to session management is described in Chapter 3.

- *GPRS Mobility Management (GMM)* supports mobility management functionality (attach, detach, and location update, described in Sections 2.5 and 2.6).

- *BSS GPRS Protocol (BSSGP)* provides the radio-related QoS and routing information required to transmit user data between a BSS and an SGSN.

Unlike GPRS, the LLC layer is not supported in UMTS. In GPRS, reliable communication between the MS and the SGSN is guaranteed by the LLC. In UMTS, the *Radio Resource Control (RRC)* protocol is responsible for reliable connection between an MS and the UTRAN. In particular, radio resources are managed by the RRC exercised between the MS and the UTRAN. The *Signaling Connection Control Part (SCCP*; see Chapter 8) is responsible for reliable connection between the UTRAN and the SGSN. On top of the SCCP, the *Radio Access Network Application Part (RANAP)* protocol supports transparent mobility management signaling transfer between the MS and the core network. These mobility management messages are not interpreted by the UTRAN. The RANAP is also responsible for serving RNC relocation (see Section 2.7), *Radio Access Bearer (RAB)* management, and so on. In [3GP02c], GMM for UMTS is also referred to as *UMTS MM (UMM)*. The MM messages are exchanged among GPRS/UMTS nodes through various interfaces described as follows:

MS and SGSN. In GPRS, the MM messages are delivered through the Gb and Um interfaces. In UMTS, the MM message transmission is performed through the Iu and the Uu interfaces. Specifically, an LLC link provides a signaling connection between the MS and the SGSN in GPRS. In UMTS, the signaling connection consists of an RRC connection between the MS and UTRAN, and an Iu connection ("one RANAP instance") between the UTRAN and the SGSN.

SGSN and other core network nodes: In both GPRS and UMTS, GSM *Mobile Application Part (MAP)* is used to interface an SGSN with the GSM nodes—for example, *Gr* for HLR, and *Gs* (the BSSAP+ protocol or *BSS Application Protocol+*) for MSC/VLR. An SGSN and a GGSN communicate using the *GPRS Tunneling Protocol (GTP)* through the *Gn* interface by using a *GTP tunnel* for packet delivery. This tunnel is identified by a *Tunnel Endpoint Identifier (TEID)*, an *Internet Protocol (IP)* address, and a UDP port number. Details of the MAP protocols and GTP are described in Chapters 4, 7, and 8.

The Gs interface merits further discussion. In both GPRS and UMTS, procedures such as attach, paging, and location update are defined separately for the CS and the PS domains. For example, LA update is performed for CS, and RA update is performed for PS. To save radio resources, execution of similar procedures for both CS and PS can be combined. Examples are combined PS/CS attach (see Section 2.5) and combined RA/LA update (see Section 2.6). Furthermore, activities such as CS paging can be performed by using the PS mechanism, so that the MS only needs to monitor a single paging channel. The above optimizations are achieved only if the Gs interface exists. Through this interface, the SGSN and the MSC/VLR can communicate to combine both PS and CS activities. The GPRS (UMTS) network is in *Network Mode I* if the Gs interface exists. Otherwise, it is in *Network Mode II*. Note that an extra network mode (Mode III) is defined for GPRS when the Gs is not present. This network mode has been removed from the UMTS specifications.

Initiated by an SGSN, a Gs association can be created between the SGSN and a VLR by storing the *SGSN number* (that is, the SGSN address) in the VLR, and storing the *VLR number* (the VLR address) in the SGSN. With this association, messages regarding CS activities can be passed between the VLR and the SGSN. We will elaborate on these activities later.

Protocols for user data transmission are defined in the *user plane*. In the early GPRS version, the user plane was called the *transmission plane*. In GPRS, the *Sub-Network Dependent Convergence Protocol (SNDCP)* carries out transmission of *Network Protocol Data Units (N-PDUs)* on top of the LLC link between the MS and the SGSN (see Chapter 3). In UMTS, the *Packet Data Convergence Protocol (PDCP)* carries out N-PDU transmission on top of the RLC connection between the MS and the UTRAN, and the GTP-U (GTP for the user plane) protocol carries out transmission of N-PDUs on top of the UDP/IP link (Iu link). Packets in user data transmission may be lost when some MM signaling procedures are executed. These procedures include attach, routing area update, and authentication.

To summarize the GPRS and UMTS architectures, in both GPRS and UMTS, IMSI is used as the common user identity, and common MAP signaling is applied to both systems as well as GSM. Unlike GPRS, the UMTS radio network parameters and radio resources are managed in the UTRAN. Like the GPRS BSS, the UTRAN does not coordinate mobility management procedures that are exercised between an MS and the core network. These procedures include location management, authentication, temporary identity management, and equipment identity check.

2.2 Concepts of Mobility Management

In order to track the MSs, the cells (i.e., BTSs/Node Bs) in the GPRS/UMTS service area are partitioned into several groups. To deliver services to an MS, the cells in the group covering the MS will page the MS to establish the radio link. Location change of an MS is detected as follows. The cells broadcast their cell identities. The MS periodically listens to the broadcast cell identity, and compares it with the cell identity stored in the MS's buffer. If the comparison indicates that the location has been changed, then the MS sends the location update message to the network.

In the CS domain, cells are partitioned into *Location Areas* (*LAs*). The LA of an MS is tracked by the VLR. In the PS domain, the cells are partitioned into *Routing Areas* (*RAs*). An RA is typically a subset of an LA. The RA of an MS is tracked by the SGSN. In GPRS, the SGSN also tracks the cell of an MS during a PS connection (that is, when packets are delivered between the MS and the SGSN). In UMTS, the cells in an RA are further partitioned into several *UTRAN RAs* (*URAs*). The URA and the cell of an MS are tracked by the UTRAN. Figure 2.3 illustrates an example of LA, RA, and URA layout. The areas controlled by the VLR, the SGSN, and the UTRAN are listed in Table 2.1.

In UMTS, the UTRAN tracking is triggered by the establishment of the RRC connection between an MS and the UTRAN, and an RRC state machine is executed (its state diagram is shown in Figure 2.4 [3GP05j, 3GP05r]).

- In the **RRC Idle Mode**, no RRC connection is established, and the MS is tracked by the SGSN at the RA level.

- When the RRC connection is established, the state moves from **RRC Idle** to **RRC Cell Connected**, and the MS is tracked by the UTRAN at the cell level.

- If no PDUs are transmitted before an *inactivity timer* expires, the state moves from **RRC Cell Connected** to **RRC URA Connected**, and the MS is tracked by the UTRAN at the URA level.

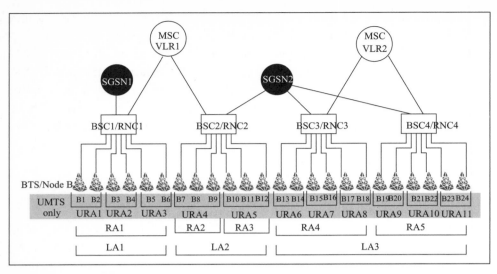

Figure 2.3   LAs, RAs, and URAs

The mobility management functions emphasizing PS-based services are listed below:

PS attach procedure allows an MS to be "known" by the PS service domain of the network. For example, after the MS is powered on, the PS attach procedure must be executed before the MS can obtain access to the PS services. Note that the term "PS attach" is used in UMTS and the term "GPRS attach" is used in GPRS. Similarly, we have the term "CS attach" for UMTS and "IMSI attach" for GPRS. For the discussion here, we use the terms PS attach and CS attach.

PS detach procedure allows the MS or the network to inform each other that the MS will not access the SGSN-based services. PS attach and detach are described in Section 2.5.

Table 2.1   Areas tracked by the network nodes

	MSC/VLR			SGSN		UTRAN
	GSM	GPRS	UMTS	GPRS	UMTS	UMTS
Cell	no	no	no	yes	no	yes
URA	–	–	no	–	no	yes
RA	–	no	no	yes	yes	no
LA	yes	yes	yes	no	no	no

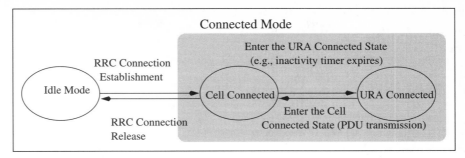

Figure 2.4 RRC State Diagram

Security procedures include authentication, user identity confidentiality (for example, P-TMSI reallocation and P-TMSI signature) and ciphering. Details of security procedures can be found in Section 9.1 and [ETS 04, 3GP04g]. Here, we elaborate on the P-TMSI signature. When the SGSN allocates a P-TMSI to an MS, it may also send the P-TMSI signature to the MS. When the next MS identity checking is performed, for example, in the attach procedure (see Step 2 in Figure 2.6), the MS sends the P-TMSI signature to the SGSN for comparison. If the comparison fails, the authentication procedure must be used by the SGSN to authenticate the MS.

GPRS ciphering is performed between the MS and the SGSN. Conversely, UMTS ciphering is performed between the UTRAN and the MS.

Location management procedures track the location of an MS. These procedures are elaborated in Section 2.6.

Tunneling of non-GSM signaling message procedures supports communication between GPRS/UMTS and non-GSM systems such as EIA/TIA IS-136. The SGSN forwards the signaling messages to the non-GSM MSC/VLR using the BSSAP+ protocol in the Gs interface.

Subscriber management procedures are used by the HLR to inform the SGSN about changes of the PS subscriber data. This procedure is needed, for example, in Step 8 of the location update procedure in Figure 2.7, Section 2.6.

Service request procedure (UMTS only) is used by the MS to establish a secure connection to the SGSN, so that the MS can send uplink signaling messages or user data. This procedure is used, for example, when the MS replies to a page from the UMTS network or when the MS attempts to request resource reservation. In GPRS, an LLC link is always established between an MS and the SGSN after the attach procedure. Therefore, the service request procedure is not needed and is not defined in GPRS.

UMTS-GPRS intersystem change procedures allow a dual-mode MS to move between GPRS and UMTS systems. Details are given in Section 2.8.

2.3   Mobility Management States

In GPRS and UMTS, an MM *Finite State Machine* (*FSM*) is exercised in both an MS and the SGSN to characterize the mobility management activities for the MS. In GPRS, the states in the machine are **IDLE**, **STANDBY**, and **READY**. For the UMTS PS service domain, these states are renamed as **PMM-DETACHED**, **PMM-IDLE**, and **PMM-CONNECTED**, respectively. The MM states are stored in the MM contexts maintained by the MS and the SGSN. Details of the MM context are given in Section 2.4. In this section, we describe the MM states and the transitions among these states. Figure 2.5 illustrates the MM state diagrams. The figure indicates that the MM state machines for both GPRS and UMTS are basically the same. The MM states are described as follows:

IDLE or **PMM-DETACHED:** The MS is not known (that is, not attached) to GPRS (UMTS/PS). That is, the MS is not reachable by the network. In this state, the MS may perform the attach procedure.

STANDBY or **PMM-IDLE:** The MS is attached to GPRS (UMTS/PS); that is, both the MS and the SGSN have established MM contexts. In this state, the MS may perform the detach and location update procedures. The SGSN may perform the paging procedure. The MS is tracked by the SGSN at the RA level (see Table 2.1).

READY or **PMM-CONNECTED:** PDUs can only be delivered in this state. In GPRS, the SGSN tracks the MS at the cell level. In UMTS, a PS signaling connection is established between the MS and the SGSN (that is, the MS is in the **RRC Connected Mode**). The SGSN tracks the MS with accuracy of the RA level, and the serving RNC is responsible for cell-level tracking. In UMTS, the serving RNC relocation (see Section 2.7) is executed in this state.

The transitions among the MM states (see Figure 2.5) are described as follows:

IDLE → READY (PMM-DETACHED → PMM-CONNECTED): This transition is triggered by the MS when the MS performs GPRS/PS attach.

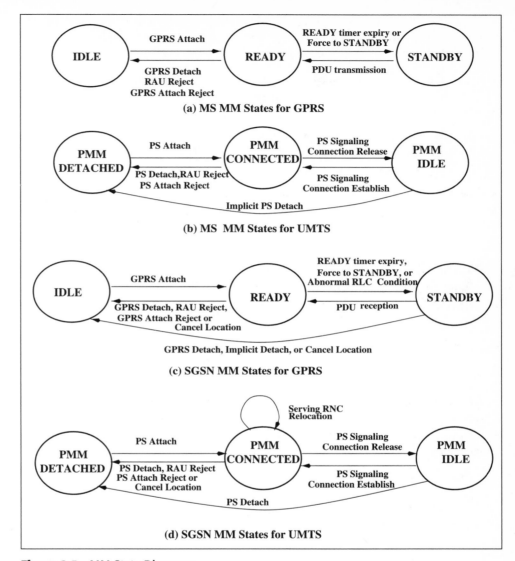

Figure 2.5   MM State Diagrams

STANDBY → IDLE (PMM-IDLE → PMM-DETACHED). This transition can be triggered by the MS or the SGSN:

■ This transition is triggered by the SGSN when tracking of the MS is lost. In this case, the SGSN performs an *implicit GPRS/PS detach.* A *mobile reachable timer* is maintained in the SGSN to monitor the periodic RA update procedure (see Section 2.6). If the SGSN does not receive the **Routing Area Update Request** message from the MS after the timer expires, the MS is considered detached. This timer is used only when the MM state is **STANDBY/PMM-IDLE**.

- This transition may also be triggered by the SGSN when the SGSN receives the **Cancel Location** (de-registration) message from the HLR. In this case, the MM and the PDP contexts are already moved to the new SGSN that serves the MS, and the contexts in the old SGSN can be deleted. Note that the MS will be associated with the new SGSN in this case.

- This transition is triggered by the MS when it performs implicit detach due to removal of the *Subscriber Identity Module (SIM)* card or the battery. This case is defined for UMTS, but not for GPRS.

STANDBY → READY (PMM-IDLE → PMM-CONNECTED). This transition is triggered by the MS. In GPRS, this transition occurs when the MS sends an LLC PDU to the SGSN, possibly in response to a page from the SGSN. In UMTS, this transition occurs when the service request procedure is executed (possibly in response to a page from the SGSN) to establish the PS signaling connection between the MS and the SGSN.

READY → STANDBY (PMM-CONNECTED → PMM-IDLE). This transition is triggered by either the SGSN or the MS. In GPRS, a READY timer is maintained in the MS and the SGSN. If no LLC PDU is transmitted before the timer expires, then this MM transition occurs. The length of the READY timer can be changed only by the SGSN. The MS is informed of the READY timer value change through messages such as **Attach Accept** and **Routing Area Update Accept**. This MM transition may also occur when the SGSN forces it, or when an abnormal RLC condition is detected during radio transmission.

In UMTS, this MM transition occurs when the PS signaling connection is released or broken (for example, RRC connection failure), or when the URA update timer at the RNC expires.

READY → IDLE (PMM-CONNECTED → PMM-DETACHED). This transition can be triggered by the MS or the SGSN:

- This transition is triggered by the MS when the MS-initiated GPRS/PS detach is performed.

- This transition is triggered by the SGSN when the network-initiated GPRS/PS detach is performed.

- This transition is triggered by the SGSN when the SGSN receives an **Cancel Location** message from the HLR, or when the SGSN rejects an RA update or an attach request from the MS.

In UMTS, the PS signaling connection is released after this transition. Specifically, both the RRC and the SCCP connections are released. In GPRS, the LLC link is removed after this transition.

2.4 MM and PDP Contexts

Two important contexts are defined in GPRS/UMTS:

- Mobility Management (MM) context provides mobility information about an MS.
- Packet Data Protocol (PDP) context provides information to support packet delivery between an MS and the network.

While an MS may be associated with several PDP contexts, it only has one MM context. The MM context is maintained in the MS and the SGSN. The PDP contexts are maintained in the MS, the SGSN, and the GGSN. This section describes the MM and PDP contexts, and shows the differences between the GPRS and UMTS contexts.

2.4.1 Contexts in SGSN

The following fields in the MM context are maintained in both GPRS SGSN and UMTS SGSN:

- *Mobile Station ISDN Number* (*MSISDN*; telephone number of the MS),
- IMSI (used to identify the MS in the GSM/GPRS/UMTS network; unlike MSISDN, IMSI is not known to the users, and is used in the network only),
- MM state (see Section 2.3),
- P-TMSI, P-TMSI signature (temporarily identities of the MS; usage of these identities are described in Section 2.5),
- *International Mobile Equipment Identity* (*IMEI*; the serial number of the handset or mobile equipment),
- routing area (see Section 2.2),
- VLR number (associated VLR address),

- MS network access capability
- new SGSN address
- selected ciphering algorithm
- subscribed charging characteristics, and
- several flags.

The following MM context fields are different in GPRS SGSN and UMTS SGSN:

Location Information. GPRS SGSN maintains cell identity (the current cell in the **READY** state, or the last known cell in the **STANDBY** or the **IDLE** state) and cell identity age (time elapsed since the last LLC PDU was received from the MS at the SGSN). These two fields are not maintained in UMTS SGSN because cell tracking is performed by the serving RNC.

UMTS SGSN maintains the last known *Service Area Code* (*SAC*) when an initial MS message was received or when the location reporting procedure was executed, and the elapsed time since the last SAC was received at the SGSN. The SAC is used to uniquely identify an area consisting of one or more cells belonging to the same location area. SAC and the location reporting procedure are used in UMTS for *Location Service* (*LCS*) and other services such as emergency calls [3GP04h]. These fields are not maintained in GPRS SGSN because the concept of SAC does not exist in GPRS.

Security Information. UMTS provides enhanced security functions over GPRS, and thus extra security parameters are maintained in the UMTS SGSN MM context. Specifically, UMTS SGSN maintains authentication vectors, CK (currently used ciphering key), IK (currently used integrity key), and KSI (key set identifier). Conversely, GPRS SGSN maintains Kc (currently used ciphering key) and CKSN (ciphering key sequence number of Kc). KSI in UMTS corresponds to CKSN in GSM, and they have the same format. The CK parameter in UMTS is equivalent to Kc in GPRS.

Radio Resource Information. GPRS SGSN maintains *radio access capability* (MS's GPRS multislot capabilities and so on) and *discontinuous reception* (*DRX*) parameters, radio priority *Short Message Service* (*SMS*; that is, the RLC/MAC radio priority level for uplink SMS transmission). DRX allows discontinuous radio transmission to save on power consumption by the MS. In UMTS, the radio resources are controlled by UTRAN and are not known to the SGSN. Thus, the above fields are not kept in the UMTS SGSN MM context.

In GPRS, if DRX mode is selected, the MS may specify the DRX parameters that indicate the delay for the network to send a page request or a channel assignment to the MS. DRX usage is independent of the MM states. However, during the GPRS attach and RA update, the GPRS MS shall not apply DRX in the **READY** state.

The following fields in a PDP context are maintained in both GPRS SGSN and UMTS SGSN:

- PDP Route Information includes PDP context identifier, PDP state, PDP type, and PDP address.

- Access Point Name (APN) Information includes the APN subscribed and the APN in use. An APN represents an external network that can be accessed by the MS (see Section 4.1).

- QoS Information includes QoS profile subscribed, QoS profile requested and QoS profile negotiated. After the subscription time, a subscriber specifies the subscribed QoS for a specific service. When the subscriber accesses this service, he/she specifies the requested QoS. Based on the available network resources, the GGSN determines the negotiated QoS for the service.

- N-PDU Information includes GTP-SND and GTP-SNU. The GTP-SND parameter is the GTP sequence number of the N-PDU to be sent from the SGSN to the MS. The GTP-SNU parameter is the GTP sequence number of the N-PDU to be sent from the SGSN to the GGSN.

- Charging Information includes the charging identifier.

- Other Routing Information includes NSAPI, TI, TEID for Gn/Gp, GGSN address in use, and VPLMN address allowed. *Network Layer Service Access Point Identifier (NSAPI)* is used by LLC (in GPRS) or RLC (in UMTS) to route the N-PDUs to appropriate higher layer protocols such as signaling, SMS, or packet data protocols. *Transaction Identifier (TI)* is used to represent NSAPI for some session management signaling messages. VPLMN specifies the GPRS/UMTS networks visited by the MS.

- Subscribed Charging Characteristics can be normal, prepaid, flat-rate, and/or hot billing. In the early GPRS/UMTS version, charging characteristics for PDP contexts are maintained in the SGSN. In the latest version, charging characteristics are included in the SGSN MM context.

The following PDP context fields are different in GPRS SGSN and UMTS SGSN:

Core Network to Radio Access Network Connection. The UMTS maintains the TEID for the Iu interface and the IP address of the RNC currently used. These two fields are not maintained in the GPRS SGSN.

Radio Resource Information. The GPRS SGSN maintains radio priority (the RLC/MAC radio priority level for uplink user data transmission). These fields are not kept in UMTS SGSN.

PDU Information. GPRS SGSN maintains the Send N-PDU number (SNDCP sequence number of the next downlink N-PDU to be sent to the MS), the Receive N-PDU number (SNDCP sequence number of the next uplink N-PDU to be received from the MS), the packet flow identifier, and the aggregate BSS QoS profile negotiated.

Conversely, UMTS SGSN maintains PDCP-SND (the next PDCP sequence number to be sent to the MS) and PDCP-SNU (the next PDCP sequence number expected from the MS).

Note that in both GPRS and UMTS, the radio resource information for SMS is kept in the MM context, while the radio resource information for user data is maintained in the PDP context. The reason is the following: The PDP context is defined for data transfer in the user plane. Conversely, the MM context is defined for mobility management signaling in the control plane. SMS is delivered through the control plane by using a common channel, which is more efficient than delivery through the user plane. Furthermore, through the control plane, the same SMS transfer procedure is used for both the CS and the PS domains [3GP02c]. Thus, the radio resource information for SMS is kept in the MM context.

2.4.2 Contexts in the MS

The following fields in the MM context are maintained in both GPRS MS and UMTS MS: IMSI, MM state, P-TMSI, P-TMSI signature, routing area, MS network access capability, CKSN/KSI, ciphering algorithm, and DRX parameters. The following MM context fields are different in GPRS MS and UMTS MS:

Location Information. GPRS MS maintains cell identity. In UMTS, cell tracking is not conducted at the mobility management layer between the MS and the SGSN. Thus, cell identity is not maintained in the MM context of the MS. Instead, it is maintained between the MS and the UTRAN.

Security Information. UMTS MS maintains the extra security parameter CK.

Radio Resource Information. GPRS MS maintains radio priority SMS. In UMTS, the SMS as well as signaling are delivered through dedicated control channels. Thus, the radio priority is not maintained in the UMTS MS [3GP05f]. The GPRS MS maintains the MS radio access capability (for example, multislot capability and power class) while the UMTS MS maintains UE capability (for example, power control, code resource, UE mode, and PDCP capability).

The following fields in a PDP context are maintained in both GPRS MS and UMTS MS: PDP type, PDP address, PDP state, dynamic address allowed, APN requested, NSAPI, TI, QoS profile requested, QoS profile negotiated, and a flag.

The following PDP context fields are different in a GPRS MS and a UMTS MS:

Radio Resource Information. The GPRS MS maintains radio priority. In UMTS, the radio priority for data delivery is determined by the QoS profile, and the radio priority is not kept separately in the MS.

PDU Delivery Information. GPRS MS maintains the BSS packet flow identifier, the Send N-PDU number, and the Receive N-PDU number. Conversely, UMTS MS maintains PDCP-SND and PDCP-SNU.

2.4.3 Relationship between the MM States and the Contexts

The status of an MM or a PDP context is affected by the MM states. The relationship between the MM states and the contexts is summarized in Table 2.2 and is described as follows:

Table 2.2 Relationship between the MM states and the contexts

| Context | IDLE/DETACHED | | | STANDBY/IDLE | | | READY/CONNECTED | | |
	MS	SGSN	GGSN	MS	SGSN	GGSN	MS	SGSN	GGSN
MM	×/△	×/△	–	○	○	–	○	○	–
PDP	×/△	×/△	×	×/○	×/○	×/○	×/○	×/○	×/○

△: The context is kept but is stale.   ×: The context is removed.
–: The context does not exist.   ○: The current context is maintained.

- In the **IDLE/PMM-DETACHED** state, the PDP context in the GGSN is deleted. The MM and PDP contexts in MS and SGSN may or may not be deleted. If the MM state moves from **STANDBY/PMM-IDLE** to **IDLE/MM-DETACHED** because the mobile reachable timer expires (for example, the MS temporarily moves out of the GPRS/UMTS coverage), then these two contexts shall not be deleted. In this case, the location and routing information are stale.

- In **STANDBY/PMM-IDLE**, valid MM contexts are maintained in the MS and the SGSN. In this state, the PDP context can be activated or deactivated. In UMTS, when the PDP context is activated in this state, no Iu/radio connection is established between the MS and the network because PDU delivery is not allowed in this state. In GPRS, the LLC link is connected.

- In **READY/PMM-CONNECTED**, valid MM contexts are maintained in the MS and the SGSN. As in **STANDBY/PMM-IDLE**, the MS may initiate PDP context activation and deactivation. In this state, the signaling connection is established in UMTS.

2.5   Attach and Detach

With the attach procedure, the MS informs the network of its presence. Figure 2.6 illustrates the message flow of the combined PS/CS (GPRS/IMSI) attach procedure. In each step, we point out the differences between GPRS and UMTS:

Step 1. The MS initiates the attach procedure by sending the Attach Request message to the new SGSN. In GPRS, besides the MS network access capability, the message includes parameters such as MS radio access capability. These radio-related parameters are not included in the UMTS Attach Request message. Conversely, the UMTS message includes the "follow on request" parameter to indicate whether there is pending uplink traffic that needs Iu connection after the attach procedure is completed. This field is not needed in GPRS because the Iu interface does not exist. Furthermore, as we previously mentioned, the security parameters for UMTS and GPRS are different.

When the SGSN receives the attach request at the end of Step 1, several results are possible:

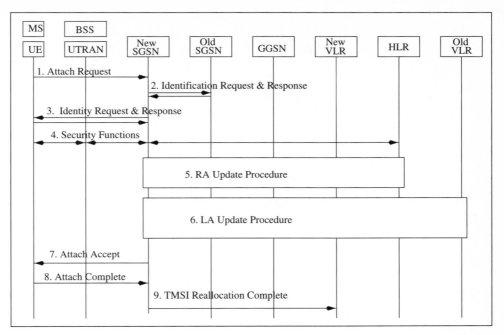

Figure 2.6 Combined PS/CS (GPRS/IMSI) Attach Procedure

- If the MS has changed SGSN since the last detach, then Step 2 is executed so that the new SGSN can obtain the MS identity (that is, IMSI) from the old SGSN.

- If the MS has not changed SGSN, then the received P-TMSI is used by the SGSN to identify the MM context of the MS. If the MM context has not been deleted since the last detach (that is, the MS is known by the new SGSN), then Steps 2–6 are skipped, and Step 7 is executed. Otherwise (the MS is not known by the old and the new SGSNs), Step 2 is skipped, and Step 3 is executed.

Step 2 (the MS is known by the old SGSN). The new SGSN sends the **Identification Request** message to the old SGSN. The P-TMSI is used to obtain the IMSI and authentication information from the old SGSN. If the old SGSN cannot find the MM context for the MS, then Step 3 is executed. Otherwise, the IMSI is returned to the new SGSN, and Step 4 is executed.

Step 3 (the MS is unknown to both the old and the new SGSNs). The new SGSN asks the MS to supply IMSI through the **Identity Request** and **Response** messages exchange.

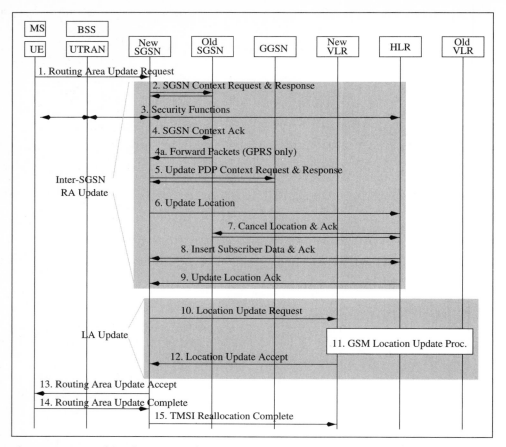

Figure 2.7 Combined RA/LA Update

Step 4. Authentication is mandatory if the MM context of the MS has been deleted since the last detach. The IMEI of the MS may be optionally checked.

Step 5. If the MS has moved from the old SGSN to the new SGSN since the last detach or if the MS is performing the first attach, then the RA update procedure is executed so that the new SGSN can obtain the current MM context of the MS. This step is the same as Steps 6–9 in Figure 2.7.

Step 6. If the Gs interface does not exist, then this step is skipped. Otherwise, the attach type in Step 1 is checked. If the attach type indicates (1) combined PS/CS attach or (2) PS attach and the MS is already CS attach, then LA update is performed. The LA update is required so that the SGSN-VLR association is established and the VLR can maintain current LA information of the MS. This step is the same as Steps 10–12 in Figure 2.7.

Step 7. For GPRS, if attach is successful, then the SGSN selects radio priority SMS and sends the **Attach Accept** message to the MS. P-TMSI is included in the message if the SGSN allocates a new P-TMSI. In UMTS, radio priority SMS is not maintained in mobility management. However, this parameter is still reserved in the UMTS **Attach Accept** message in order to support handoff between UMTS and GSM networks [3GP05f].

Steps 8 and 9. If P-TMSI or TMSI have been changed, the MS sends the **Attach Complete** message to the SGSN to acknowledge receipt of the TMSIs. The SGSN sends the **TMSI Reallocation Complete** message to the VLR.

After PS attach, the MS is in the **READY** (for GPRS) or the **PMM-CONNECTED** (for UMTS) state and MM contexts are established in the MS and the SGSN.

When PS detach is executed, the MS will not receive the SGSN-based service. The network or the MS may request detach explicitly. Conversely, implicit PS detach is executed by the network (without notifying the MS) if the mobile reachable timer expires or when the radio path is disconnected due to errors. After implicit PS detach is performed, the MS's MM context is deleted after an implementation-dependent timeout period. The PS detach procedure also inactivates the PDP contexts. The PS detach procedures are basically the same for both GPRS and UMTS.

2.6 Location Update

In location management, the MS informs the network of its location through RA and LA update procedures. The update procedures are executed in two situations:

Normal location update is performed when the MS detects that the location has been changed.

Periodic location update is exercised even if the MS does not move. That is, the MS periodically reports its "presence" to the network.

Periodic RA update enables the network to detect whether an MS is still attached to the network. A *periodic RA update timer* is maintained in both the MS and the SGSN. Every time this timer expires, the MS performs a periodic RA update. The periodic RA update timer value is set/changed by

Table 2.3   RA/LA update (the MS is not engaged in CS connection.)

NETWORK MODE	MODE I	MODE II
PS Attached	RA update	RA update
CS Attached	LA update	LA update
PS/CS Attached	RA update (periodic) combined RA update (normal)	Separate LA and RA updates

the SGSN, and is sent to the MS through the **RA Update Accept** or the **Attach Accept** messages when the MS visits an RA. This value cannot be changed before the MS leaves the RA.

RA update is periodically performed for a PS-attached MS that is not CS-attached (see Table 2.3). Conversely, LA update is periodically performed for a CS-attached MS that is not PS-attached. For a PS/CS attached MS, two cases are considered:

The MS is not engaged in a CS connection (see Table 2.3). For a PS/CS attached MS in Network Mode I, periodic RA update is performed, and LA update must not be performed. In this case, the VLR relies on SGSN to receive periodic RA updates. If the SGSN detects that the MS is lost, the SGSN detaches the MS, and notifies the VLR of this detach by the **IMSI Detach Indication** message. For normal location update, combined RA/LA update is performed when the MS changes locations.

In Network Mode II, the RA update (to the SGSN) and LA update (to the VLR) are performed separately. In this case, LA update is always performed before RA update.

The MS is engaged in a CS connection. During a CS connection, the network knows that the MS is attached, and no periodic location update is performed. In terms of normal location update, two cases are considered (see Table 2.4):

- **Class A MS (GPRS) or PS/CS MS (UMTS).** During a CS connection, RA update is exercised when the MS changes RAs. LA update is not performed when the MS changes LAs. Suppose that only inter-RA crossings occur during a CS connection, then at the end of the CS connection, no action is required. For Network Mode I, if there are inter-SGSN or inter-LA crossings during a CS connection, then at the end of the CS connection, a combined RA/LA update is executed to modify the SGSN-VLR association. For Network Mode II, if there are inter-LA crossings during the CS connection, then at the end of the CS connection, an LA update is performed.

Table 2.4 RA/LA update (the MS is engaged in CS connection.)

MS MODE	CLASS A (PS/CS)		
Movement Type	Inter-RA	Inter-SGSN	Inter-LA
During CS Connection	RA update	RA update	No action
Connection Release (Mode I)	No action	Combined RA/LA update	Combined RA/LA update
Connection Release (Mode II)	No action	No action	LA update
MS MODE	**CLASS B (GPRS ONLY)**		
Movement Type	Inter-RA	Inter-SGSN	Inter-LA
During CS Connection	No action	No action	No action
Connection Release (Mode I)	RA update	Combined RA/LA update	Combined RA/LA update
Connection Release (Mode II)	RA update	RA update	LA update

■ **Class B MS (GPRS only).** During a CS connection, the MS does not execute any RA/LA updates.

For Network Mode I, at the end of the CS connection, RA update is performed if inter RA crossings occur in the CS connection, and combined RA/LA update is performed if inter-SGSN or inter-LA crossings occur in the CS connection.

For Network Mode II, at the end of the CS connection, RA update is performed if inter-RA or inter-SGSN crossings occur in the CS connection. LA update is performed if inter-LA crossings occur in the CS connection.

Figure 2.7 illustrates the message flow of the combined RA/LA update. In each step, we point out the differences between GPRS and UMTS:

Step 1. The MS sends the **Routing Area Update Request** message to the new SGSN. This message is not ciphered so that the new SGSN can process the message. For both GPRS and UMTS, the update type can be RA update, periodic RA update, combined RA/LA update, or combined RA/LA update with IMSI attach. In this message, the "follow on request" parameter is used in UMTS to indicate whether the Iu connection should be kept for pending uplink traffic. This parameter does not exist in GPRS.

In GPRS, before the BSS passes the message to the SGSN, it adds the cell global identity information (including cell, RA and LA identities). In

UMTS, the RNC adds the routing area identity information (including RA and LA identities).

For inter-SGSN update, Steps 2–9 are executed. Otherwise (intra-SGSN update), these steps are skipped.

Step 2. To obtain the MM and the PDP contexts of the MS, the new SGSN sends the **SGSN Context Request** message to the old SGSN. Basically, the old SGSN validates the old P-TMSI signature, and returns the MM and the PDP contexts of the MS using the **SGSN Context Response** message. The old SGSN starts a timer. The MM context in the old SGSN is deleted when both of the following conditions are satisfied:

- The timer expires.

- The old SGSN receives the **Cancel Location** message from the HLR.

This timer mechanism ensures that if the MS initiates another inter-SGSN routing area update before the current update procedure is completed, the old SGSN still keeps the MM context. In GPRS, the old SGSN stops assigning SNDCP N-PDU numbers to downlink N-PDUs received. The old SGSN will forward buffered packets to the new SGSN at Step 4a. In UMTS, packet forwarding is not performed between the SGSNs. Also, the *Temporary Logical Link Identity (TLLI)* included in the GPRS **SGSN Context Request** message is not found in the UMTS message.

Step 3. If the old P-TMSI signature checking at Step 2 fails, a security function involving the MS, the BSS/UTRAN, the new SGSN, and the HLR is performed. If this security procedure also fails, then the old SGSN continues as though the **SGSN Context Request** message were never received and this procedure exits. Otherwise (security check succeeds), Step 4 is executed.

Step 4. The new SGSN sends the **SGSN Context Acknowledge** message to the old SGSN, which invalidates the SGSN-VLR association in the old MM context. In GPRS, this message includes the address of the new SGSN, which is used to inform the old SGSN that the new SGSN is ready to receive the buffered packets to be forwarded from the old SGSN. The new SGSN address is not included in the UMTS **SGSN Context Acknowledge** message.

Step 4a (GPRS only). The old SGSN then tunnels the buffered N-PDUs to the new SGSN. Note that no packets are forwarded from the old SGSN to the new SGSN in UMTS.

Step 5. The new SGSN sends the **Update PDP Context Request** message to the corresponding GGSNs. With this message, the GGSN PDP contexts are modified. The GGSNs return the **Update PDP Context Response** messages.

Step 6. The SGSN issues the **Update Location** message to inform the HLR that the SGSN for the MS has been changed.

Step 7. The HLR and the old SGSN exchange the **Cancel Location** message pair. The MM and the PDP contexts in the old SGSN are not deleted until the timer described in Step 2 expires.

Steps 8 and 9. The HLR inserts the user profile (subscriber data) into the new SGSN. For each PDP context, the new SGSN checks whether the context is new, active, or inactive. If the PDP context is active, then extra tasks are performed by the SGSN. For example, the SGSN checks whether the received "QoS subscribed" value is the same as the value of the QoS negotiated parameter. If not, the SGSN should initiate the PDP context modification procedure to adjust the QoS parameters of the context.

Steps 10–12 are executed if the new SGSN detects that the LA has been changed or the update type in Step 1 indicates combined RA/LA update with IMSI (CS) attach.

Step 10 (LA Update). Through a table lookup technique, the SGSN translates the *Routing Area Identifier (RAI)* into the VLR number and sends the **Location Update Request** message to the VLR. The VLR creates or updates the SGSN-VLR association by storing the SGSN number.

Step 11. The standard GSM location update procedure is performed. The details can be found in Section 9.2.

Step 12. The new VLR allocates a new TMSI and responds with **Location Update Accept** to the SGSN. TMSI allocation is optional if the VLR is not changed.

Step 13. The new SGSN sends the **Routing Area Update Accept** message to the MS. In GPRS, the new SGSN also confirms that all mobile-originated N-PDUs successfully transferred before the start of the update procedure.

Step 14. The MS sends the **Routing Area Update Complete** message to the new SGSN to confirm the reallocation of the TMSI. In GPRS, the MS also confirms all received mobile-terminated N-PDUs before the RA update procedure started. This information is used by the new SGSN to check whether the packets forwarded from the old SGSN have been received by the MS. If so, these redundant packets are discarded.

Step 15. If the MS receives a new TMSI, then it sends the **TMSI Realloca-tion Complete** message to the VLR.

In terms of RA update, the major differences between UMTS and GPRS are as follows:

- In GPRS, packet forwarding is performed between old and new SGSNs during RA update. In UMTS, packet forwarding is handled at the RNC level, and the SGSN is not involved.

- In the RA update, the UMTS MS may determine whether the Iu connection should be maintained (see Step 1 in Figure 2.7), which is not needed in GPRS.

Note that for a pure intra-SGSN RA update, Steps 2–12, 14, and 15 in Figure 2.7 are not executed. For a pure inter-SGSN RA update, Steps 10–12 and 15 are not executed.

2.7   Serving RNC Relocation

Like GPRS, packets are routed between the MS and the GGSN in UMTS. An example of the routing path is illustrated in Figure 2.8 (a). In this figure, the MS communicates with two Node Bs (B1 and B2). In WCDMA [3GP04e], an MS is allowed to transmit signals through multiple radio paths connected to different Node Bs, and the signals are merged in a network node (RNC1 in Figure 2.8 (a)). In a packet routing path between the core network and the MS, the RNC that directly connects to the SGSN is called the *Serving RNC (SRNC)*. In Figure 2.8 (a), RNC1 is the serving RNC. If the MS moves during packet transmission, the packet routing path may be changed. In Figure 2.8, when the MS moves toward Node B3, the radio link between the MS and B1 is removed due to radio path loss, and the radio link between B3 and the MS is added (Figure 2.8 (b)). In this case, B3 is connected to RNC2, and an Iur link between RNC1 and RNC2 is established so that the signal received by B3 can be forwarded to RNC1 through RNC2. RNC1 then combines the signals from B2 and B3, and forwards it to SGSN1. In this case, RNC1 is the SRNC, and RNC2 is called the *Drift RNC (DRNC)*. The DRNC transparently routes the data through the Iub and the Iur interfaces, and only performs Layer 1 and partial Layer 2 functionality (for example, MAC for common and shared channels). Thus, in Figure 2.8 (b), the RLC connections are defined between the SRNC and the MS, and the DRNC is bypassed.

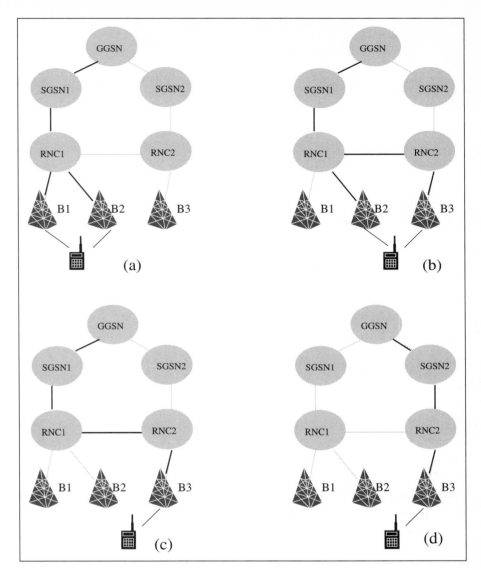

Figure 2.8 PS SRNC Relocation

Suppose that the MS moves away from B2, and the radio link between the MS and B2 is disconnected. In this case, the MS does not communicate with any Node Bs connected to RNC1. The routing path is now MS↔ B3 ↔ RNC2↔RNC1↔ SGSN1↔GGSN as shown in Figure 2.8 (c). In this case, it does not make sense to route packets between the MS and the core network through RNC1.

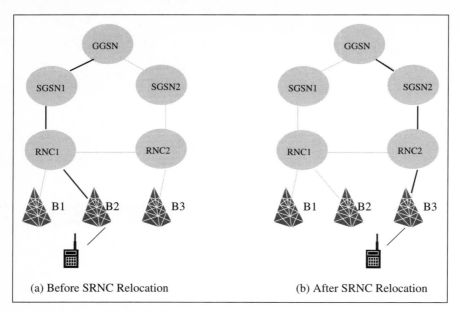

(a) Before SRNC Relocation   (b) After SRNC Relocation

Figure 2.9 Combined Hard Handoff with SRNS Relocation

SRNC relocation may be performed to remove RNC1 from the routing path. Suppose that RNC2 connects to SGSN2. Then, after RNC relocation, the packets are routed to the GGSN through RNC2 and SGSN2 (Figure 2.8 (d)). At this point, RNC2 becomes the serving RNC. SRNC relocation may also be executed when hard handoff occurs. As shown in Figure 2.9 (a), before the relocation, the communication path is GGSN $\leftrightarrow$ SGSN1 $\leftrightarrow$ RNC1 $\leftrightarrow$ B2 $\leftrightarrow$ MS, and the MS is not connected to any Node Bs of RNC2. During hard handoff and SRNC relocation, the radio link connected to the MS is switched from B2 to B3. After the relocation, the communication path is GGSN $\leftrightarrow$ SGSN2 $\leftrightarrow$ RNC2 $\leftrightarrow$ B3 $\leftrightarrow$ MS.

The SRNC relocation procedures for PS and CS services are different. Figure 2.10 illustrates the CS connection before and after SRNC relocation. Before relocation, the call path is MSC1 $\leftrightarrow$ RNC1 $\leftrightarrow$ RNC2 $\leftrightarrow$ MS. After relocation, the call path is MSC1 $\leftrightarrow$ MSC2 $\leftrightarrow$ RNC2 $\leftrightarrow$ MS, and MSC1 becomes the *anchor* MSC.

The SRNC relocation procedure for PS is illustrated in Figure 2.11. This procedure is only performed for an MS in the **PMM-CONNECTED** state. The details are given here:

Step 1. RNC1 determines that RNC2 is the target for relocation. RNC1 informs SGSN1 of this decision through the **Relocation Required** message.

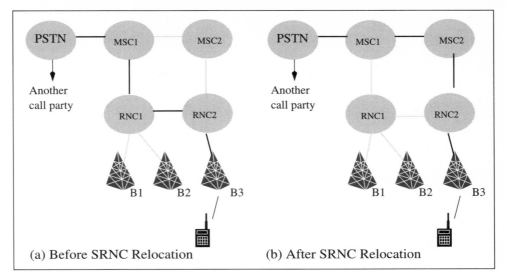

Figure 2.10 CS SRNC Relocation

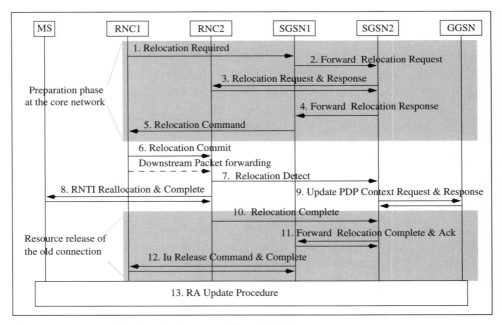

Figure 2.11 SRNC Relocation Message Flow for PS

Step 2. If both RNC2 and RNC1 are connected to SGSN1 (intra-SGSN relocation), Steps 2–4 are skipped, and the relocation procedure proceeds to Step 5. Otherwise, it is an inter-SGSN SRNC relocation. In this case, suppose that RNC2 connects to SGSN2. Then SGSN1 sends the MM and the PDP contexts of the MS to SGSN2 by using the **Forward Relocation Request** message.

Step 3. SGSN2 and RNC2 exchange the **Relocation Request** and **Response** message pair to establish the Iu user plane transport bearers between SGSN2 and RNC2, and exchange routing information required for packet delivery.

Step 4. SGSN2 sends the **Forward Relocation Response** message to SGSN1. The message indicates that SGSN2 and RNC2 are ready to receive the downstream packets buffered in RNC1 (that is, the packets that have not been acknowledged by the MS).

Step 5. SGSN1 sends the **Relocation Command** message to RNC1. This message instructs RNC1 to forward the buffered downstream packets to RNC2.

Steps 1–5 reserve the core network resources for the new path. Before Step 6 is executed, the packets are routed through the old path.

Step 6. When RNC1 receives the **Relocation Command** message, it starts the *data-forwarding* timer. Expiration of this timer will be checked at Step 12. RNC1 sends the **Relocation Commit** message to RNC2, which provides information about the buffered packets (for example, sequence numbers) to be tunneled to RNC2. RNC1 stops exchanging packets with the MS, and forwards the buffered packets (which are sent from the GGSN) to RNC2.

Step 7. Immediately after RNC2 is successfully switched at Step 6, RNC2 sends the **Relocation Detect** message to SGSN2. The purpose of this message is to inform the SGSN2 that RNC2 is starting the SRNC operation, and the core network should switch the packet routing path from RNC1 to RNC2.

Step 8. RNC2 restarts the RLC connections. RNC2 and the MS exchange information to identify the last up-stream packets received by RNC2 and the last down-stream packets received by the MS. This is achieved by exchanging the **RNTI Reallocation** and **Complete** message pair. In the **RNTI Reallocation** message, RNC2 provides RA, LA, and possibly RRC information. Since the RA has been changed, the MS also triggers the RA update procedure at Step 13. After the MS has reconfigured itself, it sends the **RNTI Reallocation Complete** message to RNC2. Packet exchange between RNC2 and the MS can start. Two messages, **UTRAN Mobility Information** and **UTRAN Mobility Information Confirm**, can also be used by the UTRAN to allocate a new RNTI and to convey other UTRAN mobility-related information to an MS.

In Steps 6–8, the UTRAN connection point is moved from RNC1 to RNC2. In this period, packet exchange between the MS and network is stopped to ensure lossless relocation.

Step 9. After Step 7, SGSN2 switches the routing path from RNC1 to RNC2. For inter-SGSN SRNS relocation, SGSN2 and every corresponding GGSN exchange the **Update PDP Context Request** and **Response** message pair to modify the GGSN address, SGSN TEID, and QoS profile negotiated stored in the GGSN PDP context. This operation switches each GGSN connection from SGSN1 to SGSN2.

Step 10. After Step 8, RNC2 sends the **Relocation Complete** message to SGSN2. This message triggers resource release of the old Iu connection. For intra-SGSN SRNC relocation, the procedure proceeds to Step 12 to release the old Iu connection. For inter-SGSN SRNS relocation, Step 11 is executed to release the old Iu connection.

Step 11. SGSN2 instructs SGSN1 to release the old Iu connection by exchanging the **Forward Relocation Complete** and **Acknowledge** message pair.

Step 12. SGSN1 sends the **Iu Release Command** message to RNC1. When the data-forwarding timer set in Step 6 expires, RNC1 returns the **Iu Release Complete** message to SGSN1, and the old Iu connection is released.

Step 13. The RA update procedure described in Section 2.6 is triggered by the MS after Step 8 is executed.

For combined hard handoff with SRNC relocation, the message flow is similar to the one in Figure 2.11, with the following differences: The SRNC relocation procedure in Figure 2.11 is initiated by RNC1 without involving the MS. For combined hard handoff with SRNC relocation, at the beginning, RNC1 decides that the MS should be involved, and the MS reconfigures the physical channel immediately after Step 5. Thus, RNTI relocation at Step 8 is not needed. Also, in the combined procedure, the SRNC context (of RNC1) must be forwarded through the path SGSN1↔ SGSN2 ↔ RNC2.

2.8   UMTS-GPRS Intersystem Change

When a GPRS/UMTS dual-mode MS moves from a cell supporting GSM/GPRS radio technology to a cell supporting WCDMA radio technology (or vice versa), a UMTS-GPRS intersystem change may take place. To provide this feature, mechanisms should exist to derive the area identities (for LA, RA, and cell) and the routing-related information from one system to another.

In this section, we describe UMTS-GPRS intersystem change using simple examples in which the GSM/GPRS cells and the UMTS cells are connected to the same SGSN. In this case, the SGSN supports both the Gb and Iu-PS interfaces.

2.8.1   SGSN Change from UMTS to GPRS

For SGSN change from UMTS to GPRS, if the MS is in the **PMM-IDLE** state, then the normal GPRS RA update procedure is executed. If the MS makes the intersystem change decision when it is in the **PMM-CONNECTED** state, then it stops the transmission to the network, and the following steps are executed for intra-SGSN change (see Figure 2.12):

Step 1. An LLC link is established between the MS and the SGSN. The MS sends the **Routing Area Update Request** message to the SGSN through the new BSS. This step is exactly the same as Step 1 of Figure 2.7, initiated by a GPRS MS.

Step 2. The SGSN exchanges the **SRNS Context Request** and **Response** messages with the old SRNS to obtain the following information:

- GTP-SND and GTP-SNU are used to resume transmission to the GGSN.

- PDCP-SNU is used to resume transmission to the MS for lossless relocation.

The SGSN converts the UMTS PDCP sequence number into the GPRS SNDCP sequence number and saves it in the GPRS PDP context. The SRNS stops sending packets to the MS, and starts buffering the packets received from the GGSN.

Step 3. Security functions may be executed as in Step 3 of Figure 2.7. If the MS is not allowed to attach in the RA, or if subscription checking fails, then the SGSN rejects this RA update.

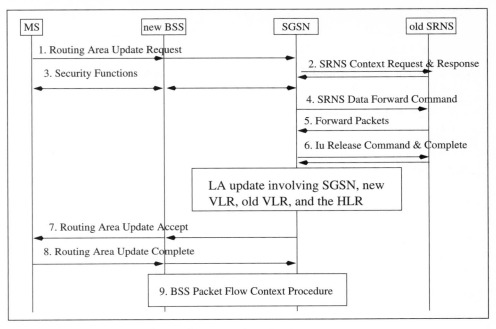

Figure 2.12 Intra-SGSN Change from UMTS to GPRS

Step 4. The SGSN is ready to receive packets. The SGSN sends the **SRNS Data Forward Command** message to the old SRNS, which instructs the SRNS to forward the buffered packets to the SGSN. The SRNS starts the *data-forwarding timer*. Before this timer expires, the Iu connection between the SRNS and the SGSN will be maintained (see Step 6).

Step 5. Packets received by the old SRNS from the SGSN but not yet sent to the MS are tunneled back from the SRNS to the SGSN.

Step 6. When the SGSN timer set at Step 4 expires, the **Iu Release Command** and **Complete** messages are exchanged to release the Iu connection. If the type parameter in the **Routing Area Update Request** message at Step 1 is combined RA/LA update (for Network Mode I), or if the LA has been changed, then the SGSN triggers LA update (see Steps 10–12 in Figure 2.7) that involves the SGSN, the new VLR, the old VLR and the HLR.

Step 7. The SGSN updates the MM and the PDP contexts. New P-TMSI and new TMSI may be allocated. The SGSN sends the **Routing Area Update Accept** message to the MS.

Step 8. The MS returns the **Routing Area Update Complete** message to the SGSN if a new P-TMSI is allocated or if the MS needs to acknowledge the packets received from the network. If a new TMSI is allocated to the MS, then the SGSN sends the **TMSI Reallocation Complete** message to the new VLR (this message is not shown in Figure 2.12).

Step 9. The SGSN and the BSS execute the BSS packet flow context procedure if no BSS packet flow context exists in the BSS. In this context, the SGSN provides the BSS with information related to ongoing user data transmission.

2.8.2 SGSN Change from GPRS to UMTS

For SGSN change from GPRS to UMTS, if the MS is in the **STANDBY** state, then the normal UMTS RA update procedure is executed. If the MS makes the intersystem change decision when it is in the **READY** state, then it stops the transmission to the network by disconnecting the LLC link. The following steps are executed for intra-SGSN change (see Figure 2.13):

Step 1. The MS establishes an RRC connection to the new SRNS, and sends the **Routing Area Update Request** to the SGSN through the SRNS.

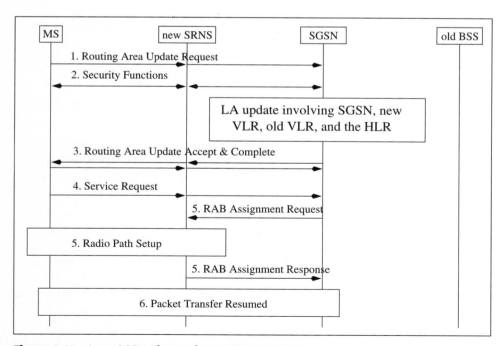

Figure 2.13 Intra-SGSN Change from GPRS to UMTS

Step 2. The SGSN stops the transmission to the old BSS. The security functions may be executed among the SGSN, the SRNS, and the MS. If the type parameter in the **Routing Area Update Request** message at Step 1 is combined RA/LA update (for Network Mode I), or if the LA has been changed, then the SGSN triggers LA update (see Steps 10–12 in Figure 2.7) that involves the SGSN, the new VLR, the old VLR and the HLR.

Step 3. The SGSN updates the MM and the PDP contexts for the MS. A new P-TMSI may be allocated. The SGSN sends the **Routing Area Update Accept** message to the MS. Reception of the new P-TMSI is acknowledged by the MS through the **Routing Area Update Complete** message. If a new TMSI is allocated to the MS, then the SGSN sends the **TMSI Reallocation Complete** message to the new VLR (this message is not shown in Figure 2.13).

Step 4. If the MS has pending uplink data or signaling, it sends the **Service Request** message to the SGSN.

Step 5. The SGSN requests the SRNS to set up the radio bearer between the SRNS and the MS. The N-PDU sequence numbers in the GPRS PDP context of the SGSN are used to derive the PDCP sequence numbers for the next packets to be delivered in the UTRAN radio bearer.

Step 6. Packet transmission is resumed among the SGSN, the SRNS, and the MS.

A major difference between the message flows in figures 2.12 and 2.13 is that packet forwarding is not required in intra-SGSN change from GPRS to UMTS. The reason is that in GPRS, the packets are buffered in SGSN. For inter-SGSN change from GPRS to UMTS, packet forwarding will occur from the old SGSN to the new SGSN. The details can be found in [3GP05q].

2.9 Concluding Remarks

Based on [Lin 01c] and the 3GPP Technical Specification (TS) 23.060 [3GP05q], this chapter describes mobility management evolution from GPRS to UMTS. In the architecture evolution from GPRS to UMTS, the radio access network UTRAN is introduced. Most radio management functions handled in the GPRS core network have been moved to UTRAN in UMTS. This architectural change results in a clean design that allows radio technology and core network technology to develop independently. The

GPRS mobility management functionality has been significantly modified to accommodate UMTS. This chapter emphasized the differences between the GPRS and the UMTS procedures.

For further reading, details of the UMTS core network architecture can be found in [3GP02c]. The UMTS protocol stacks are completely described in [3GP05q]. Details for GSM and GPRS mobility management can be found in Chapter 11 in [Lin 01b] (see also **http://liny.csie.nctu.edu.tw/supplementary**). The complete 3GPP specifications can be found at www.3gpp.org.

2.10   Questions

1. Why is RA introduced in GPRS and UMTS? Can we eliminate the concept of RA so that both SGSN and VLR track LA?

2. Consider the RRC state machine in Figure 2.4. In the **Cell Connected Mode**, how does the machine know when to move to **Idle Mode**, and when to move to **URA Connected Mode**?

3. An MS in **RRC URA Connected Mode** is tracked by the RNC when no data is actively transferred, and the probability of data transfer is quite high. An MS in **RRC Idle Mode** is tracked by the SGSN, where the probability of data transfer is quite low. The MS moves from the **RRC Connected Mode** to the **RRC Idle Mode** when the RRC-connection-release timer expires. Do optimal values of the timeout period, RA size, and URA size exist that minimize the net cost of paging and location update?

4. What is the purpose of periodic location update? How do you set the periodic RA update timer?

5. Why is P-TMSI introduced in GPRS and UMTS? Can we replace P-TMSI by TMSI?

6. We say that radio resources are managed by UTRAN nodes (that is, RNCs) instead of the core network nodes (that is, SGSNs). However, SGSN is responsible for the RAB management (RAB assignment is initiated by SGSN). Why? (Hint: See p. 90, [3GP05k].)

7. In the MM state machine, why is the transition **STANDBY → READY** (**PMM-IDLE → PMM-CONNECTED**) always initiated by the MS? Can SGSN initiate this transition?

8. The radio resource information for SMS is kept in the MM context, while the radio resource information for user data is maintained in the PDP context. Why?

9. In Network Mode II, the RA update (to the SGSN) and the LA update (to the VLR) are performed separately. In this case, why is the LA update always performed before the RA update? (Hint: The sizes of LAs are larger than those of RAs.)

10. DRX usage is independent of the MM states. However, during the GPRS attach and RA update, the GPRS MS will not apply DRX in the READY state. Why?

11. Draw the message flows for intra-SGSN RA update and inter-SGSN RA update.

12. Draw the message flow for inter-SGSN SRNC relocation.

13. Draw the message flow for combined hard handoff with SRNC relocation.

14. What is the purpose of the data-forwarding timer?

15. Describe the role of the SGSN for the items listed in Table 4.1 of Chapter 4.

## CHAPTER 3

Session Management for Serving GPRS Support Node

General Packet Radio Service (*GPRS*) provides efficient integrated Internet access for wireless networks [ETS 00b]. With GPRS, the wireless data services can be directly developed based on the Internet protocol (IP) paradigm. GPRS has also evolved into UMTS to support IP multimedia applications. Figure 3.1 (a) illustrates the GPRS architecture, where the Serving GPRS Support Nodes (SGSNs) and the Gateway GPRS Support Nodes (GGSNs) interact to establish an Internet connection to the external Packet Data Network. In this figure, the Base Station Subsystem (BSS) interacts with the SGSN and the Mobile Station (MS) for radio resource allocation. In this chapter, we describe the *session management* functions performed at the SGSN. In Chapter 4, we will elaborate on the session management functions at the GGSN. Figure 3.1 (b) shows the protocol stack of the SGSN, in which the *Service Access Points* (*SAPs*) are marked with circles. At each SAP, services are provided or received by exchanging corresponding sequences of primitives. The protocols in the figure are described as follows:

- The GPRS Tunneling Protocol (GTP) tunnels signaling messages and *Packet Data Units* (*PDUs*) between the SGSN and the GGSN [ETS 00c].

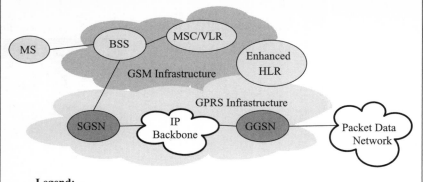

Legend:

MS: Mobile Station
BSS: Base Station System
MSC: Mobile Switching Center
SGSN: Serving GPRS Support Node

VLR: Visitor Location Register
HLR: Home Location Register
GGSN: Gateway GPRS Support Node

(a) System Architecture.

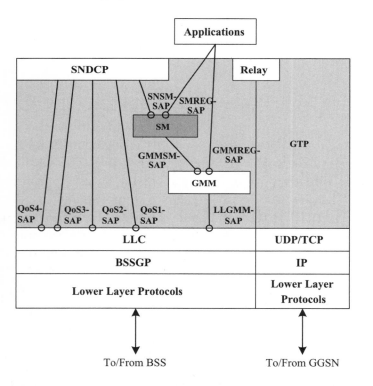

To/From BSS   To/From GGSN

Legend:

BSSGP: Base Station System GPRS Protocol
LLC: Logical Link Control
GMM: GPRS Mobility Management
SAP: Service Access Point

SM: Session Management
SNDCP: SubNetwork Dependent
          Convergence Protocol
GTP: GPRS Tunneling Protocol

(b) SGSN Protocol Architecture.

Figure 3.1 GPRS Network and SGSN Protocol Stack

- UDP/IP is used as an unreliable path to transfer signaling messages and to tunnel connectionless PDUs. On the other hand, TCP/IP is used as a reliable path to tunnel connection-oriented PDUs.

- The Relay transfers the signaling messages between the GTP and the *GPRS Mobility Management (GMM)/Session Management (SM)*. In addition, it transfers the PDUs between the GTP and the *Sub-Network Dependent Convergence Protocol (SNDCP)* [3GP04b].

- SNDCP ensures that all functions related to network-layer PDU transfer are carried out in a transparent way by the GPRS network entities. The SNDCP multiplexes the PDUs coming from different upper-layer protocol entities (for example, IP or X.25) so that the PDUs can be sent using the same service provided by the *Logical Link Control (LLC)* layer.

- The GMM supports attach, detach, security, and location management functions. It provides the GMMREG-SAP for the applications to invoke the network-initiated detach procedure, and the GMMSM-SAP for the SM to send/receive SM signaling messages to/from the LLC layer [3GP05h].

- The LLC (for session management) provides one or more ciphered logical link connections with sequence control, flow control, detection and recovery of transmission, format and operational errors. The LLC link is maintained as the MS moves between cells served by the same SGSN. When the MS moves to a new SGSN, the existing connection is released, and a new LLC connection is established with the new SGSN.

- The *BSS GPRS Protocol (BSSGP)* conveys QoS, radio, and routing-related information that is required to transmit PDUs between the BSS and the SGSN [ETS 00a]. BSSGP supports three service models: The *Relay service model* controls the transfer of the LLC frames. The *GMM service model* performs mobility management functions between an SGSN and a BSS. The *network management service model* handles functions related to BSS/SGSN node management.

The SM supports *Packet Data Protocol (PDP)* context handling of a GPRS-attached MS. When the GPRS-attached MS attempts to send or receive data, a PDP context with a specific *Quality of Service (QoS)* profile between the MS and the GGSN is established. The SM provides the SMREG-SAP for the applications to invoke the SM function. Moreover, through the SNSM-SAP, the SM informs the SNDCP about the PDP context information that is, negotiated QoS profile, the activated SNDCP *Network SAP Identifier (NSAPI)*, and the

LLC SAPI assigned for this NSAPI, for transmitting PDUs. The SNDCP maps network-level characteristics onto that of the underlying network. It supports compression/decompression and segmentation/reassembly of PDUs. The LLC supports a highly reliable ciphered logical link between the MS and the SGSN. It provides four SAPs (that is, QoS1–QoS4) for the SNDCP to carry PDUs with specific QoS requirements. It also provides the LLGMM-SAP to transfer mobility management signaling messages between the MS and the SGSN.

3.1   Session Management Functions

The *Session Management (SM)* functions consist of PDP context activation, deactivation, and modification procedures [ETS 04]. The PDP context activation procedure is used to establish a PDP context with a specific QoS profile. This procedure is initiated by the MS or is requested by the GGSN. The PDP context deactivation procedure deactivates an existing PDP context. This procedure is initiated by the MS or by the network (that is, SGSN or GGSN). The PDP context modification procedure is invoked by the SGSN in order to modify PDP context parameters (for example, QoS profile and radio priority) that were negotiated during the PDP context activation procedure or during the previously performed PDP context modification procedures.

3.1.1   PDP Context Activation

The PDP context can be activated by the MS or the GGSN. The procedures are discussed in this subsection.

The message flow of the MS-requested PDP context activation procedure is shown in Figure 3.2 (a), and is described in the following steps [ETS 04]:

Step 1. The MS sends the **Activate PDP Context Request** message with the requested PDP type, PDP address (optional), *Access Point Name (APN)*, and QoS profile to the SGSN. Note that the MS leaves the PDP address empty to request a dynamic PDP address.

Step 2. The GMM security functions may be executed for authentication (see Section 9.1) and ciphering.

Step 3. The SGSN derives the GGSN address according to the APN selection rules (see Section 4.1). If no GGSN address can be derived, then the SGSN rejects the activation request by sending the **Activate PDP Context Reject** message to the MS. If a GGSN address can be derived, the SGSN sends the **Create PDP Context Request** message with the

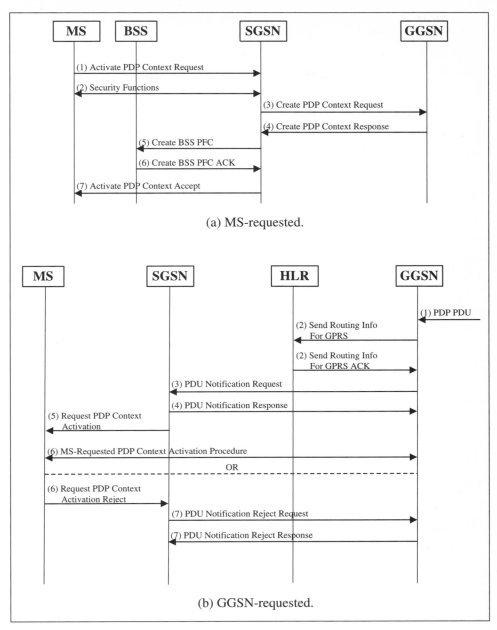

(a) MS-requested.

(b) GGSN-requested.

Figure 3.2 PDP Context Activation

requested PDP type, PDP address (optional), APN, and QoS profile to the GGSN [ETS 00c]. If the MS requests a dynamic address, then the GGSN is responsible for allocating the PDP address.

Step 4. The GGSN uses the APN to find an external network. In addition, the GGSN may restrict the negotiated QoS profile given its capabilities and the current load. The GGSN then returns the **Create PDP Context Response** message to the SGSN to indicate whether a PDP context has been created in the GGSN.

Step 5. The SGSN derives the *Aggregate BSS QoS Profile (ABQP)* from the negotiated QoS profile, and then sends the **Create BSS PFC** (Packet Flow Context) message with the requested ABQP to the BSS [ETS 00a]. Note that the ABQP defines the QoS that must be provided by the BSS for a given packet flow between the MS and the SGSN.

Step 6. The BSS may restrict the requested ABQP given its capabilities and the current load. If the PFC creation request is accepted, then the BSS returns the **Create BSS PFC Ack** message to the SGSN.

Step 7. The SGSN selects radio priority based on the negotiated QoS profile, and returns the **Activate PDP Context Accept** message to the MS. At this point, the SGSN is ready to route the PDUs between the MS and the GGSN.

The message flow of the GGSN-requested PDP context activation procedure is shown in Figure 3.2 (b), and is described as follows:

Step 1. Upon receipt of a PDP PDU, the GGSN determines whether the activation procedure has to be initiated.

Step 2. The GGSN sends the **Send Routing Info for GPRS** message to the HLR to request the routing information. If the HLR determines that the request can be served, it returns the **Send Routing Info for GPRS Ack** message with the SGSN address to the GGSN.

Steps 3 and 4. The GGSN sends the **PDU Notification Request** message with the requested PDP type and PDP address to the SGSN. The SGSN returns the **PDU Notification Response** message to inform the GGSN whether the activation will proceed or not.

Steps 5 and 6. If the activation will proceed, the SGSN sends the **Request PDP Context Activation** message to request the MS to activate the PDP context indicated with the PDP address [ETS 04]. Upon receipt of the **Request PDP Context Activation** message, the MS either activates the

PDP context with the MS-requested PDP context activation procedure or rejects the activation request by sending the **Request PDP Context Activation Reject** message to the SGSN.

Step 7. If the MS does not respond or refuses the activation request, the SGSN sends the **PDU Notification Reject Request** message with cause "MS not responding" or "MS refusing" to the GGSN [ETS 00c]. The GGSN returns the **PDU Notification Reject Response** message to the SGSN.

3.1.2 PDP Context Deactivation

The message flow of the MS-initiated PDP context deactivation procedure is shown in Figure 3.3 (a), and is described in the following steps [ETS 04]:

Step 1. The MS sends the **Deactivate PDP Context Request** message to the SGSN.

Step 2. The GMM security functions may be executed for authentication and ciphering.

Steps 3 and 4. The SGSN sends the **Delete PDP Context Request** message to the GGSN [ETS 00c]. The GGSN removes the PDP context and returns the **Delete PDP Context Response** message to the SGSN.

Step 5. The SGSN sends the **Delete BSS PFC** message to the BSS [ETS 00a].

Step 6. The BSS deletes the corresponding PFC and returns the **Delete BSS PFC Ack** message to the SGSN.

Step 7. The SGSN returns the **Deactivate PDP Context Accept** message to the MS.

The message flow of the SGSN-requested PDP context deactivation procedure is shown in Figure 3.3 (b). Steps 1–4 are the same as Steps 3–6 described in Figure 3.3 (a). In Step 5, the SGSN sends the **Deactivate PDP Context Request** message to the MS. The MS then returns the **Deactivate PDP Context Accept** message to the SGSN in Step 6.

The message flow of the GGSN-requested PDP context deactivation procedure is shown in Figure 3.3 (c). In Step 1, the GGSN sends the **Delete PDP Context Request** message to the SGSN. Steps 2–5 are the same as Steps 3–6 described in Figure 3.3 (b). In Step 6, the SGSN returns the **Delete PDP Context Response** message to the GGSN. The GGSN then releases the

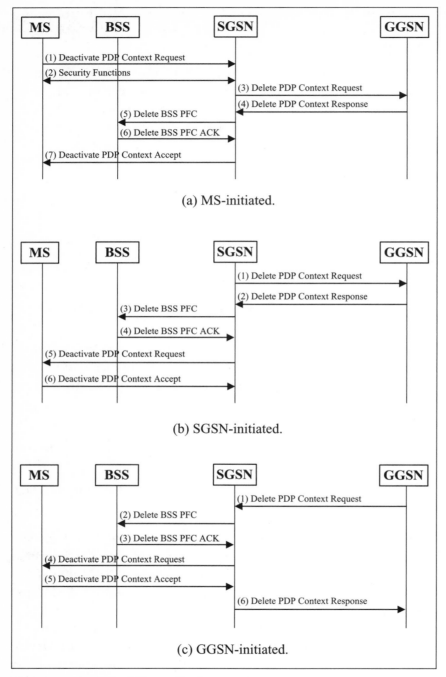

(a) MS-initiated.

(b) SGSN-initiated.

(c) GGSN-initiated.

Figure 3.3   PDP Context Deactivation

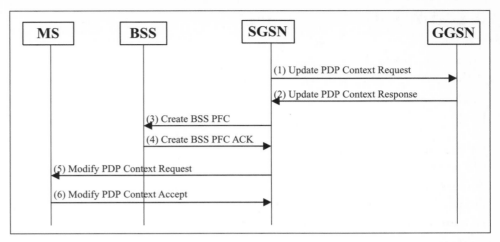

Figure 3.4   PDP Context Modification

network resources (for example, IP address and bandwidth) and makes them available for subsequent activation by other MSs.

3.1.3   PDP Context Modification

The message flow of the PDP context modification procedure is shown in Figure 3.4, and is described as follows:

Steps 1 and 2. The SGSN sends the **Update PDP Context Request** message with the QoS profile requested to the GGSN [ETS 00c]. The GGSN may restrict the requested QoS profile given its capabilities and the current load, and returns the **Update PDP Context Response** message (with the negotiated QoS profile) to the SGSN.

Steps 3 and 4. The SGSN sends the **Create BSS PFC** message with the requested ABQP to the BSS. If the PFC already exists, the BSS interprets the message as a modification request. If the PFC modification request is accepted, then the BSS returns the **Create BSS PFC Ack** message to the SGSN [ETS 00a].

Steps 5 and 6. The SGSN sends the **Modify PDP Context Request** message with the negotiated QoS profile, radio priority, and LLC SAPI to the MS [ETS 04]. The MS acknowledges by returning the **Modify PDP Context Accept** message to the SGSN. If the MS does not accept the negotiated QoS profile, it deactivates the PDP context with the MS-initiated PDP context deactivation procedure.

3.2   SM Software Architecture

Figure 3.5 shows the SM software architecture for the SGSN. The *SM-Core* is the kernel of the SM software, which consists of functions that interact with the SM components. Upon receipt of an invocation from an SM component, the SM-Core dispatches the corresponding SM components to complete the tasks triggered by that invocation. The interaction between an SM component (for example, SNSM) and a protocol entity (for example, SNDCP) takes place when primitives are sent across an SAP (for example, SNSM-SAP). The standardized primitives of an SAP allow independent implementations of the SM components, thereby achieving the modularity goal. In addition, the standardized GPRS SM primitives guarantee compatibility with various versions of the GPRS standards.

The design of the primitives for the SM components follows the primitive flow model shown in Figure 3.6. A primitive can be one of the following types:

- REQ (Request)
- IND (Indication)
- RSP (Response)
- CNF (Confirm)/REJ (Reject)

The REQ primitive is initiated by the *service user* of a protocol entity called the *initiator*. The *service provider* of the initiator then delivers the request to the corresponding protocol entity called the *responder*. When the service provider of the responder receives the request, it informs the service user by issuing the IND primitive. The service user of the responder then sends the RSP primitive to the service provider. After the service provider of the initiator receives this response, it issues the CNF primitive to confirm the request. Descriptions of the SM components are given in the following section.

3.2.1   SM REGistration

The *SM REGistration* (*SMREG*) provides the primitives at the SMREG-SAP that interfaces the SM software with the SGSN applications (such as PDP context modification and deactivation). For these primitives, the SM software is the service provider, and the applications are service users. The SMREG primitive types can be REQ or CNF, and are described as follows:

- **SMREG-PDP-DEACTIVATE:** The REQ primitive initiates a PDP context deactivation procedure. The CNF primitive confirms that the PDP context deactivation procedure has been concluded.

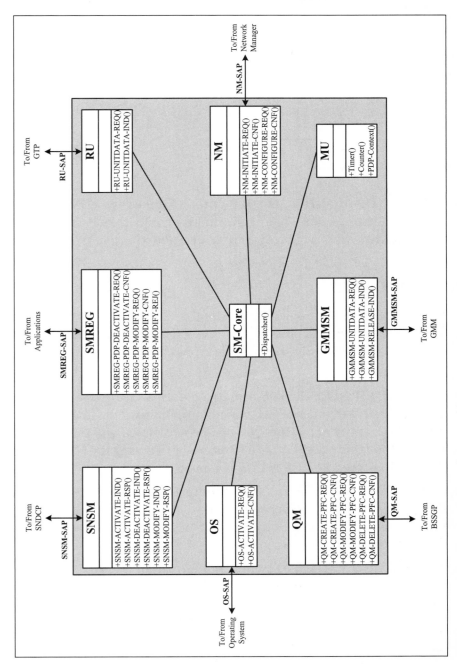

Figure 3.5 Session Management Software Architecture

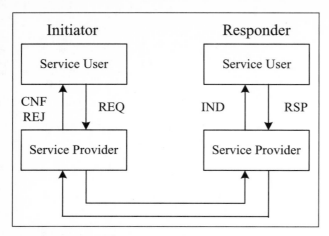

Figure 3.6   Primitive Flow Model

- **SMREG-PDP-MODIFY:** The REQ primitive initiates a PDP context modification procedure. The CNF primitive confirms that the PDP context modification procedure has been concluded.

3.2.2 SNDCP SM

The *SNDCP SM* (*SNSM*) provides the primitives at the SNSM-SAP that interfaces the SM software (that is, service provider) with the SNDCP entity (that is, service user). The SNSM primitive types can be IND or RSP, and are described as follows:

- **SNSM-ACTIVATE:** The IND primitive informs the SNDCP entity that an SNDCP NSAPI has been activated for data transfer. It also informs the SNDCP entity about the QoS profile negotiated and the LLC SAPI assigned for this NSAPI. The RSP primitive informs the SM software that the indicated SNDCP NSAPI is now in use.

- **SNSM-DEACTIVATE:** The IND primitive informs the SNDCP entity that an SNDCP NSAPI has been de-allocated and cannot be used anymore. The RSP primitive informs the SM software that the indicated SNDCP NSAPI is no longer in use.

- **SNSM-MODIFY:** The IND primitive informs the SNDCP entity about the change of the QoS profile for an SNDCP NSAPI and the LLC SAPI assigned for this NSAPI. It also indicates that an NSAPI shall be created, together with the re-negotiated QoS profile and the assigned SAPI. The RSP primitive informs the SM software that the indicated SNDCP NSAPI and the negotiated QoS profile are now in use.

3.2.3 Relay Unit

The *Relay Unit* (RU) transfers the signaling messages between the SM software (that is, service user) and the GTP entity (that is, service provider). The RU primitive types can be REQ or IND, and are described as follows:

- **RU-UNITDATA:** The REQ primitive requests the GTP entity to forward a GTP message to the GGSN. With the IND primitive, the GTP entity forwards a GTP message (sent from the GGSN) to the SM software.

3.2.4 QoS Manager

The *QoS Manager* (*QM*) acts as the admission controller and the resource manager. The QM-SAP provides an interface for the SM software (that is, service user) to invoke the BSSGP entity (that is, service provider) to create, modify, or delete the BSS PFC. The QM primitive types can be REQ or CNF, and are described as follows:

- **QM-CREATE-PFC:** The REQ primitive initiates the BSS PFC creation. The CNF primitive indicates whether the requested PFC has been created in the BSS or not.
- **QM-DELETE-PFC:** The REQ primitive initiates the BSS PFC deletion. The CNF primitive informs the SM software whether the indicated PFC has been deleted in the BSS.
- **QM-MODIFY-PFC:** The REQ primitive initiates the BSS PFC modification. The CNF primitive indicates whether the PFC modification request is accepted by the BSS or not.

3.2.5 GMM SM

The *GMM SM* (*GMMSM*) provides the primitives at the GMMSM-SAP that interfaces the SM software (that is, service user) with the GMM entity (that is, service provider). The GMMSM primitive types can be REQ or IND, and are described as follows:

- **GMMSM-UNITDATA:** The REQ primitive requests the GMM entity to forward the SM message to the MS. Upon receipt of the SM message from the MS, the GMM entity forwards it to the SM software by using the IND primitive.
- **GMMSM-RELEASE:** The IND primitive informs the SM software that the MS has been detached, and therefore the PDP contexts are not valid anymore.

3.2.6 Maintenance Unit and Operating Service

The *Maintenance Unit* (*MU*) is responsible for maintaining the SM parameters (that is, timers and retransmission counters). It is also responsible for creating and maintaining the PDP contexts. The *Operating Service* (*OS*) encapsulates all operating system dependencies of the SM software. The operating system functions (for example, memory and queue management) required by the SM software are provided by the OS-SAP. The OS primitive types can be REQ or CNF, and are described as follows:

- **OS-ACTIVATE:** The REQ primitive activates the operating system functions. The CNF primitive is used by the operating system to confirm the activation procedure.

3.2.7 Network Manager

The *Network Manager* (*NM*) manages the SM software configuration parameters. The NM primitive types can be REQ or CNF, and are described as follows:

- **NM-INITIATE:** The REQ primitive initiates the SM software. With the CNF primitive, the SM software confirms the initiation procedure.
- **NM-CONFIGURE:** The REQ primitive configures specific components of the SM software. With the CNF primitive, the SM software confirms the configuration procedure.

3.2.8 SM for UMTS

The SM design described in this section can be extended for UMTS SM software architecture. Figure 3.7 shows the SGSN protocol architecture for UMTS [3GP05g]. The SM entity interacts with the GTP entity through the Relay, uses services provided by the GMM entity through the GMMSM-SAP, and provides services to the applications and the *Radio Access Bearer Manager* (*RABM*) entity through the SMREG-SAP and the RABMSM-SAP, respectively. The Relay, GMMSM-SAP, and SMREG-SAP are identical to that for GPRS SM. Only the RABMSM-SAP is different. Thus, to extend our SM software architecture for UMTS SM, we need to develop a new RABMSM component to replace the SNSM component in the SM software. The SM-Core and the NM component need to be modified slightly to accommodate the RABMSM component. In addition, the QM component needs to be

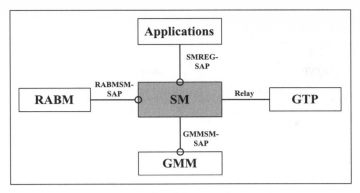

Figure 3.7   SGSN Protocol Architecture for UMTS

modified to accommodate the *Radio Access Network Application Part* (*RANAP*) entity. In UMTS, the RANAP [3GP05k], instead of BSSGP, is used to convey QoS-related information between the *Radio Network Subsystem* (*RNS*) and the SGSN. Other SM components are not affected by the introduction of the RABMSM component and the RANAP entity.

3.3   SM Software Initiation and Configuration

With the initiation procedure, the network manager initiates the global data structures of the SM software and activates operating system functions. Through the configuration procedure, the specific components of the SM software can be configured at run time via the network manager interface (that is, NM-SAP). The details are elaborated as follows.

3.3.1   Initiation

The initiation procedure is illustrated in Figure 3.8 (a), and is described in the following steps:

Step 1. The network manager sends the NM-INITIATE-REQ primitive to the NM component to activate the initiation procedure. The NM then invokes the SM-Core.

Step 2. The SM-Core instructs all SM components to initiate the global data structures.

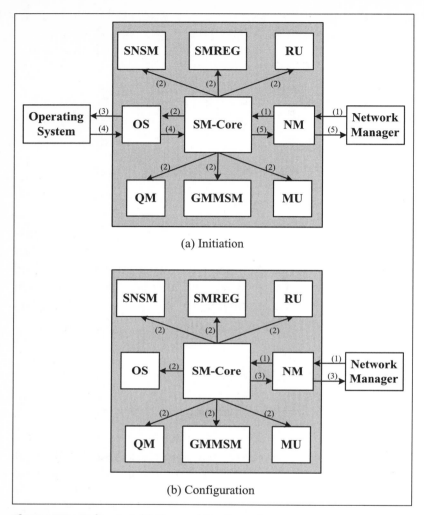

Figure 3.8   Software Initiation and Configuration

Step 3. The OS sends the OS-ACTIVATE-REQ primitive to the operating system to activate operating system functions.

Step 4. After completing the activation procedure, the operating system sends the OS-ACTIVATE-CNF primitive to the OS. The OS then informs the SM-Core of the completion of the activation procedure.

Step 5. The SM-Core dispatches the NM to send the NM-INITIATE-CNF primitive to the network manager, which indicates the completion of the initiation procedure.

3.3.2   Configuration

The configuration procedure is illustrated in Figure 3.8 (b), and is described as follows:

Step 1. To initiate the configuration procedure, the network manager sends the NM-CONFIGURE-REQ primitive to the NM component. The NM then invokes the SM-Core.

Step 2. The SM-Core instructs the related components to perform the configuration task. For example, when the network congestion situation occurs, the QM is invoked to reduce the maximum number of activated PDP contexts.

Step 3. After completing the configuration procedure, the SM-Core dispatches the NM to acknowledge the configuration procedure by sending the NM-CONFIGURE-CNF primitive to the network manager.

3.4   SM Procedures in the SGSN

Based on the SM software architecture, we show how the SGSN behaves in the PDP context activation, deactivation, and modification procedures.

3.4.1   PDP Context Activation

As shown in Figure 3.2, the PDP context activation procedure establishes a PDP context with a specific QoS profile between the MS and the GGSN. The procedure is requested by the MS or the GGSN.

Figure 3.9 illustrates the behavior of the SGSN in the MS-requested PDP context activation procedure. Each step is described as follows:

Step 1. The GMM entity sends the GMMSM-UNITDATA-IND primitive with the **Activate PDP Context Request** message (sent from the MS) to the GMMSM (see Step 1 in Figure 3.2 (a)). The GMMSM then informs the SM-Core that the MS is requesting the establishment of a PDP context.

Step 2a. The SM-Core derives the GGSN address according to the APN selection rules. If no GGSN address can be derived, the SM-Core rejects the PDP context activation by dispatching the GMMSM to send the GMMSM-UNITDATA-REQ primitive with the **Activate PDP Context Reject** message to the GMM entity. In this case, Steps 2b–6 are skipped.

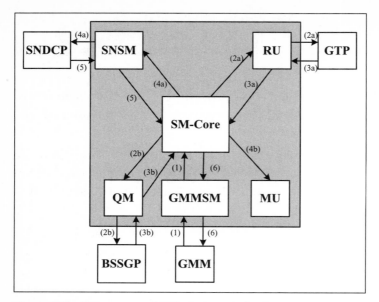

Figure 3.9 MS-requested PDP Context Activation

If a GGSN address can be derived, then the SM-core dispatches the RU to send the RU-UNITDATA-REQ primitive to the GTP entity. The GTP entity forwards the **Create PDP Context Request** message to the GGSN (see Step 3 in Figure 3.2 (a)).

Step 2b. The SM-Core dispatches the QM to perform admission control and resource allocation, to derive the ABQP, and then to send the QM-CREATE-PFC-REQ primitive to the BSSGP entity. The BSSGP entity forwards the **Create BSS PFC** message with the ABQP to the BSS (see Step 5 in Figure 3.2 (a)).

Step 3a. The GTP entity sends the RU-UNITDATA-IND primitive with the **Create PDP Context Response** message (sent from the GGSN) to the RU (see Step 4 in Figure 3.2 (a)). The RU then informs the SM-Core whether the PDP context activation request is accepted or not. If the PDP context activation request is not accepted, then the SM-Core dispatches the GMMSM to send the GMMSM-UNITDATA-REQ primitive with the **Activate PDP Context Reject** message to the GMM entity, and Steps 4b–6 are skipped.

Step 3b. The BSSGP entity sends the QM-CREATE-PFC-CNF primitive with the **Create BSS PFC Ack** message (sent from the BSS) to the QM (see Step 6 in Figure 3.2 (a)). The QM then informs the SM-Core that the PFC creation request is accepted. If the PFC creation request is not accepted (that is, a **Create BSS PFC Nack** message is received), then the SM-Core dispatches the GMMSM to send the GMMSM-UNITDATA-REQ primitive with the **Activate PDP Context Reject** message to the GMM entity, and Steps 4b–6 are skipped.

Step 4a. If the creation of the GGSN PDP context and the BSS PFC are accepted, then the SM-Core dispatches the SNSM to send the SNSM-ACTIVATE-IND primitive to inform the SNDCP entity that an SNDCP NSAPI has been activated for data transfer. If only the creation of the GGSN PDP context is successful, then the SM-Core dispatches the RU to send the RU-UNITDATA-REQ primitive to the GTP entity. The GTP entity forwards the **Delete PDP Context Request** message to the GGSN, which instructs the GGSN to release the reserved resources.

If only the creation of the BSS PFC is successful, then the SM-Core dispatches the QM to send the QM-DELETE-PFC-REQ primitive to the BSSGP entity. The BSSGP entity forwards the **Delete BSS PFC** message to the BSS, which requests the BSS to perform resource de-allocation.

Step 4b. The SM-Core dispatches the MU to create the PDP context.

Step 5. Upon receipt of the SNSM-ACTIVATE-IND primitive from the SNSM, the SNDCP entity sends the SNSM-ACTIVATE-RSP to the SNSM to inform the SM-Core that the indicated SNDCP NSAPI is now in use.

Step 6. The SM-Core dispatches the GMMSM to send the GMMSM-UNITDATA-REQ primitive to the GMM entity. The GMM entity forwards the **Activate PDP Context Accept** message to the MS (see Step 7 in Figure 3.2 (a)).

Figure 3.10 illustrates the actions of the SGSN in the GGSN-requested PDP context activation procedure. The steps are described as follows [ETS 04]:

Step 1. When the SGSN receives the **PDU Notification Request** message from the GGSN (see Step 3 in Figure 3.2 (b)), the GTP entity sends the RU-UNITDATA-IND primitive with the received message to the RU. The RU then informs the SM-Core that the PDP context activation is requested by the GGSN.

Step 2. If the SGSN has no information about that MS, the SM-Core dispatches the RU to send the RU-UNITDATA-REQ primitive to the GTP entity. The GTP entity forwards the **PDU Notification Response** message to the GGSN with cause "IMSI not known" or "MS detached," and the procedure exits.

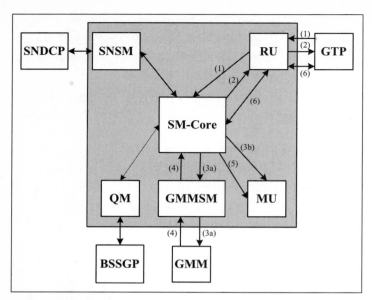

Figure 3.10   GGSN-requested PDP Context Activation

If the SGSN has information about that MS, the cause is "Activation proceeds" (see Step 4 in Figure 3.2 (b)) and Steps 3a–6 are executed.

Step 3a. The SM-Core dispatches the GMMSM to send the GMMSM-UNITDATA-REQ primitive to the GMM entity. The GMM entity forwards the **Request PDP Context Activation** message to the MS. This message requests the MS to activate the indicated PDP context (see Step 5 in Figure 3.2 (b)).

Step 3b. The SM-Core dispatches the MU to start the timer T3385 and increment the retransmission counter.

Step 4. When the SGSN receives the **Activate PDP Context Request** message or the **Request PDP Context Activation Reject** message from the MS (see Step 6 in Figure 3.2 (b)), the GMM entity sends the GMMSM-UNITDATA-IND primitive with the received message to the GMMSM. The GMMSM then informs the SM-Core whether the PDP context activation request is accepted or not.

Step 5. The SM-Core dispatches the MU to stop the timer T3385. If the message at Step 4 is **Activate PDP Context Request**, then Steps 2a–6 described in Figure 3.9 are executed. Otherwise, Step 6 is executed. Note that if the message indicated at Step 4 is not received within the T3385 timeout period, then Step 3a is repeated up to four times. On the fifth expiry of T3385, the SM-Core aborts the activation procedure.

Step 6. The SM-Core dispatches the RU to send the RU-UNITDATA-REQ primitive to the GTP entity. The GTP entity forwards the **PDU Notification Reject Request** message to the GGSN.

When the SGSN receives the **PDU Notification Reject Response** message from the GGSN (see Step 7 in Figure 3.2 (b)), the GTP entity sends the RU-UNITDATA-IND primitive with the received message to the RU. The RU then informs the SM-Core.

3.4.2   PDP Context Deactivation

As shown in Figure 3.3, the PDP context deactivation procedure deactivates an existing PDP context between the MS and the GGSN. The procedure is initiated by the MS or the network (that is, the SGSN or the GGSN).

The actions of the SGSN in the MS-initiated PDP context deactivation procedure are illustrated in Figure 3.11, and the steps are described as follows:

Step 1. When the SGSN receives the **Deactivate PDP Context Request** message from the MS (see Step 1 in Figure 3.3 (a)), the GMM entity sends the GMMSM-UNITDATA-IND primitive with the received message to the GMMSM. The GMMSM then informs the SM-Core that the MS requests deactivation of a PDP context. Note that if the GMM entity sends the GMMSM-RELEASE-IND primitive to the GMMSM to inform the SM-Core that the MS has been detached, then Step 2d is skipped.

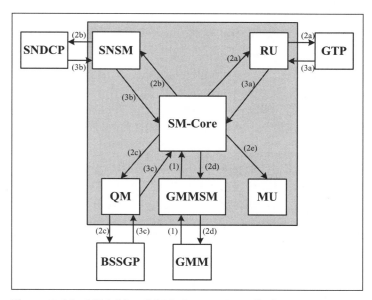

Figure 3.11   MS-initiated PDP Context Deactivation

Step 2. The SM-Core dispatches the following related components to perform the deactivation procedure:

Step 2a. The RU sends the RU-UNITDATA-REQ primitive to the GTP entity. The GTP entity forwards the **Delete PDP Context Request** message to the GGSN (see Step 3 in Figure 3.3 (a)).

Step 2b. The SNSM sends the SNSM-DEACTIVATE-IND primitive to inform the SNDCP entity that an SNDCP NSAPI has been de-allocated and cannot be used anymore.

Step 2c. The QM performs resource de-allocation and sends the QM-DELETE-PFC-REQ primitive to the BSSGP entity. The BSSGP entity forwards the **Delete BSS PFC** message to the BSS (see Step 5 in Figure 3.3 (a)).

Step 2d. The GMMSM sends the GMMSM-UNITDATA-REQ primitive to the GMM entity that forwards the **Deactivate PDP Context Accept** message to the MS (see Step 7 in Figure 3.3 (a)).

Step 2e. The MU changes the PDP context state to **INACTIVE**. The PDP context is deleted after an implementation-dependent timeout period. The PDP context exists in one of two states: **INACTIVE** or **ACTIVE**. The **INACTIVE** state indicates that the data service for a certain PDP address of the MS is not activated. In the **ACTIVE** state, the PDP context for the PDP address in use is activated in the MS, the SGSN, and the GGSN. See Section 6.2.2 for more details.

Step 3. The SM-Core is informed for actions triggered by the GTP, SNDCP, and the BSSGP entities.

Step 3a. When the SGSN receives the **Delete PDP Context Response** message from the GGSN (see Step 4 in Figure 3.3 (a)), the GTP entity sends the RU-UNITDATA-IND primitive with the received message to the RU. The RU then informs the SM-Core that the PDP context is deleted.

Step 3b. Upon receipt of the SNSM-DEACTIVATE-IND primitive from the SNSM, the SNDCP entity sends the SNSM-DEACTIVATE-RSP to the SNSM. The SNSM informs the SM-Core that the indicated SNDCP NSAPI is no longer in use.

Step 3c. When the SGSN receives the **Delete BSS PFC Ack** message from the BSS (see Step 6 in Figure 3.3 (a)), the BSSGP entity sends the QM-DELETE-PFC-CNF primitive with the received message to the QM. The QM then informs the SM-Core that the PFC is deleted.

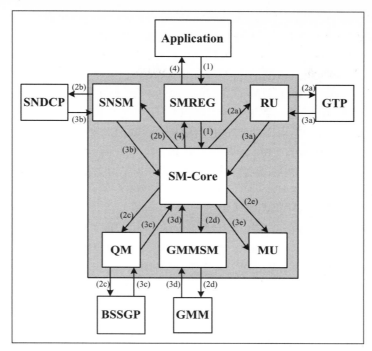

Figure 3.12   SGSN-initiated PDP Context Deactivation

The behavior of the SGSN in the SGSN-initiated PDP context deactivation procedure is shown in Figure 3.12, and is described in the following steps [ETS 04]:

Step 1. The application sends the SMREG-PDP-DEACTIVATE-REQ primitive to the SMREG. The SMREG initiates the PDP context deactivation procedure by sending an invocation to the SM-Core.

Step 2. The SM-Core dispatches the related components to perform the deactivation procedure. Steps 2a–2c are exactly the same as Steps 2a–2c described in Figure 3.11.

Step 2d. The GMMSM sends the GMMSM-UNITDATA-REQ primitive to the GMM entity. The GMM entity forwards the **Deactivate PDP Context Request** message to the MS to deactivate the indicated PDP context (see Step 5 in Figure 3.3 (b)).

Step 2e. The MU starts the timer T3395 and increments the retransmission counter. In addition, the MU changes the PDP context state into **INACTIVE**. The PDP context is deleted after an implementation-dependent timeout period.

Step 3. Steps 3a–3c are the same as Steps 3a–3c described for Figure 3.11.

Step 3d. When the SGSN receives the **Deactivate PDP Context Accept** message from the MS (see Step 6 in Figure 3.3 (b)), the GMM entity sends the GMMSM-UNITDATA-IND primitive with the received message to the GMMSM. The GMMSM then informs the SM-Core that the PDP context deactivation request is accepted.

Step 3e. The SM-Core dispatches the MU to stop the timer T3395. Note that if the message indicated at Step 3d is not received within the T3395 timeout period, then Step 2d is repeated up to four times. On the fifth expiry of T3395, the SM-Core dispatches the MU to erase the PDP context-related data for that MS.

Step 4. The SM-Core dispatches the SMREG to send the SMREG-PDP-DEACTIVATE-CNF primitive to inform the application that the PDP context deactivation procedure has been concluded.

Figure 3.13 shows the actions of the SGSN in the GGSN-initiated PDP context deactivation procedure. The steps are described as follows:

Step 1. When the SGSN receives the **Delete PDP Context Request** message from the GGSN (see Step 1 in Figure 3.3 (c)), the GTP entity sends the RU-UNITDATA-IND primitive with the received message to the RU.

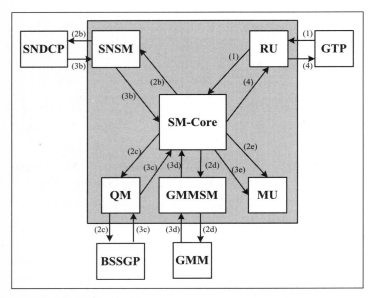

Figure 3.13   GGSN-initiated PDP Context Deactivation

The RU then informs the SM-Core that the PDP context deactivation is requested by the GGSN.

Step 2. Steps 2b–2e are the same as Steps 2b–2e for Figure 3.12.

Step 3. Steps 3b–3e are the same as Steps 3b–3e for Figure 3.12.

Step 4. The SM-Core dispatches the RU to send the RU-UNITDATA-REQ primitive to the GTP entity. The GTP entity forwards the **Delete PDP Context Response** message to the GGSN (see Step 6 in Figure 3.3 (c)).

3.4.3   PDP Context Modification

As shown in Figure 3.4, the PDP context modification procedure changes the QoS profile and radio priority negotiated during the PDP context activation procedure or during the previously-executed PDP context modification procedure. The procedure can be initiated by the SGSN at any time when a PDP context is active. Figure 3.14 illustrates the behavior of the SGSN in the SGSN-initiated PDP context modification procedure. The steps are described as follows [ETS 04]:

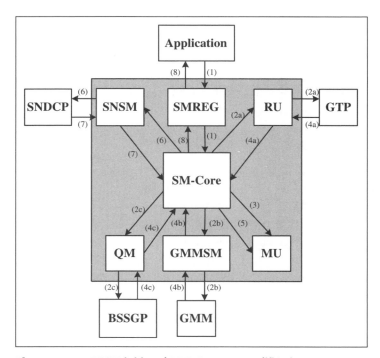

Figure 3.14   SGSN-initiated PDP Context Modification

Step 1. The application sends the SMREG-PDP-MODIFY-REQ primitive to the SMREG to initiate the modification procedure. The SMREG then invokes the SM-Core.

Step 2. The SM-Core dispatches the related components to perform the modification procedure.

> **Step 2a.** The RU sends the RU-UNITDATA-REQ primitive to the GTP entity. The GTP entity forwards the **Update PDP Context Request** message to the GGSN (see Step 1 in Figure 3.4).

> **Step 2b.** The GMMSM sends the GMMSM-UNITDATA-REQ primitive to the GMM entity. The GMM entity forwards the **Modify PDP Context Request** message to the MS (see Step 5 in Figure 3.4).

> **Step 2c.** The QM performs resource allocation, derives the ABQP, and then sends the QM-CREATE-PFC-REQ primitive to the BSSGP entity. The BSSGP entity forwards the **Create BSS PFC** message with the ABQP to the BSS (see Step 3 in Figure 3.4).

Step 3. The MU starts the timer T3386 and increments the retransmission counter.

Step 4. The SM-Core receives responses from the GGSN, the MS, and the BSS.

> **Step 4a.** When the SGSN receives the **Update PDP Context Response** message from the GGSN (see Step 2 in Figure 3.4), the GTP entity sends the RU-UNITDATA-IND primitive with the received message to the RU. The RU then informs the SM-Core whether the PDP context update request is accepted or not. If the PDP context update request is not accepted, then the SGSN-initiated PDP context deactivation is executed, and Steps 6–7 are skipped.

> **Step 4b.** When the SGSN receives the **Modify PDP Context Accept** message from the MS (see Step 6 in Figure 3.4), the GMM entity sends the GMMSM-UNITDATA-IND primitive with the received message to the GMMSM. The GMMSM then informs the SM-Core that the PDP context modification request is accepted. If the PDP context modification request is not accepted (that is, a **Deactivate PDP Context Request** message is sent from the MS), then the MS-initiated PDP context deactivation procedure is executed, and Steps 6–7 are skipped.

> **Step 4c.** When the SGSN receives the **Create BSS PFC Ack** message from the BSS (see Step 4 in Figure 3.4), the BSSGP entity sends the QM-CREATE-PFC-CNF primitive with the received message to the QM. The QM then informs the SM-Core that the PFC modification

request is accepted. If the PFC modification request is not accepted (that is, a **Create BSS PFC Nack** message is received), then the SGSN-initiated PDP context deactivation is executed, and Steps 6–7 are skipped.

Step 5. The SM-Core invokes the MU to stop the timer T3386. Note that if the message indicated at Step 4b is not received within the T3386 timeout period, then Step 2b is repeated up to four times. On the fifth expiry of T3386, the SGSN-initiated PDP context deactivation procedure is executed, and Steps 6–7 are skipped.

Step 6. The SM-Core dispatches the SNSM to send the SNSM-MODIFY-IND primitive to inform the SNDCP entity about the change of the QoS profile.

Step 7. Upon receipt of the SNSM-MODIFY-IND primitive from the SNSM, the SNDCP entity sends the SNSM-MODIFY-RSP to the SNSM. The SNSM informs the SM-Core that the negotiated QoS profile is now in use.

Step 8. The SM-Core dispatches the SMREG to send the SMREG-PDP-MODIFY-CNF or the SMREG-PDP-MODIFY-REJ primitive. The SMREG informs the application that the request for PDP context modification has been accepted or rejected.

3.5 Concluding Remarks

In this chapter, we presented a modular, compatible, and flexible software framework to implement GPRS session management functions. These functions can be implemented in modules and are compatible with the current and upcoming GPRS standards. In addition, this framework enables specific modules to be configured at run time. Based on the proposed software architecture, we described the behavior of the SGSN in the PDP context activation, deactivation, and modification procedures. We also described the design of the SM software architecture for UMTS.

For a complete description of the SGSN protocol stacks and SAPs, the readers are referred to Chapter 18 in [Lin 01b] and [ETS 00b, ETS 99].

3.6 Questions

1. Describe the relationship among the SGSN protocols, including GTP, TCP/UDP/IP, Relay, SNDCP, GMM, LLC, and BSSGP.

2. Describe the PDP context activation procedures. In the GGSN-initiated procedure, how is the MS routing information derived? Can SGSN activate PDP context? Why or why not?

3. Can a PDP context modification procedure be initiated by the MS or the GGSN?

4. Describe the steps of the SGSN-requested and the GGSN-requested PDP context deactivation procedures.

5. Describe the SGSN SM software architecture. Map the SM components in this architecture with the SGSN protocols in Figure 3.1 (b).

6. Describe the primitive flow model. What are the primitive types? What are the roles of the initiator and the responder?

7. The SM is not responsible for mobility. Why is GMMSM required in the SGSN SM software architecture?

8. Describe the SGSN protocol architecture for UMTS. What are the differences between this architecture and the GPRS version?

9. Based on the description in Section 3.2.8, describe the SM activities at the UMTS SGSN due to MS-initiated PDP context activation.

10. Describe the timers T3385, T3386, and T3395. How do you set the timeout periods of these timers, and should the timeout periods be the same for all of these timers?

Session Management for Gateway GPRS Support Node

In UMTS, the *Gateway GPRS Support Node* (*GGSN*) acts as a gateway to control user data sessions and data packet transfer between the UMTS network and the external PDN. Table 4.1 shows the four UMTS meta-functions defined in UMTS [3GP05q]. In this table, the functions implemented in the GGSN are described as follows:

Network access control (item 1 in Table 4.1) enables a user to access the UMTS services, which include the following GGSN functions:

- *Admission control* (item 1.3 in Table 4.1). The GGSN admits the incoming service requests and assigns the IP address for the MS depending on the requested *Access Point Name* (*APN*) and its available resources (in terms of bandwidth and QoS). APN and IP address allocation are described in Section 4.1 and 4.2. Additionally, the GGSN coordinates with the SGSN and the UTRAN to determine the end-to-end resources allocated for a user. We elaborate on the GGSN QoS mechanism in Section 4.5.

- *Message screening* (item 1.4 in Table 4.1). This function filters out the unauthorized and unsolicited messages. In 3GPP TS 23.060, message screening is defined in the GGSN. In most GPRS products,

Table 4.1 Functions of UMTS network elements

1. NETWORK ACCESS CONTROL	MS	UTRAN	SGSN	GGSN	HLR
1.1 Registration	–	–	–	–	v
1.2 Authentication and authorization	v	–	v	–	v
1.3 Admission control	v	v	v	v	–
1.4 Message screening	–	–	–	v	–
1.5 Packet terminal adaptation	v	–	–	–	–
1.6 Charging data collection	–	–	v	v	–
2. PACKET ROUTING AND TRANSFER	**MS**	**UTRAN**	**SGSN**	**GGSN**	**HLR**
2.1 Relay	v	v	v	v	–
2.2 Routing	v	v	v	v	–
2.3 Address translation and mapping	v	v	v	v	–
2.4 Encapsulation	v	v	v	v	–
2.5 Tunneling	–	v	v	v	–
2.6 Compression	v	v	–	–	–
2.7 Ciphering	v	v	–	–	v
3. Mobility management	v	–	v	v	v
4. Radio resource management	v	v	–	–	–

however, this function is implemented in a firewall server. Feature interactions between the message screening function and new services such as VoIP and video streaming may cause unexpected results. The traffic of these services use specific *User Datagram Protocol* (UDP) ports, which might be filtered out when enabling the message screening function at the GGSN. Therefore, the message screening function must be carefully configured to resolve the unexpected feature interaction issues.

■ *Charging data collection* (item 1.6 in Table 4.1). The GGSN generates *Call Detail Records* (*CDRs*) when a predefined event (for example, PDP context deactivation) occurs. The CDRs are included in the charging packets, which are sent to the corresponding charging gateway (see Chapter 7).

Several UMTS network access control functions are not performed at the GGSN. For example, the registration function (item 1.1 in Table 4.1) associates a user with his or her user profile (subscriber data) and routing information stored in the HLR. The authentication and authorization functions (item 1.2 in Table 4.1) are exercised between the SGSN and the MS to authenticate the user and authorize the requested services (see Section 9.1). The packet terminal adaptation function (item 1.5 in Table 4.1) is performed by the MS to adapt user data packets to be delivered across the UMTS network.

Packet routing and transfer (item 2 in Table 4.1) determines the route to transfer the user data packets within and between the UMTS networks. Within the UMTS backbone network, packets are delivered between the GGSN and the SGSN by the *GPRS Tunneling Protocol* (*GTP*) in the Gn interface [3GP05d, 3GP05q]. The following functions are defined in the GGSN:

- Relay (item 2.1 in Table 4.1): The GGSN relays the data packets between the SGSN (GTP packets) and the external PDN (IP packets). The GGSN exercises the tunneling technique to transfer the data packets between the UMTS and the external PDN. Details of the tunneling techniques are described in Section 4.4.

- Routing (item 2.2 in Table 4.1): The GGSN utilizes the APN to route packets to the external PDN. Details of APN are elaborated in Section 4.1.

- Address translation and mapping (item 2.3 in Table 4.1): The address translation function converts an address from one address type (e.g., GTP *Tunnel Endpoint Identifier* (TEID) or IP address) to another. The address mapping function translates a network address to another network address of the same type. These functions are performed by the MS, the SGSN, the GGSN, and the UTRAN.

- Encapsulation (item 2.4 in Table 4.1): The GGSN encapsulates the incoming data packets from an external PDN into GTP packets by adding the GTP headers (including the SGSN address, sequence number, etc.) to the user data packets. The GGSN decapsulates the GTP packets by removing the GTP headers to reveal the original data packets, and relays them to the external PDN.

- Tunneling (item 2.5 in Table 4.1): Through GTP, this function transfers the encapsulated data packets within and between the UMTS networks. GTP tunneling is performed by the GGSN, the SGSN, and the UTRAN. GTP-related issues are not discussed in this chapter, but can be found in Chapter 7 and [3GP05d].

Other packet routing and transfer functions such as compression (item 2.6 in Table 4.1) and ciphering (item 2.7 in Table 4.1) are only performed by the UTRAN and the MS.

Mobility management (item 3 in Table 4.1) keeps track of the current location of an MS. The GGSN involves this function only when there exist data sessions (PDP contexts) between the MS and the GGSN. When the MS moves from one SGSN to another, the GGSN updates the SGSN address and the TEID in the PDP context (i.e., the GTP link is switched from the old SGSN to the new SGSN).

Radio resource management (item 4 in Table 4.1) does not involve the GGSN. This function, which is performed by the MS and the UTRAN, is concerned with the allocation and maintenance of radio links.

The remainder of this chapter focuses on the packet routing and transfer meta-function for the GGSN. Specifically, we describe how an IP connection is established in the GGSN.

4.1   APN Allocation

An *Access Point Name* (*APN*) is used in UMTS/GPRS as a reference point for an external PDN that supports the services to be accessed by an MS. The APN information is permanently distributed and maintained in the HLR, the GGSN, and the *Domain Name Server* (*DNS*). A set of APN labels is defined in the HLR. Each mobile user can subscribe to one or more APNs from this set. The labels of these subscribed APNs are then stored in the MS at subscription time. Among the subscribed APNs, there is one default APN. If a user attempts to access a service without specifying the APN, then the default APN is used. Additionally, the HLR may also define a wildcard APN "*", which allows an MS to access any unsubscribed APNs. For each APN, the DNS keeps an IP address list of the GGSNs associated with this APN label. Figure 4.1 illustrates an example of various APN configurations. In this figure, the DNS (see Figure 4.1 (11)) is connected to the SGSNs (see Figure 4.1 (12)) and GGSNs (see Figure 4.1 (13)). When an SGSN performs an APN query, the DNS will convert the APN label to the IP address of a GGSN in this list. In a typical UMTS network, all SGSNs are fully connected to all GGSNs. During a GPRS session, when a user moves from the old SGSN to the new SGSN, the new SGSN will still connect to the original GGSN, and the same IP connectivity is maintained. Note that the *Mobile IP* (*MIP*) mechanism can be utilized to provide IP-level mobility when the user changes the GGSN (see Chapter 6).

Every GGSN maintains the configurations for a subset of the APNs defined in the HLR. Each configuration includes the IP address allocation method (static or dynamic), the IP address type (IPv4 or IPv6), and the destinated

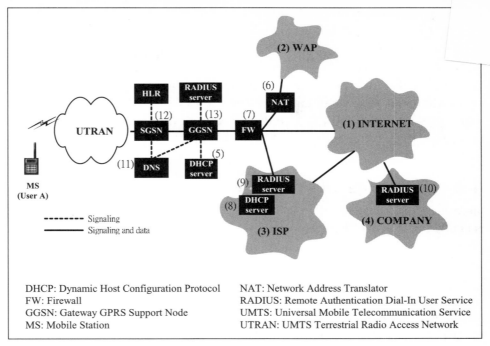

Figure 4.1 APN Configuration Examples

external PDN (which offers the service through this APN). In Figure 4.1, the HLR defines four APNs:

- The default APN is INTERNET. The mobile operator uses this APN to provide the Internet services (see Figure 4.1 (1)).

- The WAP (Wireless Application Protocol) APN is used to provide the WAP services (see Figure 4.1 (2)). The GGSN may transfer the MS's *Mobile Station ISDN Number* (*MSISDN*) to the corresponding WAP content server for the accounting purpose. With this GGSN feature, the WAP content server can provide customized personal services based on the received MSISDN.

- The ISP APN provides the Internet services offered by an *Internet Service Provider* (*ISP*) other than the mobile operator (see Figure 4.1 (3)). This APN typically is not offered in a UMTS network due to business considerations. We include this APN in Figure 4.1 for demonstration purposes.

- The COMPANY APN provides the mobile office services (see Figure 4.1 (4)), which allows a corporate user to access the intranet services of his or her company through the UMTS network.

Suppose that a user subscribes to the INTERNET and the WAP APNs in Figure 4.1. If the user requests a service through the COMPANY APN, the SGSN will reject the request. On the other hand, if the HLR defines a wild-card APN for the user, then the request will be accepted. Clearly, using the wildcard APN may cause security problems. For example, with the wild-card, an unauthorized user can access the intranet of a corporation through the COMPANY APN, and retrieve data or attack the database server of the company.

4.2   IP Address Allocation

Based on the APN setting specified in 3GPP TS 29.060 [3GP05d], there are two access modes for IP address allocation in the GGSN: *transparent* and *non-transparent*. In the transparent access mode, the mobile operator acts as an Internet service provider, and an MS is given an IP address from the opera-tor's IP address space. The IP address can be allocated statically at subscrip-tion time or dynamically at the activation of the PDP context. In Figure 4.1, the transparent access mode is exercised if the requested APN is INTERNET.

In the non-transparent access mode, the mobile operator only provides a user the access channel to an Internet service provider (if the requested APN is ISP in Figure 4.1) or a company (if the requested APN is COMPANY in Figure 4.1). The IP address pool is owned by the Internet service provider or the corporation, and the IP address for an MS is dynamically allocated. Therefore, in this access mode, the UMTS operator only serves as the access service provider.

The IP addresses can be allocated by either the GGSN, a *Dynamic Host Configuration Protocol (DHCP)* server, or a *Remote Authentication Dial-In User Service (RADIUS)* server. In the transparent access mode, the GGSN may allocate the IP address for a user by using its own address pool. This address pool is maintained by the UMTS operator. In the current implemen-tation, IPv6 addresses can be allocated only with this alternative.

In either the transparent or the non-transparent access modes, the GGSN may negotiate with a DHCP server to allocate an IP address from the address pool maintained by this DHCP server. Alternatively, the IP address of an MS may be assigned by a RADIUS server, with the IP address pool maintained by this RADIUS server.

We use the example in Figure 4.1 to illustrate the setup of the four APN configurations. Consider a mobile network with 1 million GPRS subscribers. Assume that 20% of users are expected to simultaneously access the mobile data services. The average number of PDP contexts activated by each active user is typically dimensioned to 2. Each PDP context is assigned an IP address. One PDP context is used for the WAP service, and the MS keeps an allocated IP address to browse the WAP service. Another PDP context is used for the wireless Internet service, whereby the notebook/PDA connecting to the MS keeps another allocated IP address to access the services provided by the APNs INTERNET, ISP, or COMPANY. Thus, the dimensioned network capacity is 200,000 simultaneous active GPRS users and 400,000 PDP contexts activated by these users. Note that a typical GGSN is designed to support 300,000 simultaneous users. Since we estimate 200,000 active users in this mobile network, 2 GGSNs are expected to be deployed in the network (one for operation and one for standby; or both operating in the load-sharing mode). The configurations of these APNs are listed in Table 4.2. In our example, the IP addresses for the INTERNET APN are dynamically allocated from the GGSN's IP address pool. When an MS requests an IP address through the GGSN, no authentication is required. In this scenario, the MS has already been authenticated by the standard UMTS authentication procedure in mobility management described in Section 9.1. Table 4.2 assumes that the address type is IPv6. The IP addresses range from 2002:1234:5678:1111:2222:3333:4444:0001 to 2002:1234:5678:1111:2222:3333:4444:fffe, and the maximum number of supported PDP contexts is 200,000. Note that the maximum number of PDP

Table 4.2 Four APN configuration examples

APN LABEL	INTERNET	WAP	ISP	COMPANY
GGSN access mode	Transparent	Transparent	Non-transparent	Non-transparent
IP address allocator	GGSN	DHCP server	DHCP server	RADIUS server
IP address type	IPv6	IPv4	IPv4	IPv4
DHCP server's IP address	–	192.168.30.1	140.113.214.1	–
RADIUS server's IP address	–	–	140.113.214.2	192.168.70.1
Starting MS IP address in the IP address pool	2002:1234:5678: 1111:2222:3333: 4444:0001	10.144.1.1	168.100.1.1	10.100.1.1
Maximum number of active PDP contexts	160,000	200,000	36,000	4,000

contexts and simultaneous active users is limited by the capacity of the GGSN, which is planned based on the user's requirement. The size of the IP address pool is determined by the number of simultaneous active users, which is assumed to be 20% of total subscribed users in our example.

For the WAP APN, the IP addresses are dynamically allocated from a DHCP's IP address pool (see Figure 4.1 (5)). Similar to the INTERNET APN, authentication for the MS is performed in UMTS mobility management, and no extra authentication procedure is required when the MS requests the IP address. Due to a shortage of IPv4 address spaces, each MS is allocated a private IP address. The *Network Address Translator* (*NAT*) server (see Figure 4.1 (6)) will perform address translation between the private IP address realm and the public IP address realm. In most GGSN products, the NAT function is implemented in the firewall server (see Figure 4.1 (7)). The IP addresses range from 10.144.1.1 to 10.144.15.254, and the maximum number of the PDP contexts is 20,000.

For the ISP APN, the IP address of a user is allocated from a DHCP server's IP address pool (see Figure 4.1 (8)). Before the allocation, the user is authenticated by a separate RADIUS server (see Figure 4.1 (9)). Both the RADIUS server and the DHCP server are owned by the Internet service provider. The MS is allocated a public IP address. The connection between the UMTS network and the external ISP PDN can be established by using either dedicated leased lines or a *Virtual Private Network* (*VPN*). VPN interworking is implemented by the tunneling technologies described in the next section. In our example, the IP addresses range from 168.100.1.1 to 168.100.15.254, and the maximum number of the PDP contexts is 10,000.

For the COMPANY APN, the user is authenticated by a RADIUS server (see Figure 4.1 (10)), and is allocated an IP address from the RADIUS server's IP address pool. The RADIUS server is maintained by the corporation. In the existing implementations, an IP address in a company intranet is allocated from a private IP address pool. Since the IP addresses range from 10.100.1.1 to 10.100.1.254 in Table 4.2, the maximum number of PDP contexts is 1,024.

4.3   PDP Context Activation

For a subscribed service, the corresponding APNs are recorded in a list in the MS at subscription time. When the MS attempts to access this service, it initiates the PDP context activation procedure defined in Section 3.1.1. During this procedure, the MS specifies a requested APN from the APN list.

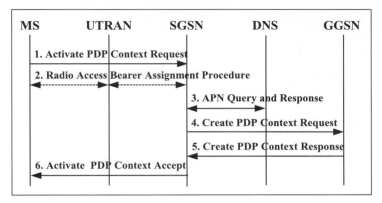

Figure 4.2 PDP Context Activation Procedure

Then the SGSN uses this requested APN to select a GGSN. If the user does not specify any requested APN in the activation procedure, the default APN is chosen by the SGSN. Recall the PDP context activation in Section 3.1.1, with the focus on the APN allocation. Figure 4.2 shows the message flow of the PDP context activation procedure. Note that the MS must attach to the UMTS network before it can activate a PDP context. In the attach procedure, the corresponding SGSN obtains the subscriber data (user profile) of the MS from the HLR, which is used in Step 3 of the PDP context activation procedure described here:

Step 1. The MS specifies the APN in the **Activate PDP Context Request** message and sends it to the SGSN.

Step 2. The SGSN negotiates with the UTRAN to allocate the radio bearer bandwidth for the data session.

Step 3. The SGSN checks whether the requested APN (obtained from the **Activate PDP Context Request** message sent by the MS) is specified in the APN list of the subscriber data for the MS. If not, the default APN is used. Then the SGSN creates the PDP context for the user, and sends the requested APN to the DNS server. The DNS server uses this APN to derive the GGSN's IP address.

Step 4. Based on the GGSN's IP address obtained from the DNS, the SGSN sends the **Create PDP Context Request** message to the GGSN to establish a GTP tunnel between the SGSN and the GGSN, which will be used as the packet routing path between the GGSN and the MS.

Step 5. The GGSN creates a PDP context for the MS. This PDP context records the requested APN, PDP type, MSISDN, IP address, and so on [3GP05q]. The GGSN allocates an IP address for the MS by using either transparent or non-transparent access mode, and determines the tunneling mechanism to the destinated external PDN (described in Section 4.4).

Step 6. Finally, the SGSN informs the MS that the session is set up.

The preceding procedure assumes IPv4 IP address allocation. For IPv6, the IP address allocation is different. Support of public IP addresses is a major difference for UMTS address allocation between IPv4 and IPv6. For IPv4, the MS is typically allocated a private address because of limited IPv4 address space. For IPv6, the MS is always allocated a public address.

At Step 5 of the PDP context activation procedure, the GGSN allocates a complete IP address for IPv4. For IPv6, there are two alternatives for dynamic address allocation [3GP05q, Tho98]:

- *Stateless* address allocation
- *Stateful* address allocation

Like IPv4, the *stateful* IPv6 address is allocated by the DHCP server at Step 5. On the other hand, in *stateless address auto-configuration,* the GGSN allocates a part of the IPv6 address called *link-local address* for the MS by using its own IPv6 address pool at Step 5. Then the MS generates the public IP address by combining the link-local address and a *network-prefix* address. The details are shown in Figure 4.3. The MS first obtains the link-local address in the PDP context activation procedure (Step 1 in Figure 4.3, which is Steps 1–6 in Figure 4.2). Then the MS activates the IPv6 address auto-configuration by sending the **Router Solicitation** message to the GGSN (Step 2 in Figure 4.3). The GGSN replies with the **Router Advertisement**

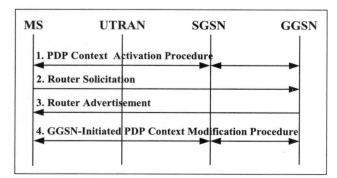

Figure 4.3 IPv6 Stateless Address Auto-configuration Procedure

message (Step 3 in Figure 4.3), which includes the network-prefix address. After the MS has received the **Router Advertisement** message, it obtains the IPv6 address by concatenating the link-local address and the network-prefix address. Then the GGSN updates the IPv6 address of the PDP contexts in the SGSN and the MS (Step 4 in Figure 4.3). To avoid conflict of link-local address assignment, the GGSN exercises neighbor discovery with other GGSNs. Note that in traditional IPv6 stateless address allocation, neighbor discovery is conducted by the mobile host. In UMTS, neighbor discovery is exercised by the GGSNs. Also note that the existing UMTS core network is developed based on the IPv4 transport network. Therefore, IPv6 packets are carried on top of the IPv4-based GTP tunnel, and are invisible to the UMTS core network.

4.4 Tunneling between UMTS and External PDN

The GGSN interworks the external data network through the *Gi interface*. The interworking mechanisms may be different for various APN configurations. Consider the example in Figure 4.1. For the INTERNET and the WAP APNs, the GGSN connects to the external PDN directly through Ethernet or leased lines. For the ISP APN, the external PDN can be connected to the GGSN either through leased lines or VPN. If the Internet service provider connects to the GGSN through VPN, then tunneling is required. For the COM-PANY APN, tunneling is always required for interworking between the GGSN and the corporation intranet. For each corporation, two tunnels are manually provisioned between the GGSNs and the VPN gateway in our example. Three tunneling methods commonly used in the existing PDNs have been proposed for UMTS:

IP-in-IP tunneling: Figure 4.4 shows the protocol stack of IP-in-IP tunneling [Sim 95]. In this method, a user IP packet (see Figure 4.4 (2)) is encapsulated within a tunnel IP packet (see Figure 4.4 (1)). The source and the destination IP addresses in the tunnel IP packet are used to identify the endpoints of the tunnel, i.e., the GGSN and the VPN gateway in the external PDN. The time-to-live and type-of-service parameters of the tunnel IP packet are the same as that of the encapsulated user IP packet. IP-in-IP tunneling incurs the smallest overhead compared with the two tunneling methods described next.

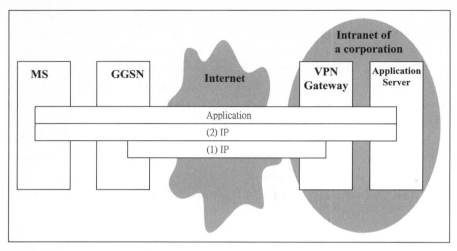

Figure 4.4   IP-in-IP Tunneling

Generic Routing Encapsulation (GRE) tunneling: GRE   supports multiprotocol tunneling whereby the GRE packets can carry user packets of various protocols, such as IP and *Point-to-Point Protocol* (*PPP*) [Han94]. The GRE tunneling that uses PPP to carry user packets (which can be IP packets) is also called *Point-to-Point Tunneling Protocol* (*PPTP*) [Ham99]. PPTP uses an enhanced GRE mechanism to provide a flow and congestion-controlled encapsulated datagram service for carrying PPP packets Figure 4.5 shows an example in which

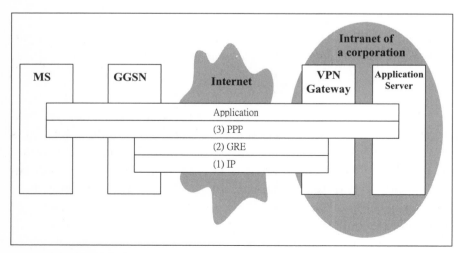

Figure 4.5   GRE (PPTP) Tunneling

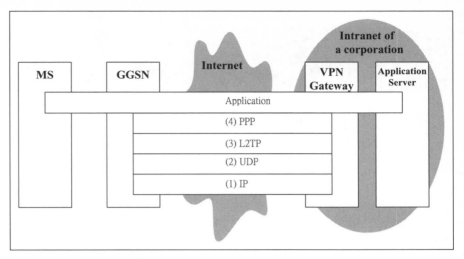

Figure 4.6   L2TP Tunneling

a user PPP packet (see Figure 4.5 (3)) is encapsulated within a GRE packet. The GRE packet (see Figure 4.5 (2)) is transfered over the IP (see Figure 4.5 (1)). Compared with IP-in-IP tunneling, GRE tunneling requires an extra IP layer, and therefore more overhead is expected.

Layer Two Tunneling Protocol (L2TP) tunneling: L2TP (see Figure 4.6 (3)) supports PPP sessions (see Figure 4.6 (4)) over UDP or other lower layer protocols such as *Frame Relay* (*FR*) and *Asynchronous Transfer Mode* (ATM) [IET99b]. The user packets (which can be IP packets) are transferred over the PPP session. For the example in Figure 4.6, an L2TP packet is encapsulated within a UDP packet (see Figure 4.6 (2)). The PPP session is established on a per-session basis when the user PDP context is created. The user data packets are encapsulated within the PPP packets.

Each of the preceding three methods can be used with *IP Security Protocol* (*IPsec*) to provide end-to-end protection for packet delivery. The IPsec protection is established between the firewall server and the VPN gateway, with the firewall function implemented within the GGSN or in a separate firewall server colocated with the GGSN. Table 4.3 summarizes the characteristics of the three tunneling methods. Note that if an MS supports both PPP and IP, then all three tunneling methods can be used to provide data sessions to the MS. IP-in-IP tunneling has better network efficiency because of small protocol overhead. However, if the user applications can only be carried

Table 4.3   Characteristics of the tunneling methods

TUNNELING METHOD	OVERHEAD	MULTIPROTOCOL SUPPORT	TRANSPORT NETWORK PROTOCOL SUPPORT	MS PROTOCOL SUPPORT
IP-in-IP	low	no	IP	IP
GRE(PPTP)	medium	yes	IP	PPP/IP
L2TP	high	yes	IP/UDP, FR, ATM	PPP/IP

over the PPP, GRE tunneling must be used. Furthermore, if the lower layer transport network is FR or ATM, then L2TP tunneling must be selected.

4.5   Quality of Service

UMTS defines four *Quality of Service* (*QoS*) classes for user data traffic: *conversational, streaming, interactive*, and *background* [3GP05q, Lin 01b]. The conversational and streaming classes support real-time traffic for services such as voice and streaming video. The interactive and background classes support non real-time traffic for services such as web browsing and email. Each class defines parameters, including maximum bit rate, guaranteed bit rate, bit error ratio, transfer delay, etc.

Table 4.4 shows major QoS parameters for VoIP and Internet access services. In this table, VoIP is a conversational-class service, with the maximum bit rate of 16 Kbps. Internet access is an interactive-class service, with the maximum bit rate of 128 Kbps. The transfer delay (100 ms) for VoIP is significantly lower and more stringent than that of (unguaranteed) Internet access. On the other hand, the Internet access traffic requires a lower bit error rate than that of VoIP.

Table 4.4   Major QoS parameters for VoIP and internet access services

QoS PARAMETER	VoIP (CONVERSATIONAL CLASS)	INTERNET ACCESS (INTERACTIVE CLASS)
Maximum bit rate	16 Kbps	128 Kbps
Guaranteed bit rate	12.2 Kbps	100 Kbps
Bit error ratio	10^{-4}	10^{-6}
Transfer delay	100 ms	unguaranteed

Table 4.5 Five conceptual end-to-end IP QoS models

SCENARIO	1	2	3	4	5
MS	–	DS	DS/RSVP	RSVP	SBLP
GGSN	DS	DS	DS	DS/RSVP	DS/SBLP
External PDN	DS	DS	DS	DS/RSVP	DS
Remote host	DS	DS	DS/RSVP	DS/RSVP	DS/SBLP

DS: Diffserv
RSVP: Resource Reservation Protocol
SBLP: Service-Based Local Policy.

Table 4.5 shows five scenarios for the end-to-end IP QoS conceptual model specified in 3GPP TS 23.207 [3GP04f]. The end-to-end QoS for packet-switched service is negotiated among the MS, the GGSN, and the remote host located in the external PDN. 3GPP TS 23.207 assumes that the external PDN supports the *Differentiated Services* (*Diffserv*) QoS mechanism, and the GGSN is required to perform the Diffserv edge function in all scenarios. Within a UMTS network path MS↔UTRAN↔SGSN↔GGSN, the IP QoS is translated and maintained by the UMTS QoS mechanism, with the QoS parameters set in the PDP contexts:

- In Scenario 1, GGSN performs the Diffserv edge function to support the IP QoS mechanism (between the GGSN and the remote host).

- Scenario 2 is similar to Scenario 1 except that the MS also supports the Diffserv edge function to control the IP QoS over the external PDN. In this scenario, the GGSN Diffserv edge function may overwrite the IP QoS setup from the MS.

- In Scenarios 3 and 4, the end-to-end QoS is controlled by the *Resource Reservation Protocol* (*RSVP*) signaling performed in the MS and the remote host.

- In Scenario 5, the MS and the GGSN support the *Service-Based Local Policy* (*SBLP*) QoS mechanism to provision 3GPP R5 IP multimedia service (see Chapter 15).

In the preceding scenarios, either Diffserv marking or RSVP signaling is carried through the PDP context transparently over the UMTS network, and the GGSN performs interworking between the UMTS QoS and the IP QoS of the external PDN.

Figure 4.7 shows a possible GGSN QoS architectural design for Scenarios 1, 2 and 3, whereby the GGSN exercises the UMTS PDP context and Diffserv

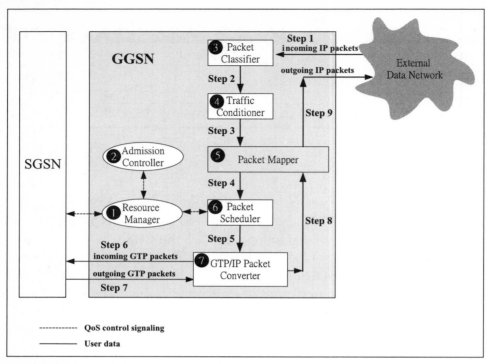

Figure 4.7 GGSN QoS Architecture

edge function. In this figure, the dashed lines represent signaling links, and the solid lines represent data links. The QoS management in the GGSN includes the following functions:

Resource Manager (see Figure 4.7 (1)) is responsible for the bearer management and resource monitor functions. The *bearer management* function interrogates the Admission Controller to determine whether the GGSN supports the specific requested service, and determines whether the resources are available. It allocates the resources requested for each individual bearer service. The negotiated resources are then specified in the PDP context. The available resources are tracked by the *resource monitor* function.

Admission Controller (see Figure 4.7 (2)) determines whether a new or modified request of a PDP context can be accepted based on the available resource information provided by the Resource Manager.

Packet Classifier (see Figure 4.7 (3)) maps each incoming user packet to the corresponding PDP context. The Packet Classifier may enable multiple PDP contexts to share one IP address using the *Traffic Flow Template (TFT)* technique. TFT employs IP header fields and higher level headers (UDP/TCP) to differentiate PDP contexts. For example, the user may activate a PDP context with interactive class for web browsing, and then activate the secondary PDP context with conversational class for VoIP. These two PDP contexts share one IP address, but are managed with different QoS classes. With TFT, UMTS can efficiently manage IP address allocation. GPRS can also simultaneously support multiple PDP contexts for a mobile user, but each PDP context will be assigned an individual IP address. Therefore, without TFT, the IP address space will be an issue when the number of mobile users becomes large.

Traffic Conditioner (see Figure 4.7 (4)) provides conformance of the QoS profile negotiated for the data traffic of a service. The incoming data packets from the external PDN may result in bursts that are not conformant with the negotiated QoS. If the incoming user data traffic exceeds the maximum bit rate specified in the corresponding PDP context, these data packets will be queued into a buffer to be transferred later. The queued packets may be dropped due to network congestion.

Packet Mapper (see Figure 4.7 (5)) marks each incoming data packet with a specific QoS indication related to the bearer service, and translates the QoS parameters of the outgoing data packet into those of the external PDN. For example, if the external PDN supports the *Differentiated Services Control Point (DSCP)* mechanism [Nic98, Hei99, Jac99], the conversational class packets are marked as *Expedited Forward (EF)* codepoint, which specifies the priority for delivery over the external PDN. The mapping between the UMTS QoS classes and the DSCP codepoints is given in Table 4.6.

Table 4.6 Mapping between UMTS QoS classes and the DSCP codepoints

QoS CLASS	DSCP CODEPOINT	DELIVERY PRIORITY
Conversational	Expedited Forward	1 (high)
Streaming	Assured Forward class 1	2
Interactive	Assured Forward class 2	3
Background	Best Forward	4 (low)

Packet Scheduler (see Figure 4.7 (6)) delivers the incoming data packets based on the priority specified for the QoS classes. The Packet Scheduler also checks the available resources with the Resource Manager. If the available resources cannot support delivery of all packets, the Packet Scheduler queues the packets, and transfers the high-priority packets first when the resources are available.

GTP/IP Packet Converter (see Figure 4.7 (7)) encapsulates the incoming IP packets into the GTP packets, and decapsulates the outgoing GTP packets into the IP packets.

Note that the Resource Manager and the Admission Controller are involved in PDP context activation. The Packet Classifier, the Traffic Conditioner, the Packet Mapper, the Packet Scheduler, and the GTP/IP Packet Converter are involved in packet delivery. Consider the QoS processing for an incoming IP packet from the external PDN to the SGSN, which is illustrated in the nine steps of Figure 4.7. Assume that the external PDN supports the DSCP QoS mechanism, and the user activates two PDP contexts. The primary PDP context (PDP-1) is used for the VoIP service (the conversational class). The secondary PDP context (PDP-2) is used for the Internet access service (the interactive class). Both PDP contexts share one IP address of the MS. Suppose that the IP address is 140.150.220.110. PDP-1 utilizes the UDP port 8010, and PDP-2 utilizes the TCP port 80.

Step 1. The incoming IP packet from the external PDN is categorized to the corresponding PDP context by the Packet Classifier. The Packet Classifier first checks whether the destination IP address of the incoming IP packet can be found in any PDP contexts created in the GGSN database. If not, the IP packet is filtered out. In our example, since the IP address is mapped to multiple PDP contexts (PDP-1 and PDP-2), the Packet Classifier exercises the TFT to determine the corresponding PDP context. That is, if the IP address of a data packet is 140.150.220.110 and the UDP port number is 8010, then the data packet is classified as the traffic for the PDP-1. If the TCP port number is 80, then this data packet will be classified as the traffic for the PDP-2.

Step 2. Based on the maximum bandwidth parameter set in the PDP context, the Traffic Conditioner determines whether the traffic volume of the incoming data packets with specific PDP context exceeds the allocated bandwidth. For example, the maximum allocated bandwidth is 128 Kbps for PDP-2 (Internet access; see Table 4.4). The exceeded packets are queued when the traffic volume of PDP-2 exceeds 128 Kbps.

Steps 3 and 4. For each data packet, the Packet Mapper specifies the corresponding UMTS QoS parameters, such as bit error ratio and transfer delay. Based on the specified QoS, the Packet Scheduler checks the available resources, and determines the delivery priority of the IP packet. In our example, the data traffic of PDP-1 has higher delivery priority than that of PDP-2.

Steps 5 and 6. The GTP/IP Packet Converter encapsulates the IP packet into a GTP packet. Then the GTP packet is transferred to the SGSN through the Gn interface.

The QoS processing for an outgoing GTP packet from the SGSN to the external PDN is described as follows:

Step 7. The GTP/IP Packet Converter decapsulates a GTP packet into an IP packet by removing the GTP header.

Step 8. The Packet Mapper marks the IP packet with the corresponding codepoint of the external PDN. For the data traffic of the VoIP service (PDP-1), the IP packet is labeled with the EF codepoint. For the data traffic of the Internet access service (PDP-2), the IP packet is labeled with the Best Forward (BF) codepoint.

Step 9. Finally, the IP packet is transfered to the external PDN through the Gi interface. The external PDN transports the IP packet according to the QoS specified in the GGSN.

4.6 Concluding Remarks

Based on [Che 05], this chapter described the GGSN session management functions for an IP connection, which includes APN and IP address allocation, tunneling technologies, and QoS management. In GPRS/UMTS, the GGSN serves as a gateway node to support IP connection toward the external PDNs. We described how the IP address can be dynamically allocated by the GGSN, a DHCP server, or a RADIUS server, and how the GGSN can interwork with the external PDN through either leased lines or VPN. For VPN interworking, tunneling is required to connect the GGSN with the external PDNs. We showed the tunneling alternatives, including IP-in-IP tunneling,

GRE tunneling, and L2TP tunneling. Finally, we elaborated on the QoS processing of a user data packet at the GGSN.

4.7 Questions

1. Describe the meta-functions performed in the GGSN. Why is mobility management performed in the GGSN? Why is radio resource management not performed in the GGSN?

2. What is APN? In UMTS, which network nodes maintain the APN information?

3. When an MS is moving within a UMTS network, is it required to use Mobile IP to maintain IP connectivity?

4. How is IP address allocation performed in UMTS? What are the roles of DHCP and RADIUS servers?

5. In the PDP context activation procedure, how is the IP address of the GGSN derived?

6. Map the steps for the PDP context activation procedure in Section 4.3 to that in Section 3.1.1.

7. Describe how an IPv6 address is allocated in UMTS. Why does the GGSN need to exercise neighbor discovery?

8. Describe and compare the three tunneling techniques used between UMTS and external PDN.

9. Describe four classes of UMTS QoS. How is UMTS QoS mapped to the IP QoS?

10. What is the traffic flow template (TFT) technique? How is IP address allocation affected if TFT is not available?

## CHAPTER 5

Serving Radio Network Controller Relocation for UMTS

In the UMTS all-IP network, the IP packets are routed between the User Equipment (UE or mobile station) and the GGSN. Through the Packet Data Protocol (PDP) context activation procedure, a PDP context is created to establish the routing path for IP packet delivery. Besides the packet routing information (e.g., the UE's IP address), the PDP context also contains the QoS profiles and other parameters. Due to the characteristics of WCDMA, multiple radio paths (for delivering the same IP packets) may exist between the UE and two or more Node Bs. An example of multiple routing paths is illustrated in Figure 5.1 (a). In this figure, an IP-based GPRS Tunneling Protocol (GTP) connection is established between the GGSN and RNC1. The UE connects to two Node Bs (B1 and B2). Node B1 is connected to RNC1, and Node B2 is connected to RNC2. An *Iur* link between RNC1 and RNC2 is established so that the signal (i.e., IP packets) sent from the UE to Node B2 can be forwarded to RNC1 through RNC2. RNC1 then combines the signals from Node B1 and B2, and forwards them to SGSN1. Similarly, the packets sent from the GGSN to RNC1 will be forwarded to both Node B1 and RNC2 (and then Node B2). As described in Section 2.7, RNC1 is called the *Serving RNC* (*SRNC*). RNC2, called the *Drift RNC* (*DRNC*), transparently routes the packets through the Iub (between Node B and the RNC) and Iur (between two RNCs) interfaces. Suppose that the UE moves from Node B1

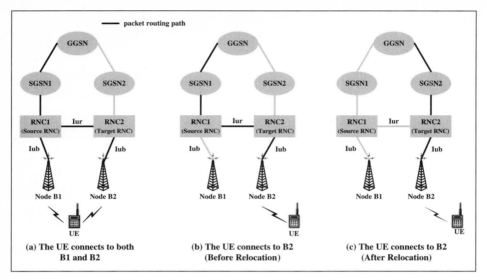

Figure 5.1   SRNC Relocation

toward Node B2, and the radio link between the UE and Node B1 is disconnected. In this case, the routing path will be

UE↔Node B2↔RNC2↔RNC1 ↔SGSN1↔GGSN

as shown in Figure 5.1 (b). In this scenario, it does not make sense to route
packets between the UE and the core network through RNC1. Therefore,
SRNC relocation may be performed to remove RNC1 from the routing path.
After SRNC relocation, the packets are routed to the GGSN directly through
RNC2 and SGSN2 (see Figure 5.1 (c)), and RNC2 becomes the SRNC.

In Section 2.7 (see also 3GPP TS 23.060 [3GP05q]), a lossless SRNC relocation procedure was proposed for non-real-time data services. In this
approach, at the beginning of SRNC relocation, the source RNC (RNC1 in
Figure 5.1 (b)) first stops transmitting downlink IP packets to the UE. Then
it forwards the next packets to the target RNC (RNC2 in Figure 5.1 (b)) via
a GTP tunnel between the two RNCs. The target RNC stores all IP packets
forwarded from the source RNC. After taking over the SRNC role, the target
RNC restarts the downlink data transmission to the UE. In this approach,
no packet is lost during the SRNC switching period. Unfortunately, this approach does not support real-time data transmission because the IP data
traffic will be suspended for a long time during SRNC switching. In order to
support real-time multimedia services, 3GPP TR 25.936 [3GP02b] proposes
SRNC Duplication (*SD*) and *Core Network Bi-casting* (*CNB*). These two

approaches duplicate data packets during SRNC relocation, which may not efficiently utilize system resources. In this chapter, we describe a new approach called *Fast SRNC Relocation (FSR)* to provide real-time SRNC switching without packet duplication.

5.1 SRNC Duplication

In Figure 5.1 (b), the target RNC is the DRNC, which is connected to the source RNC via the Iur interface. After SRNC relocation, the SRNC role is moved from the source RNC to the target RNC, and the IP packets for the UE are directly routed through SGSN2 and the target RNC (see Figure 5.1 (c)). Figure 5.2 shows the four stages of the SRNC duplication (SD) procedure. Stage I (Figure 5.2 (a)) initiates SRNC relocation. In this stage, the user IP packets are delivered through the old path

GGSN↔SGSN1↔source RNC↔ target RNC↔UE

The following steps are executed:

Steps 1 and 2. When Node B of the source RNC no longer connects to the UE, the source RNC initiates SRNC relocation. Specifically, the source RNC sends the **Relocation Required** message with the identification (Id) of the target RNC to SGSN1.

Step 3. Based on the Id of the target RNC, SGSN1 determines whether the SRNC relocation is intra-SGSN SRNC relocation or inter-SGSN SRNC relocation. Assume that it is inter-SGSN SRNC relocation. By sending the **Forward Relocation Request** message, SGSN1 requests SGSN2 to allocate the resources (described in Step 4) for the UE.

Step 4. SGSN2 sends the **Relocation Request** message with the Radio Access Bearer (RAB) parameters to the target RNC. The RAB parameters include the traffic class (e.g., conversational, streaming, interactive or background), traffic handling priority, maximum and guaranteed bit rates, and so on [3GP05k]. After all necessary resources for the RAB are successfully allocated, the target RNC sends the **Relocation Request Acknowledge** message to SGSN2.

In Stage II (see Figure 5.2 (b)), a forwarding path

source RNC→target RNC→UE

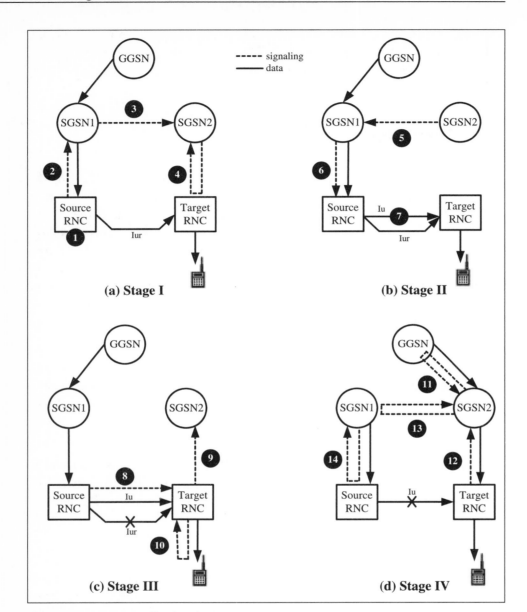

Figure 5.2   SRNC Duplication (SD)

for downlink packet delivery is created between the source and the target RNCs through the Iu interface. The source RNC duplicates the packets and forwards these packets to the target RNC. Thus, the downlink packets are simultaneously transmitted through both the old path (via the Iur interface) and the forwarding path (via the Iu interface) between the source RNC and the target RNC. Note that 3G TR 25.936 [3GP02b] did not clearly describe whether an Iu link can be directly established between two RNCs. If not, an indirect path

source RNC→SGSN1→SGSN2→target RNC

is required. We assume a direct link between the source and target RNCs. This assumption favors the SD approach. The following steps are executed in Stage II:

Steps 5 and 6. SGSN2 sends the **Forward Relocation Response** message to SGSN1, which indicates that all resources (e.g., RAB) are allocated. SGSN1 forwards this information to the source RNC through the **Relocation Command** message.

Step 7. Upon receipt of the **Relocation Command** message, the source RNC duplicates the downlink packets and transmits the duplicated packets to the target RNC through the forwarding path (via the Iu interface at the IP layer). The forwarded packets are discarded at the target RNC before it becomes the SRNC (i.e., before the target RNC receives the **Relocation Commit** message at Step 8).

In Stage III (see Figure 5.2 (c)), the Iur link between the source RNC and the target RNC (i.e., the old path) is disconnected. The downlink packets arriving at the source RNC are forwarded to the target RNC through the Iu link (i.e., the forwarding path). A data-forwarding timer is maintained in the source RNC. When the timer expires, the forwarding operation at the source RNC is stopped. The following steps are executed in Stage III:

Step 8. With the **Relocation Commit** message, the source RNC transfers the *Serving Radio Network Subsystem (SRNS)* context (e.g., QoS profile for the RAB) to the target RNC.

Step 9. Upon receipt of the **Relocation Commit** message, the target RNC sends the **Relocation Detect** message to SGSN2, which indicates that the target RNC will become the SRNC.

Step 10. At the same time, the target RNC sends the **RAN Mobility Information** message to the UE. This message triggers the UE to send the uplink IP packets to the target RNC. After the UE has reconfigured itself, it replies with the **RAN Mobility Information Confirm** message to the target RNC.

In Stage IV (see Figure 5.2 (d)), the packet routing path is switched from the old path to the new path

GGSN↔SGSN2↔target RNC↔UE

At this stage, the target RNC becomes the SRNC. The source RNC forwards the downlink packets to the target RNC until the data-forwarding timer expires. The remaining steps are executed in Stage IV:

Step 11. SGSN2 sends the **Update PDP Context Request** message to the GGSN. Based on the received message, the GGSN updates the corresponding PDP context and returns the **Update PDP Context Response** message to SGSN2. Then the downlink packet routing path is switched from the old path to the new path. At this moment, the target RNC receives the downlink packets from two paths (i.e., the forwarding and new paths), and transmits them to the UE. Since the transmission delays for these two paths are not the same, the packets arriving at the target RNC may not be in sequence, which results in out-of-order delivery.

Steps 12 and 13. By sending the **Relocation Complete** message to SGSN2, the target RNC indicates the completion of the relocation procedure. Then SGSN2 exchanges this information with SGSN1 using the **Forward Relocation Complete** and **Forward Relocation Complete Acknowledge** message pair.

Step 14. Finally, SGSN1 sends the **Iu Release Command** message to request the source RNC to release the Iu connection in the forwarding path. When the data-forwarding timer expires, the source RNC replies with the **Iu Release Complete** message.

5.2 Core Network Bi-casting

Figure 5.3 shows the 4 stages of the Core Network Bi-casting (CNB) procedure when the communicating UE moves from the source RNC to the target RNC. Stage I (Steps 1–4, Figure 5.3 (a)) is the same as Stage I in SD, which requests the target RNC to allocate the necessary resources for relocation.

In Stage II (see Figure 5.3 (b)), the downlink packets are duplicated at the GGSN, and are sent to the target RNC through both the old path

GGSN→SGSN1→Source RNC →Target RNC

and the new path

GGSN→SGSN2→Target RNC

The following steps are executed:

Step 5. Upon receipt of the **Relocation Request Acknowledge** message at Step 4, SGSN2 sends the **Update PDP Context Request** message that

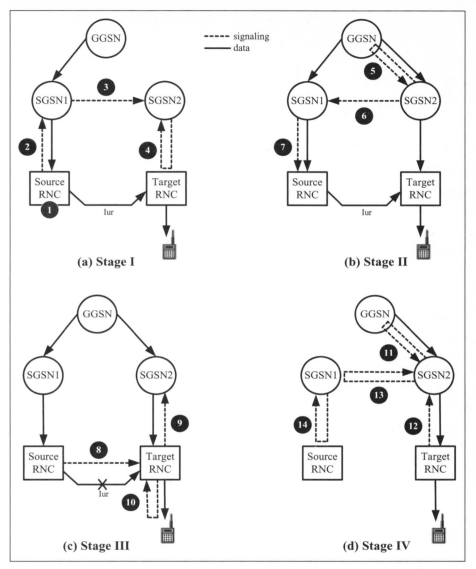

Figure 5.3 Core Network Bi-casting (CNB)

requests the GGSN to bi-cast the downlink packets. The GGSN starts to perform bi-casting and replies to SGSN2 with the message **Update PDP Context Response**. At this moment, the downlink packets are simultaneously transmitted to the target RNC through the old and new paths. Since the target RNC has not taken the SRNC role (i.e., the target RNC has not received the **Relocation Commit** message), the packets routed through the new path are discarded at the target RNC.

Steps 6 and 7. These steps inform the source RNC that all necessary resources are allocated, similar to Steps 5 and 6 in the SD approach.

In Stage III (see Figure 5.3 (c)), the Iur link between the source RNC and the target RNC is disconnected, and the downlink packets arriving at the source RNC are discarded:

Steps 8–10. These steps move the SRNC role from the source RNC to the target RNC, similar to Steps 8–10 in the SD approach.

In Stage IV (see Figure 5.3 (d)), the GGSN is informed to stop downlink packet bi-casting. The target RNC takes the SRNC role to transmit the downlink packets to the UE:

Step 11. Through the **Update PDP Context Request** message, SGSN2 informs the GGSN to stop downlink packet bi-casting. Then the GGSN removes the GTP tunnel between the GGSN and SGSN1, and replies to SGSN2 with the **Update PDP Context Response** message.

Steps 12 and 13. With the **Relocation Complete** message, the target RNC informs SGSN2 that the relocation procedure has been successfully performed. Then SGSN2 exchanges this information with SGSN1 using the **Forward Relocation Complete** and **Forward Relocation Complete Acknowledge** message pair.

Step 14. Finally, SGSN1 and the source RNC exchange the **Iu Release Command** and **Iu Release Complete** message pair to release the Iu connection in the old path.

5.3 Fast SRNC Relocation

This section describes the Fast SRNC Relocation (FSR) approach and compares this approach with SD and CNB. As shown in Figure 5.1 (b), the UE is connected to the source RNC and SGSN1 before SRNC relocation. After relocation, the data packets for the UE are directly routed through the target RNC and SGSN2 as shown in Figure 5.1 (c). Figure 5.4 illustrates the four stages of the FSR procedure.

Stage I (see Figure 5.4 (a)) initiates SRNC relocation. In this stage, the routing path of downlink packets is

GGSN→SGSN1→source RNC →target RNC→UE

The following steps are executed in Stage I:

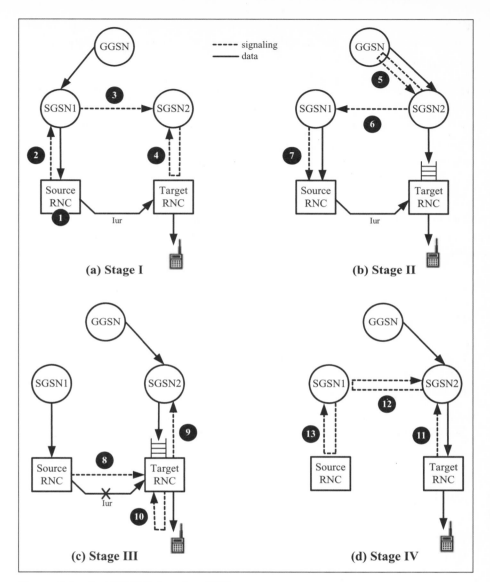

Figure 5.4 Fast SRNC Relocation (FSR)

Steps 1 and 2. When Node B of the source RNC no longer connects to the UE, the source RNC initiates SRNC relocation and sends the Id of the target RNC to SGSN1 through the **Relocation Required** message.

Step 3. Based on the Id of the target RNC, SGSN1 determines that it is inter-SGSN SRNC relocation. SGSN1 requests SGSN2 to allocate the resources for the UE through the **Forward Relocation Request** message.

Step 4. SGSN2 and the target RNC exchange the **Relocation Request** and **Relocation Request Acknowledge** message pair to allocate the necessary resources for the UE.

In Stage II (see Figure 5.4 (b)), the GGSN routes the downlink packets to the old path before receiving the **Update PDP Context Request** message (see Step 5 in Figure 5.4 (b)). The packets delivered through the old path are called *old packets*. After the GGSN has received the **Update PDP Context Request** message, the downlink packets are routed to the new path as follows:

GGSN→SGSN2→target RNC

The packets delivered by the new path are called *new packets*. The new packets arriving at the target RNC are buffered until the target RNC takes over the SRNC role. The following steps are executed in Stage II:

Step 5. Upon receipt of the **Relocation Request Acknowledge** message, SGSN2 sends the **Update PDP Context Request** message to the GGSN. Based on the received message, the GGSN updates the corresponding PDP context fields and returns the **Update PDP Context Response** message to SGSN2. Then the downlink packet routing path is switched from the old path to the new path. At this stage, the "new" downlink packets arriving at the target RNC are buffered.

Steps 6–7. SGSN2 sends the **Forward Relocation Response** message to SGSN1 to indicate that all resources for the UE are allocated. SGSN1 forwards this information to the source RNC through the **Relocation Command** message.

In Stage III (see Figure 5.4 (c)), the Iur link between the source RNC and the target RNC is disconnected. The "old" downlink packets arriving at the source RNC later than the **Relocation Command** message (see Step 7 in Figure 5.4 (b)) are dropped. In this stage, Steps 8–10 switch the SRNC role from the source RNC to the target RNC:

Step 8. With the **Relocation Commit** message, the SRNC context of the UE is transferred from the source RNC to the target RNC.

Steps 9 and 10. The target RNC sends the **Relocation Detect** message to SGSN2. At the same time, the target RNC sends the **RAN Mobility Information** message to the UE, which triggers the UE to send the uplink IP packets through the new path:

UE→target RNC→SGSN2→GGSN

By executing Steps 11 and 12 at Stage IV (see Figure 5.4 (d)), the target RNC informs the source RNC that SRNC relocation has been successfully performed. Then the source RNC releases the system resources for the UE.

Steps 11 and 12. The target RNC sends the **Relocation Complete** message to SGSN2, which indicates that SRNC relocation has been successfully performed. Then SGSN2 exchanges this information with SGSN1 through the **Forward Relocation Complete** and **Forward Relocation Complete Acknowledge** message pair.

Step 13. Finally, SGSN1 and the source RNC exchange the **Iu Release Command** and **Iu Release Complete** message pair to release the Iu connection in the old path.

5.4 Comparison of the Relocation Mechanisms

Based on the preceding discussions, Table 5.1 compares FSR with SD and CNB. The following issues are addressed:

Packet Duplication: During SRNC relocation, IP packets are duplicated at the source RNC in SD. Similarly, IP packets are duplicated at the GGSN in CNB. Packet duplication will significantly consume system resources. However, packet duplication is not needed in the FSR approach.

Packet Loss: Packet loss may occur in these 3 approaches either at the source RNC or at the target RNC. For SD and FSR, the data packets arriving at the source RNC may be lost. In SD, the "old" packets are dropped at the source RNC when the data-forwarding timer expires (see Step 13 in Figure 5.2 (d)). In FSR, the "old" packets are dropped if they arrive at

Table 5.1 Comparing FSR with SD and CNB

	FSR	SD	CNB
Packet Duplication	No	Yes	Yes
Packet Loss at Source RNC	Yes	Yes	No
Packet Loss at Target RNC	No	Yes	Yes
Packet Buffering	Yes	No	No
Out-of-order Delivery	No	Yes	No
Extra Signaling	No	No	Yes

the source RNC later than the **Relocation Command** message (see Step 7 in Figure 5.4 (b)) does.

For SD and CNB, the data packets may be lost at the target RNC. In SD, the target RNC discards the forwarded packets from the source RNC if these packets arrive at the target RNC earlier than the **Relocation Commit** message does (see Step 7 in Figure 5.2 (b)). In CNB, the duplicated packets may be lost at the target RNC because the packets from the new path are dropped before the target RNC becomes the SRNC (see Step 5 in Figure 5.3 (b)). On the other hand, since the packet buffer mechanism is implemented in FSR, the packets are not lost at the target RNC.

Packet Buffering: To avoid packet loss at the target RNC, the packet buffer mechanism is implemented in FSR, which is not found in both SD and CNB.

Out-of-order Delivery: In SD, two paths (i.e., the forwarding and new paths) are utilized to simultaneously transmit the downlink packets (see Step 11 in Figure 5.2 (d)). Since the transmission delays for these two paths are not the same, the packets arriving at the target RNC may not be in sequence, which results in out-of-order delivery. This problem does not exist in FSR and CNB because the target RNC in these two approaches only processes the packets from one path (either the old path or the new path) at any time, and the out-of-order packets are discarded (see Step 5 in Figure 5.3 (b)).

Extra Signaling: The SD approach follows the standard SRNC relocation procedure proposed in 3GPP TS 23.060 [3GP05q] (see Section 2.7). The FSR approach reorders the steps of the 3GPP TS 23.060 SRNC relocation procedure. Both approaches do not introduce any extra signaling cost. Conversely, CNB exchanges additional **Update PDP Context Request** and **Update PDP Context Response** message pairs (see Step 5 in Figure 5.3) between the GGSN and SGSN2, which incurs extra signaling cost. Note that all three approaches can be implemented in the GGSN, the SGSN, and the RNC without introducing new message types to the existing 3GPP specifications.

In conclusion, SD and CNB require packet duplication that will double the network traffic load during SRNC relocation. For the SD approach, it is not clear if the Iu link in the forwarding path can be directly established between two RNCs. If not, an indirect path

source RNC→SGSN1→ SGSN2→target RNC

is required. Also, it is not clear if the target RNC will be informed to stop receiving the forwarded packets when the data-forwarding timer expires. Packet duplication is avoided in FSR. Note that packets may be lost during SRNC relocation for these three approaches. Packet loss cannot be avoided in SRNC relocation if you want to support real-time applications. Our study indicates that the effect of packet loss for FSR can be ignored [Pan04a].

5.5 Concluding Remarks

In 3GPP TR 25.936, SRNC duplication (SD) and core network bi-casting (CNB) were described to support real-time multimedia services in the UMTS all-IP network. Both approaches require packet duplication during SRNC relocation, which significantly consumes system resources. This chapter proposed the fast SRNC relocation (FSR) approach, which provides real-time SRNC switching without packet duplication. In FSR, a packet buffer mechanism is implemented to avoid packet loss at the target RNC. In [Pan04a] (see also question 1 in Section 5.6), we developed an analytic model to investigate the performance of FSR. Our performance study indicated that packet loss at the source RNC can be ignored in FSR. Furthermore, the expected number of buffered packets at the target RNC is small, which does not result in long packet delay. FSR can be implemented in the GGSN, the SGSN, and the RNC without introducing new message types to the existing 3GPP specifications. As a final remark, the FSR approach is a ROC patent (No. 194136).

5.6 Questions

1. The FSR approach can be modeled as follows. The routing path of the downlink packets for the UE is switched from the old path

    GGSN→ SGSN1→ source RNC→target RNC

    to the new path

    GGSN→SGSN2→target RNC

    after the GGSN receives the **Update PDP Context Request** message (see Step 5 in Figure 5.4 (b)). The packets delivered through the old path are lost if these packets arrive at the source RNC later than the **Relocation Command** message does (see Step 7 in Figure 5.4 (b)). Therefore, an important performance measure is the expected number of lost packets

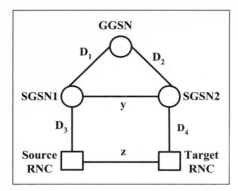

Figure 5.5   The Transmission Delays

$E[N_L]$ during SRNC relocation. Furthermore, the packets transmitted through the new path are buffered at the target RNC if they arrive at the target RNC earlier than the **Relocation Commit** message does (see Step 8 in Figure 5.4 (c)). Hence, another important performance measure is the expected number of buffered packets $E[N_B]$ during SRNC relocation.

Figure 5.5 denotes the transmission delays among the network nodes, which are represented by the random variables described here:

- D_1: the transmission delay between the GGSN and SGSN1

- D_2: the transmission delay between the GGSN and SGSN2. Without loss of generality, we assume that D_1 and D_2 have the same distribution.

- D_3: the transmission delay between SGSN1 and the source RNC

- D_4: the transmission delay between SGSN2 and the target RNC. Without loss of generality, we assume that D_3 and D_4 have the same distribution.

- y: the transmission delay between SGSN1 and SGSN2

- z: the transmission delay between the source RNC and the target RNC

Based on the preceding random variables, please derive analytically the expected numbers of lost and buffered packets for FSR. (Hint: See the solution in http://liny.csie.nctu.edu.tw/supplementary)

Expected Number of Lost Packets. Consider the timing diagram in Figure 5.6. Suppose that the GGSN receives the **Update PDP Context Request** message from SGSN2 at time τ_0 (see Step 5 in

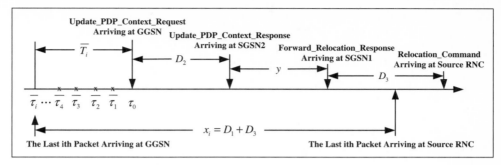

Figure 5.6 Timing Diagram for Computing the Number of Lost Packets

Figure 5.4 (b)). Tracing back from τ_0, the previous ith packet was sent from the GGSN to SGSN1 at time $\bar{\tau}_i$. Following the assumption widely used in the literature [Ros96], we assume that the inter-packet arrivals are a Poisson stream, and the inter-packet arrival times $\bar{\tau}_{i-1} - \bar{\tau}_i$ are exponentially distributed with the arrival rate λ_a. If the arrival of the **Update PDP Context Request** message from SGSN2 to the GGSN at time $\bar{\tau}_0$ is a random observer, then from the residual life theorem (see question 2 in Section 9.6) and the memoryless property of the exponential distribution, $\bar{\tau}_0 - \bar{\tau}_1$ has the exponential distribution with mean $\frac{1}{\lambda_a}$. Therefore, $\bar{T}_i = \bar{\tau}_0 - \bar{\tau}_i$ has an Erlang distribution with the density function

$$f_{\bar{T}_i}(t) = \left[\frac{(\lambda_a t)^{i-1}}{(i-1)!} \right] \lambda_a e^{-\lambda_a t}. \tag{5.1}$$

For $i \geq 1$, the transmission delay for the previous ith packet through the path

GGSN $\rightarrow$ SGSN1 $\rightarrow$ source RNC

can be represented by the random variable $x_i = D_1 + D_3$. The transmission delay for the signaling messages **Update PDP Context Response** (see Step 5 in Figure 5.4 (b)), **Forward Relocation Response** (see Step 6 in Figure 5.4 (b)), and **Relocation Command** (see Step 7 in Figure 5.4 (b)) through the path

GGSN$\rightarrow$SGSN2$\rightarrow$SGSN1$\rightarrow$source RNC

can be represented by the random variable $x + y$, where $x = D_2 + D_3$ is identical to the random variable x_i. The intervals x and y have general distributions determined by the layout and transmission property of the UMTS all-IP core network. We assume that both x and y have mixed-Erlang density functions

$$f_x(t) = \sum_{j=1}^{J} \alpha_{x,j} \left[\frac{(\lambda_{x,j} t)^{m_{x,j}-1}}{(m_{x,j}-1)!} \right] \lambda_{x,j} e^{-\lambda_{x,j} t} \qquad \text{where } \sum_{j=1}^{J} \alpha_{x,j} = 1$$

$$(5.2)$$

and

$$f_y(t) = \sum_{l=1}^{L} \alpha_{y,l} \left[\frac{(\lambda_{y,l} t)^{m_{y,l}-1}}{(m_{y,l}-1)!} \right] \lambda_{y,l} e^{-\lambda_{y,l} t} \qquad \text{where } \sum_{l=1}^{L} \alpha_{y,l} = 1$$

$$(5.3)$$

In (5.2) and (5.3), J, L, $\alpha_{x,j}$ and $\alpha_{y,l}$ determine the shapes and scales of the distributions. The mixed-Erlang distribution is selected because this distribution has been proven as a good approximation to many other distributions, as well as measured data [Kel79].

The previous ith packet is lost if it arrives at the source RNC later than the **Relocation Command** message does. Let N_L be the number of the lost packets, and define

$$\delta_i = \begin{cases} 1, & \text{if the previous } i\text{th packet is lost (i.e., } \bar{T}_i + x + y < x_i) \\ 0, & \text{otherwise} \end{cases}$$

$$(5.4)$$

Then $N_L = \sum_{i=1}^{\infty} \delta_i$, and

$$E[N_L] = E \left[\sum_{i=1}^{\infty} \delta_i \right]$$

$$= \sum_{i=1}^{\infty} \Pr[\text{the previous } i\text{th packet is lost}]$$

$$= \sum_{i=1}^{\infty} \Pr[\bar{T}_i + x + y < x_i] \qquad (5.5)$$

Please derive (5.5) by using (5.2)–(5.4).

Expected Number of Buffered Packets. Consider the timing diagram in Figure 5.7. Suppose that at time τ_0, the **Update PDP Context Response** message is sent from the GGSN to SGSN2 (Step 5 in Figure 5.4 (b)). SGSN2 receives this message at time t_1, and issues the **Forward Relocation Response** message to SGSN1 (Step 6 in Figure 5.4 (b)). SGSN1 receives the message at t_2 and sends the **Relocation Command** message to the source RNC (Step 7 in

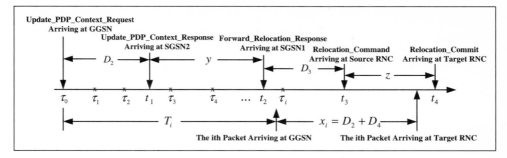

Figure 5.7 Timing Diagram for Computing the Number of Buffered Packets

Figure 5.4 (b)). The source RNC receives the message at time t_3, and transfers SRNS contexts to the target RNC by using the **Relocation Commit** message (Step 8 in Figure 5.4 (c)). The message arrives at the target RNC at time t_4. The transmission delay $t_4 - \tau_0$ can be represented by the random variable $D_2 + y + D_3 + z = x + y + z$. During this period, several packets may have been sent from the GGSN to the target RNC through SGSN2. We assume that z has a mixed-Erlang distribution with density function $f_z(t)$ and Laplace transform $f_z^*(s)$, where for $\sum_{p=1}^{P} \alpha_{z,p} = 1$, we have

$$f_z(t) = \sum_{p=1}^{P} \alpha_{z,p} \left[\frac{(\lambda_{z,p}t)^{m_{z,p}-1}}{(m_{z,p} - 1)!} \right] \lambda_{z,p} e^{-\lambda_{z,p}t} \quad \text{and}$$

$$f_z^*(s) = \sum_{p=1}^{P} \alpha_{z,p} \left(\frac{\lambda_{z,p}}{s + \lambda_{z,p}} \right)^{m_{z,p}} \tag{5.6}$$

The ith packet was sent from the GGSN at time $\tau_i = \tau_0 + T_i$ through the new path

GGSN$\rightarrow$ SGSN2$\rightarrow$target RNC

and its transmission delay can be represented by the random variable $x_i = D_2 + D_4$. Note that T_i has the same distribution as $\bar{T}_i$.

Suppose that N_B packets arrive at the target RNC during the period $[\tau_0, t_4]$ (i.e., during the transition of routing path switching). These packets must be buffered in the target RNC. From the preceding discussion, the expected number $E[N_B]$ of buffered packets at the target RNC is

$$E[N_B] = \sum_{i=1}^{\infty} \Pr[\text{the } i\text{th packet is queued in the buffer}]$$

$$= \sum_{i=1}^{\infty} \Pr[T_i + x_i < x + y + z] \qquad (5.7)$$

Please derive (5.7) by using (5.6).

2. What is serving RNC? What is drift RNC?

3. Section 2.7 described lossless SRNC relocation. However, it does not support real-time applications. Explain why.

4. Describe SRNC duplication (SD), core network bi-casting (CNB), and fast SRNC relocation (FSR).

5. Compare SD, CNB, and FSR in terms of packet duplication, packet loss, packet buffering, out-of-order delivery, and extra signaling needed.

UMTS and cdma2000 Mobile Core Networks

Universal Mobile Telecommunications System (UMTS) and cdma2000 are two major standards for the third generation (3G) mobile telecommunication [Lin 01b]. Evolving from the existing 2G networks, construction of an effective 3G network is critical for provisioning advanced mobile services. This chapter describes mobility management and session management for UMTS and cdma2000, and compares their design guidelines. These guidelines are also followed by beyond 3G all-IP core networks.

The mobility management functions are used to keep track of the current location of a mobile user. There are three types of mobility: radio mobility, core network mobility, and IP mobility:

- *Radio mobility* supports handoff (i.e., radio link switching) of a mobile user during conversation.

- *Core network mobility* (or link-layer mobility) provides tunnel-related management for packet re-routing in the core network due to user movement.

- *IP mobility* allows the mobile user to change the access point of IP connectivity without losing ongoing application sessions.

The network nodes such as *Gateway GPRS Support Node* (*GGSN*) and *Home Agent* (*HA*) hide the mobility of a mobile user from the external network so that the corresponding user (who communicates with the mobile user) will not be notified.

Session management maintains the routing path for a communication session between a mobile user and the mobile core network, which provides packet routing functions, including IP address assignment, QoS setting, and so on.

The UMTS network architecture is introduced in Chapter 2 (see Figure 2.1). In this architecture, *Mobile Application Part* (*MAP*) provides mobility management interfaces between the Serving GPRS Support Node (SGSN) and the GSM network nodes—for example, Gr for HLR and Gs for MSC/VLR. SGSNs and GGSNs communicate by using the GPRS Tunneling Protocol (GTP) through the Gn interface.

Figure 6.1 shows the cdma2000 architecture [3GP02d]. In this figure, the *Base Station Controller* (*BSC*) connects to the core network through the *Selection and Distribution Unit* (*SDU*). Like UMTS, cdma2000 supports both the Packet Switched (PS) and the Circuit Switched (CS) service domains. The SDU distributes the circuit-switched traffic (e.g., voice) to the MSC via interfaces A1, A2, and A5:

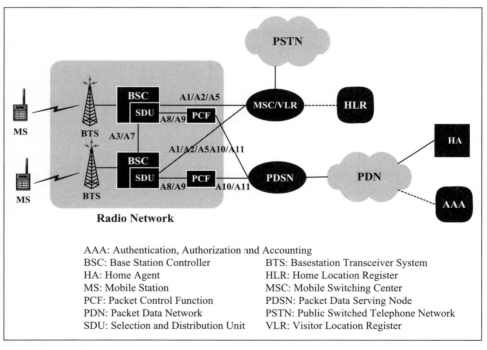

Figure 6.1 cdma2000 Architecture

- The A1 interface supports call control and mobility management between the MSC and the BSC.

- The A2 and A5 interfaces support voice traffic and circuit-switched data traffic between the BSC and the MSC, respectively.

The MSC, VLR, and HLR functions are basically the same as those in UMTS, and will not be re-elaborated. The SDU distributes packet-switched traffic to the *Packet Control Function (PCF)* and then to the *Packet Data Serving Node (PDSN)* through the following interfaces.

- Interfaces A8 and A9 support packet-switched data and signaling between the PCF and the SDU, respectively.

- Interfaces A10 and A11 (i.e., the R-P interface) support packet-switched data (which is delivered in sequence) and signaling between the PCF and the PDSN.

The *Generic Routing Encapsulation (GRE)* tunnel with standard IP QoS is used for data routing in A1 [3GP01]. A11 utilizes the *Mobile IP (MIP)* [Per 02] based messages to convey the signaling information between the PCF and PDSN [3GP01, Per 98]. The R-P interface (i.e., A10/A11 interface) also supports PCF handoff (inter-or intra-PDSN), described in Section 6.2.

Connecting to one or more BSCs, a PDSN establishes, maintains, and terminates link-layer sessions to the MSs. The PDSN supports packet compression and packet filtering (see the *PPP* description in Section 6.1.1) before the packets are delivered through the air interface. The PDSN provides IP functionality to the mobile network, which routes IP datagrams to the PDN with differentiated service support. A PDSN interacts with the *Authentication, Authorization and Accounting (AAA)* [3GP03c] to provide IP authentication, authorization, and accounting support for packet data services (note that user and device authentication functions are not provided in the PDSN). The PDSN may act as a MIP *Foreign Agent (FA)* in the mobile network, which provides mobility management mechanisms with the MIP HA. The interfaces among the PDN nodes (i.e., PDSN, HA, AAA) follow the *Internet Engineering Task Force (IETF)* standards. In cdma2000, user and device authentication functions are handled by the MSC/VLR/HLR. In UMTS, the SGSN directly interacts with the HLR to exercise user, device, and service authentication.

The preceding discussion indicates that network architectures of mobility and session management are basically the same for both UMTS and cdma2000. However, the protocols exercised on these network architectures are different. For example, cdma2000 uses IETF protocols (for example, MIP and PPP) to support mobility and session management mechanisms. Conversely, the UMTS protocols include SS7-based MAP and IP-based GTP. Moreover, one of the key 3G goals is to provide independence between the radio and core networks. As shown in the following sections, UMTS provides a clear demarcation between the radio access network and the core network. Conversely, this goal is only partially achieved in cdma2000, where the MSC is still involved in radio resource allocation for a packet session.

6.1   UMTS and cdma2000 Protocol Stacks

In this section, we briefly discuss the lower layer protocols that support mobility and session management and user data transport. Figures 6.2 and 6.3 illustrate the control planes and the user planes for cdma2000 and UMTS networks, respectively.

The control plane carries out tasks for mobility management, session management, and short message service. The mobility and session tasks in cdma2000 are based on the same lower layer protocols (for example, IP-based protocols; cf. Figures 6.2 (a) and 6.3 (a)) for user data transportation except that the user data flow bypasses the link access control layer. An advantage of the cdma2000 approach is that the same lower layer protocols can support both control and user planes.

In UMTS, the lower layer protocols supporting these tasks in the control plane (see Figure 6.2 (b)) are different from the lower layer protocols in the user plane (see Figure 6.3 (b)). The UMTS control plane protocols such as *Radio Resource Control* (*RRC*) and *Radio Access Network Application Part* (*RANAP*) are utilized for signaling. Specifically, the signaling path between an MS and an SGSN consists of an RRC connection between MS and *UMTS Terrestrial Radio Access Network* (*UTRAN*), and an Iu connection ("one RANAP instance") between UTRAN and SGSN. The RRC protocol is responsible for a reliable connection between MS and UTRAN, that is, radio resources are managed by RRC exercised between MS and UTRAN. *Signaling Connection Control Part* (*SCCP*) is responsible for a reliable connection between UTRAN and SGSN. On top of SCCP, the RANAP protocol supports transparent Non-Access Stratum (for example, mobility management and session management) signaling transfer between MS and the core network, which are not interpreted by the UTRAN. By using these signaling protocols, efficient radio resource management, session management, and mobility management can be achieved. In cdma2000, the task for radio resource management involves the MSC. That is, the MSC instructs the BSC

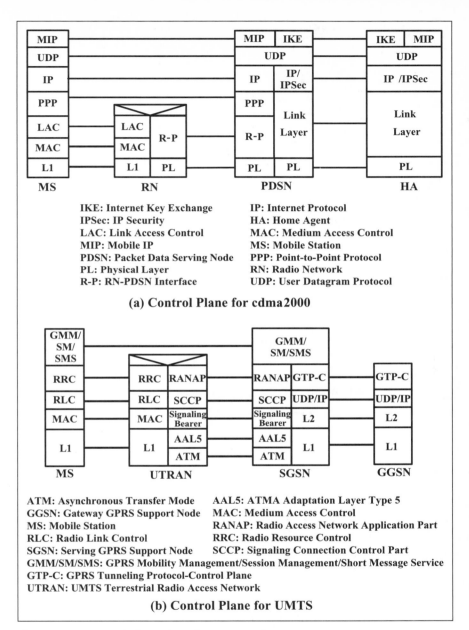

Figure 6.2   Control Planes for cdma2000 and UMTS

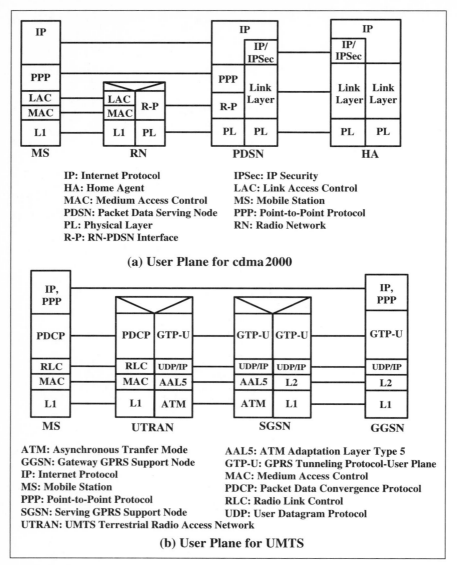

IP: Internet Protocol
HA: Home Agent
MAC: Medium Access Control
PDSN: Packet Data Serving Node
PL: Physical Layer
R-P: RN-PDSN Interface

IPSec: IP Security
LAC: Link Access Control
MS: Mobile Station
PPP: Point-to-Point Protocol
RN: Radio Network

(a) User Plane for cdma 2000

ATM: Asynchronous Tranfer Mode
GGSN: Gateway GPRS Support Node
IP: Internet Protocol
MS: Mobile Station
PPP: Point-to-Point Protocol
SGSN: Serving GPRS Support Node
UTRAN: UMTS Terrestrial Radio Access Network

AAL5: ATM Adaptation Layer Type 5
GTP-U: GPRS Tunneling Protocol-User Plane
MAC: Medium Access Control
PDCP: Packet Data Convergence Protocol
RLC: Radio Link Control
UDP: User Datagram Protocol

(b) User Plane for UMTS

Figure 6.3 User Planes for cdma2000 and UMTS

o resource through the A1 interface. The details for further
cdma2000 specifications [3GP01]. In UMTS, the UTRAN
transfer procedure for handling mobility and session man-
es exchanged between the MS and the SGSN. The direct
re relays these messages between the MS and the SGSN
e message formats without interpreting the contents of the
ma2000, these messages are delivered using the PPP session

(between an MS and the PDSN); the radio network is not involved in format translation of these messages.

In UMTS, the PS domain services are supported by the *Packet Data Convergence Protocol (PDCP)* in the user plane. The PDCP contains compression methods that are needed to provide better spectral efficiency for IP packet transmission over the radio. In cdma2000, the header and payload compression mechanism is provided by PPP between an MS and the PDSN.

Both the UMTS *Radio Link Control (RLC)* protocol (see Figures 6.2 (b) and 6.3 (b)) and cdma2000 *Link Access Control (LAC)* protocol (see Figures 6.2 (a) and 6.3 (a)) provide segmentation and retransmission services for user and control data. In addition, the cdma2000 LAC protocol supports the authentication functionality for wireless access, which is equivalent to GPRS transport layer authentication in UMTS. In the remainder of this section, we elaborate on the protocols regarding IP support and tunneling.

6.1.1   Point-to-Point Protocol

In both control and user planes for cdma2000, Point-to-Point Protocol (PPP) is carried over the LAC/MAC; and R-P tunnels (elaborated in Section 6.1.2) are utilized to establish a connection between an MS and the PDSN. The PPP provides a standard method for transporting multi-protocol datagrams (for example, IP) over point-to-point links [Sim 94]. The PPP encapsulates network-layer datagrams over a serial communications link, and enables two network nodes to negotiate particular types of network layer protocols (such as IP) to be used during a session. In cdma2000, a PPP connection is equivalent to a packet data session, which is comparable to the UMTS PDP context described in Section 6.2.2. After establishing the PPP connection, the MS registers with the MIP HA to indicate its presence and to acquire a temporary home address (which is an IP address). The details are given in Section 6.3.

In the UMTS control plane, no PPP/IP connection is established between MS and SGSN. Instead, signaling is carried over the RRC and Iu connections as previously described. The UMTS user plane provides two alternatives for IP services. For the scenario in which IP is supported by non-PPP lower layer protocols (specifically, PDCP in the MS and GTP-U in GGSN), the IP address of an MS is either permanently (statically) allocated or dynamically assigned by the GGSN. Alternatively, the MS's IP address can be assigned by the network nodes outside the GGSN, where the IP protocol is supported over PPP. This alternative is utilized when the MIP is introduced to UMTS for global network roaming. In this case, the MIP HA is responsible for the MS's IP address assignment. That is, the MS sends the MIP registration message to the MIP HA via the PPP connection before obtaining a temporary home address.

6.1.2   Tunneling Protocols

In both UMTS and cdma2000, tunneling is used to support various types of mobility. Figure 6.4 shows the tunneling approaches used in UMTS and cdma2000. In UMTS, the PDCP/RLC/MAC links between the MS and the UTRAN support radio link mobility (that is, handoff). The GTP tunnels between the UTRAN and the SGSN and between the SGSN and the GGSN are used to provide RNC relocation (see Section 2.7 and Chapter 5) and core network mobility, respectively. Conversely, cdma2000 utilizes MIP, R-P, and LAC/MAC connections to support IP mobility, core network mobility, and radio mobility, respectively.

We focus on the comparison of core network tunneling implementations for UMTS and cdma2000. GTP is utilized in both the user and control planes of UMTS.

- In the user plane, *GTP User Plane* (*GTP-U*) provides services for carrying user data packets between SGSNs and/or GGSNs, and between an SGSN and an RNC.

- In the control plane, *GTP Control Plane* (*GTP-C*) provides GTP-U tunnel-related management and MS mobility-related management between SGSNs and GGSNs.

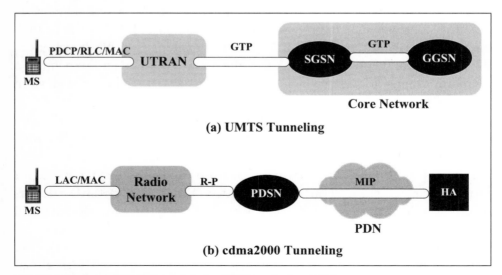

Figure 6.4   Tunneling Approaches for UMTS and cdma2000

Conversely, cdma2000 utilizes the MIP-based tunneling protocol to maintain the R-P connection between PCF and PDSN. We compare the following two aspects of the UMTS GTP and cdma2000 MIP-based tunneling protocols:

Tunnel Establishment: In UMTS, the SGSN initiates tunnel establishment through the **Create PDP Context Request** message to the GGSN. In UMTS, every UMTS node (RNC, GGSN, or SGSN) is assigned an IP address. A UMTS GTP tunnel is identified with a *Tunnel Endpoint Identifier (TEID)*, an IP address (that is, the IP address of the receiving node), and a UDP port number. The TEID is locally assigned by the receiving side of a GTP tunnel, which unambiguously identifies the tunnel endpoint. The TEID values are exchanged between the tunnel endpoints using GTP-C messages.

In cdma2000, PCFs and PDSNs are assigned unique IP addresses. The PCF uses the **MIP Registration Request** message to initiate R-P connection establishment. In this **MIP Registration Request** message, the care-of-address and home agent fields are set to the IP addresses of the PCF and PDSN, respectively. For each packet data bearer, the PCF assigns a *PCF Session Identifier (PSI)*, which is set in the session-specific extension of the **MIP Registration Request** message. The PSI, PCF-IP-Address, and PDSN-IP-Address form a unique identification for each R-P connection.

Tunnel Release: The UMTS GTP tunnel release procedure is initiated by either the SGSN or the GGSN that issues the **Delete PDP Context Request** message. Release of an R-P connection in the cdma2000 network is controlled by the PCF. The PCF initiates the R-P connection release procedure by sending the **MIP Registration Request** message to the PDSN, with the lifetime field set to zero (elaborated in Section 6.2.1). If the PDSN initiates release of an R-P connection, the **MIP Registration Update** message is sent from the PDSN to the PCF. Then the PCF sends the accounting-related information to the PDSN through the normal R-P connection release procedure.

6.2 Mobility and Session Management Mechanisms

As we previously mentioned, network architectures of mobility and session management are basically the same for both UMTS and cdma2000. However, the protocols exercised on these network architectures are different.

Table 6.1   Comparison of MM and SM between UMTS and cdma2000

	UMTS	cdma2000
Protocol (MM)	Mobile Application Part	Mobile IP
Involved Network Nodes (MM)	Serving GPRS Support Node and Home Location Register	Packet Data Serving Node and Home Agent
Protocol (SM)	GPRS Tunneling Protocol	Point-to-Point Protocol
Involved Network Nodes (SM)	Serving GPRS Support Node and Gateway GPRS Support Node	Packet Data Serving Node

Table 6.1 compares the mobility and session management mechanisms of UMTS and cdma2000. These mechanisms are implemented in layer 3 protocols of the control plane; specifically GPRS Mobility Management/Session Management/Short Message Service (GMM/SM/SMS) for UMTS (see Figure 6.2 (b)), and MIP (over UDP) and PPP for cdma2000 (see Figure 6.2 (a)).

6.2.1   Mobility Management

In UMTS, the base stations covered by an SGSN are partitioned into several *Routing Areas* (*RAs*). When the MS roams into a new RA, location update is performed. Mobility management in UMTS is carried out by the *Mobility Management* (*MM*) *Finite State Machine* (*FSM*). The MM FSMs executed in the MS and the SGSN are not the same. The details are given in Section 2.3. In this section, we only provide common parts of the MS and the SGSN MM state machines (see Figure 6.5), which give necessary details to compare

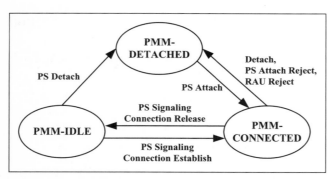

Figure 6.5   UMTS Mobility Management State Diagram (Common Parts in both the MS and the SGSN)

UMTS with cdma2000. By simplifying the description in Chapter 2, Figure 6.5 redefines the UMTS MM states:

- In the **PMM-DETACHED** state, the MS has not yet performed a PS attach and is not known to the core network nodes (for example, SGSN and HLR). When the MS attaches to the PS service domain (for example, the MS powers on and performs a PS attach), the FSM moves into **PMM-CONNECTED**.

- In the **PMM-IDLE** state, the PS signaling connection does not exist; that is, the Iu connection and RRC connection are released. The MS is attached to the PS service domain, and its location is known by the SGSN with the accuracy of an RA area; that is, the RA of an MS is tracked by the SGSN. In this state, the MS can be reached by paging, and no packet arrives at the MS. The FSM moves into **PMM-CONNECTED** if the packets are exchanged between the SGSN and the MS. The FSM moves into **PMM-DETACHED** if the PS detach procedure is executed.

- In the **PMM-CONNECTED** state, the PS signaling connection is established. Packets can only be delivered in this state. The SGSN tracks the MS with accuracy of the RA level, and the RNC is responsible for cell-level tracking. The FSM moves into **PMM-IDLE** if, for example, no packet is delivered for a period of time. The FSM moves into **PMM-DETACHED** if the PS detach procedure is executed.

In both the **PMM-IDLE** and the **PMM-CONNECTED** states, two types of RA update are performed:

- *Normal RA update* is performed when the RA of an MS has been changed.

- *Periodic RA update* is periodically initiated by an MS to report its "presence" to the network even if the MS does not move.

A *Periodic RA Update Timer (PRUT)* is maintained in the MS. The length of PRUT is sent from the SGSN to the MS in the **Routing Area Update Accept** or **Attach Accept** message. The timer is unique with an RA. Upon expiration of PRUT, the MS starts a periodic routing area update procedure. Corresponding to the PRUT, a *Mobile Reachable Timer (MRT)* is maintained in the SGSN. MRT is slightly longer than PRUT, and both of them are stopped in the **PMM-CONNECTED** state and started in the **PMM-IDLE** state.

In cdma2000, the MIP protocol is utilized for mobility management. Each PCF corresponds to a packet zone. An FA (a PDSN) usually covers one or more packet zones. Location update is performed when the MS roams into a new PCF area. When a change of packet zone is detected in the same FA coverage (that is, intra-PDSN movement), the MS registers by issuing the **Origination Message** to the BSC. Then the BSC establishes an A8/A9 connection to the new PCF. When an MS moves into the coverage area of a new FA (that is, inter-PDSN movement), the MIP registration is activated. Thus, in addition to sending the **Origination Message** to the BSC, the MS performs MIP registration to the HA by issuing the **Registration Request** message to its FA. A lifetime value is included in this message. If the registration request is accepted, the HA may grant or change the lifetime value. The negotiated lifetime is included in the **Registration Reply** message sent to the MS. The MS must perform the re-registration operation before the lifetime expires. Details of cdma2000 mobility management procedures are elaborated in Section 6.3.

To detach an MS from the cdma2000 network, the MS simply issues the **Registration Request** with a zero lifetime period, which is equivalent to the **Detach** message in UMTS. As previously mentioned, the MS initiates re-registration before the lifetime expires, which performs the same function as periodic RA update in UMTS.

6.2.2   Session Management

Both UMTS and cdma2000 adopt a similar pipe concept for session management. However, different approaches are used to implement the packet data session between an MS and a gateway node (that is, GGSN in UMTS and PDSN in cdma2000):

- A PPP connection represents a session pipe between the MS and the PDSN in cdma2000.

- A *Packet Data Protocol* (*PDP*) context represents a session pipe between the MS and the GGSN in UMTS, which consists of a PDCP tunnel between the MS and the RNC, a GTP-U tunnel between the RNC and the GGSN, and a GTP-U tunnel between the SGSN and the GGSN.

In UMTS, the PDP context provides information (specifically, PDP state for session management) to support packet delivery between an MS and the core network. That is, to support packet routing in a data communication session, PDP context activation is required, and the PDP contexts must be created in the MS, the GGSN, and the SGSN.

Some properties of the PDP context (such as QoS profile) are not found in the PPP connection of cdma2000. Figure 6.6 illustrates the PDP state

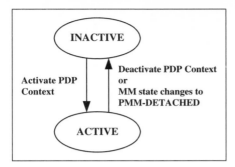

Figure 6.6 UMTS PDP State Diagram

diagram for the FSM of UMTS session management. Two states are defined in the diagram:

- In the **INACTIVE** state, the PDP context contains no routing or mapping information related to that PDP address. Therefore, no data can be transferred. The FSM moves to **ACTIVE** when the PDP context is activated.

- In the **ACTIVE** state, the PDP context for the PDP address is activated in the MS, the SGSN, and the GGSN. Specifically, the PDP context contains mapping and routing information about delivering PDP *Protocol Data Units (PDUs)* between the MS and the GGSN for that particular PDP address. The FSM moves to **INACTIVE** when the PDP context is deactivated or when the MM FSM moves to **PMM-DETACHED**.

In cdma2000, a session pipe is established between the MS and the PDSN through a PPP connection. Session management for cdma2000 is similar to that for UMTS. The session states are either "open" or "close". In the "open" state, a PPP connection is established between the MS and the PDSN. Conversely, the logical pipe between the MS and the PDSN does not exist in the "close" state. Figure 6.7 shows the cdma2000 packet data service state diagram. Three states are defined in this diagram. The relationship between the transmission links (i.e., A8/A10 connections and PPP link) and the packet data service states is elaborated as follows:

- In the **NULL/INACTIVE** state, the MS detaches from the PS service domain (e.g., powers off) and the packet services are released. That is, the physical traffic channel, the A8/A10 connections, and the PPP link are released or are not established. The FSM moves to the **AC-TIVE/CONNECTED** state when the MS attaches (for example, powers on) and the packet data services are requested.

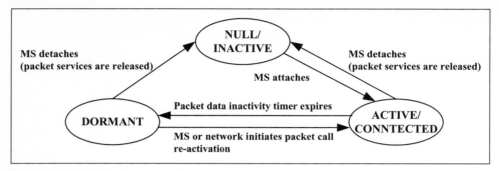

Figure 6.7 cdma2000 Packet Data Service State Transitions

- In the **ACTIVE/CONNECTED** state, a physical traffic channel exists between the MS and the BSC. The PPP link between the MS and the PDSN, the A8 connection between the BSC and the PCF, and the A10 connection between the PCF and the PDSN are maintained.

    The BSC for the MS maintains a *packet data inactivity timer*. This timer is reset when a non-idle data frame is sent or received. If the timer expires, the traffic channel is disconnected and the FSM moves to the **DORMANT** state. When the MS detaches, the FSM moves to the **NULL/INACTIVE** state.

- In the **DORMANT** state, a PPP link between the MS and the PDSN, and an A10 connection between the PCF and the PDSN are maintained. No physical traffic channel exists between the MS and the BSC, and the A8 connection between the BSC and the PCF is released. When a packet is targeted to the MS in the **DORMANT** state, the PDSN transmits this packet to the PCF through the A10 connection. Then the PCF establishes an A8 connection to the BSC. The BSC communicates with the MSC for packet session setup admission. After the MSC has approved the session request, the BSC pages the MS and establishes the traffic radio channel with the MS. At this point, the FSM moves to the **ACTIVE/CONNECTED** state. The **DORMANT** to **ACTIVE/CONNECTED** state transition may also occur when the MS initiates the packet call re-activation procedure (for example, a packet data call reconnection is requested by the MS). When the MS detaches, the FSM moves to the **NULL/INACTIVE** state.

From the preceding descriptions, it is clear that the A10/A11 connection and PPP link are maintained during the **ACTIVE/CONNECTED** and the **DORMANT** states. Conversely, the A8/A9 connection and the traffic radio channel are only maintained in the **ACTIVE/CONNECTED** state. In the

DORMANT state, we do not expect immediate packet transmission, and the connection between the MS and the PCF is not established so that the radio resources are not wasted.

6.2.3   Remarks on Mobility and Session Management

Mobility management and session management are closely related. In a cdma2000 network, if the MS moves across a PCF zone, the old A10/A11 connection is released and the new A10/A11 connection is established. The the PPP connection remains the same. If the MS moves across a PDSN area, then both the PPP and A10/A11 connections are re-established between the MS and the PDSN through the new PCF. If the MS moves to a new PCF zone during a communication session, then the A8/A9 connection also needs to be switched from the old PCF to the new PCF.

In UMTS, the attach procedure may or may not be followed by PDP context activation. That is, the session management procedure is independent of the mobility management procedure. This approach provides flexibility for PS domain services. For example, an MS can retrieve the PS domain SMS without activating the PDP context. In cdma2000, a MIP registration is always followed by an access registration. The advantage of this approach is that the session establishment time can be reduced if a communication session is required immediately after the attach.

As a final remark, UMTS supports multiple PDP contexts (that is, sessions) simultaneously for a communicating MS. On the other hand, in cdma2000, only one PPP connection can be supported at a time between an MS and the PDSN.

6.3   IP Mobility

IP mobility solutions (such as MIP-based tunneling) are different from the link-layer mobility solutions (such as GPRS tunneling). The link-layer mobility management function is used to manage tunnels between the core network nodes (for example, between the SGSN and the GGSN, or between the PCF and the PDSN). On the other hand, the IP mobility mechanism allows the mobile station to change its point of IP connectivity without losing ongoing sessions.

In cdma2000, two types of IP mobility are supported:

- *Simple IP*
- *Mobile IP*

In simple IP, IP address mobility is supported within a PDSN, but is not supported when the MS moves to a new PDSN. The IP address is dynamically assigned by the local PDSN. *Challenge Handshake Authentication Protocol (CHAP)* or *Password Authentication Protocol (PAP)* [Sim 96] are used for authentication between PDSN and the *Remote Authentication Dial-In User Service (RADIUS)* server. In mobile IP, both intra-PDSN and inter-PDSN IP address mobility are supported. In this case, the HA serves as the home router of an MS, and is responsible for location information maintenance and IP datagram tunneling of the MS. The HA interacts with the AAA entities to provide security association to the FA (a PDSN). Three kinds of AAAs are defined in cdma2000:

- The *visited AAA* forwards an AAA request from the PDSN to the home or the broker AAA based on the MS's *Network Access Identifier (NAI)*.

- The *broker AAA* provides secure AAA message delivery between the visited and the home AAAs.

- The *home AAA* authenticates the PDSN request based on the NAI. The home AAA also supports QoS services based on the AAA profile that contains the differentiated service policy and the HLR record.

FA challenge for authentication is performed through the MN-AAA Challenge Extension procedure [Per 00]. Note that in MIP, neither CHAP nor PAP are used because these protocols result in longer initial setup time due to additional RADIUS traversal.

In MIP, the IP address of an MS is statically assigned or dynamically assigned by the HA. This persistent IP address is maintained when the MS moves around PDSNs. The location update procedure is illustrated in the following steps (see Figure 6.8):

Step 1. When an MS moves into a new PDSN (MIP FA), a PPP connection is first established between the MS and the PDSN. Then the *agent discovery* procedure is exercised. Specifically, the MS sends the **Agent Solicitation** message and the PDSN broadcasts the **Agent Advertisement** message. Unlike the standard MIP environment, the MS is known by the PDSN/FA in cdma2000 during the PPP connection setup. Thus, the **Agent Solicitation** message can be eliminated for efficiency.

Step 2. Upon receipt of **Agent Advertisement**, the MS initiates the registration procedure by sending the **MIP Registration Request** to the PDSN.

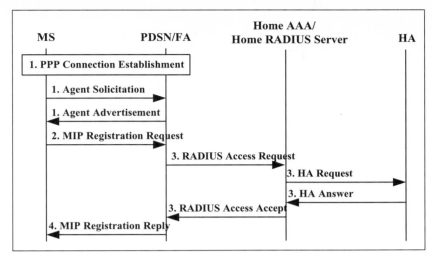

Figure 6.8 cdma2000 MIP Location Update Procedure

Step 3. The PDSN then stops broadcasting agent advertisement, and issues the **RADIUS Access Request** to the home RADIUS server (via
broker servers if required) to perform FA challenge for authentication.
If the authentication succeeds, the home RADIUS server sends the **HA
Request** message to the HA, which includes the MIP registration request of the MS. The HA validates the registration request and sends
a response back to the home RADIUS server. Then the home RADIUS
server acknowledges the PDSN by issuing the **RADIUS Access Accept**
message.

Step 4. The PDSN sends **MIP Registration Reply** to the MS, and the location update procedure is completed.

The GTP mechanism of UMTS/GPRS provides IP mobility within a GGSN
area. IP mobility for inter-GGSN areas is not supported. Without the MIP-
like solution, an MS must stay within a fixed GGSN area as long as the PDP
context is activated. Based on 3GPP Specification 23.923 [3GP00b], MIP is
used to provide IP mobility for inter-UMTS/GPRS networks through a three-
stage evolution:

- Stage 1 enables mobile users to roam between wireless LAN and UMTS
  by using MIP.

- Stage 2 supports IP mobility between two UMTS networks where the
  GGSN is changed during a session.

■ Stage 3 provides the same mechanism as Stage II except that the SGSN
and the GGSN are combined into one node, which is similar to the
cdma2000 solution in Figure 6.8.

Details for the three stages are described as follows. In Stage 1, the current GPRS bearer transport is maintained to handle mobility within UMTS
networks. MIP enables a user to roam between wireless LAN and UMTS
without losing an ongoing session (e.g., TCP), where the GGSN acts as a
MIP FA. The Access Point Name (APN) is used to find the desired GGSN,
and the MS stays with this GGSN as long as the PDP context is activated.
Since MIP is implemented in the application level, all MIP signaling messages
are transported over the UMTS/GPRS user plane. Figure 6.9 shows the MIP
registration procedure in Stage 1, and the steps are described as follows:

Step 1.1. The standard PS attach procedure is performed between the
MS and the SGSN. Then the MS sends the **Activate PDP Context Request** message to the SGSN. The APN "MIPv4FA" for MIP registration
is included in this message.

Step 1.2. After receiving the **Activate PDP Context Request** message, the
SGSN selects a suitable GGSN based on the APN. The selected GGSN
must be equipped with the MIP FA capability. Then the SGSN and the

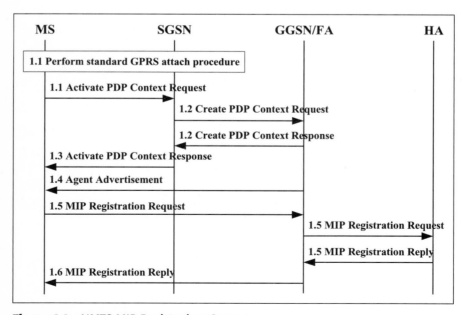

Figure 6.9   UMTS MIP Registration: Stage 1

GGSN exchange the **Create PDP Context Request** and **Response** message pair to set up the MS's PDP context. Note that the GGSN is not responsible for IP address assignment, and the IP address field of the PDP context is not filled at this point.

Step 1.3. The **Activate PDP Context Response** message is sent from the SGSN to the MS to indicate that creation of the PDP context was successful. At this point, the bearer-level connection is established.

Step 1.4. The GGSN (MIP FA) periodically broadcasts the **Agent Advertisement** message to announce its presence. When the MS receives this message, Step 1.5 is executed.

Step 1.5. The **MIP Registration Request** message is sent from the MS to the GGSN across the UMTS/GPRS backbone through the user plane (that is, the bearer-level connection constructed in Steps 1.1–1.3). The GGSN forwards the request to the HA. The HA assigns a home (IP) address to the MS and includes this address in the **MIP Registration Reply** message. Then the message is sent from the HA to the GGSN.

Step 1.6. The GGSN receives the **MIP Registration Reply** message from the HA and extracts the needed information (per example, the home IP address of the MS) in this message. Then the **MIP Registration Reply** message is forwarded to the MS to indicate that the MIP registration procedure is complete.

Steps 1.1–1.3 establish GPRS bearer-level connection. Then, on top of this connection, Steps 1.4–1.6 perform application-level registration (that is, MIP registration). Also note that the IP address of the MS is assigned by the HA instead of the GGSN.

In Stage 2, efficient packet re-routing is supported. The core network is the same as that in Stage 1. During a session, the GGSN in the communication path may be changed in two situations:

- After inter-SGSN handoff, the GGSN is changed to optimize the route.
- The GGSN may also be changed for load balancing purposes.

Figure 6.10 illustrates the scenario for changing GGSN after the change of SGSN. When changing GGSN, two tunnels are maintained between the new SGSN and the old GGSN, and between the new SGSN and the new GGSN (see Figure 6.10 (c)). This will reduce the possibility of packet loss. Figure 6.11 shows the procedure for changing SGSNs and GGSNs in Stage 2.

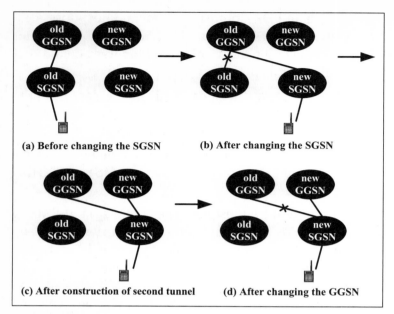

Figure 6.10   UMTS Scenario for Changing SGSNs and GGSNs: Stage 2

Step 2.1. After an inter-SGSN handoff, the new SGSN decides to change to an "optimal" GGSN. The **Create PDP Context Request** and **Response** message pair is exchanged between the new SGSN and the new GGSN to create a new PDP context.

Step 2.2. The HA, the GGSN (MIP FA), and the MS perform the standard MIP registration procedure as described at Steps 1.4–1.6 in Figure 6.9.

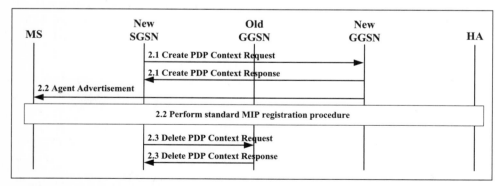

Figure 6.11   UMTS Handoff in Stage 2

Step 2.3. After successful creation of the new PDP context, a timer is set in the new SGSN. When the timer expires, the new SGSN instructs the old GGSN to delete the old PDP context by sending the **Delete PDP Context Request** message. The old GGSN deletes the PDP context and responds to the new SGSN with the **Delete PDP Context Response** message. At this time, the GGSN handoff procedure is complete.

In Stage 3, the SGSN and the GGSN will be combined into one node called *Internet GPRS Support Node (IGSN)*, and MIP is utilized to handle inter-IGSN handoff. The IGSN main functionality includes support of UMTS mobility management across UTRANs, interaction with HLR, and provision of FA functionality. Details of the IGSN procedures are similar to those in Figure 6.9 except that the SGSN and the GGSN are merged into an IGSN, which is similar to the WGSN design described in Chapter 14.

In cdma2000, the Internet AAA functionality is utilized to provide authentication, authorization, and accounting. In UMTS, if the mobile operator and the *Internet Service Provider (ISP)* are different, then AAA is also required when MIP is introduced to provide inter-system IP mobility. In IETF, the AAA working group has incorporated requirements provided by the mobile IP working group. Figure 6.12 illustrates the network architecture for AAA support of MIP in UMTS. In this figure, the MS roams to a visited UMTS network. After the MS has successfully performed the UMTS-based authentication with the HLR in the home UMTS network, the AAA mechanism (between the visited AAA and the home AAA) is initiated to authenticate the MS. Then the MS performs MIP registration to the HA in the home ISP to update the location of the MS.

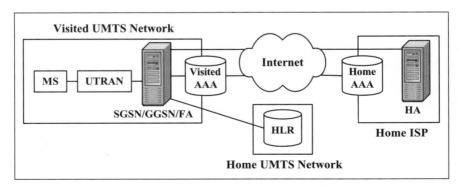

Figure 6.12 Network Architecture for AAA Support in UMTS

6.4   UMTS and cdma2000 Interworking

This section describes an IP-level mechanism for UMTS-cdma2000 interworking. Several countries (such as Taiwan, Japan, and Korea) have issued both UMTS and cdma2000 licenses to provide 3G services. In this case, UMTS-cdma2000 interworking is essential to support global roaming between these two systems. To achieve complete interoperability between UMTS and cdma2000, the interworking issues at the radio, the core network, and the IP-levels must be addressed. We focus on the IP-level interworking that supports roaming between the UMTS and the cdma2000 networks.

Based on the UMTS and cdma2000 mobility management mechanisms, we describe an architecture and algorithms for UMTS-cdma2000 interworking to provide real-time IP multimedia services. Figure 6.13 shows the architecture that interconnects UMTS and cdma2000. In this figure, the UMTS network connects to the IP network through the GGSN that acts as a MIP FA. Following the PS attach and PDP context activation procedures described in Section 6.3, the MS performs the MIP registration to its HA via the GGSN (MIP FA). The HA maintains the MS's location information (per example, the GGSN/MIP FA address) and tunnels the IP packets to the MS. In the cdma2000 network, the PDSN is a MIP FA, which communicates with the HA through the IP network as described in Section 6.3. Both the UMTS and cdma2000 networks connect to the IP multimedia network through the MIP HA. That is, when a terminal in the IP multimedia network originates

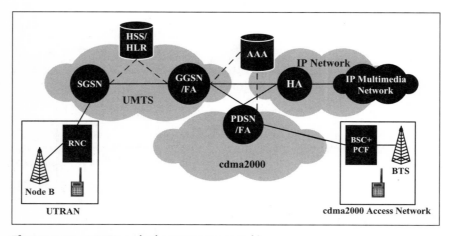

Figure 6.13   UMTS and cdma2000 Interworking

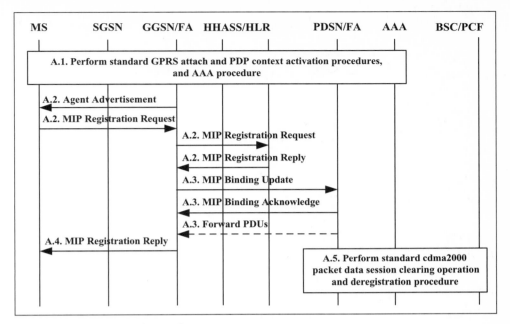

Figure 6.14 Roaming from cdma2000 to UMTS

a multimedia call to the UMTS or cdma2000 MS by using protocols such as H.323 [ITU03] or *Session Initiation Protocol* (*SIP*; see Chapter 12), the voice packets are first delivered to the MIP HA. Then the MIP HA forwards these packets to the MS through the MIP FA (a GGSN or a PDSN).

When an MS in communication roams between cdma2000 and UMTS systems, inter-system roaming occurs. The inter-system roaming message flow from cdma2000 to UMTS (see Figure 6.14) is described as follows:

Step A.1. The MS performs the standard attach and PDP context activation procedures to register with the UMTS network and set up the bearer communication. This step is the same as Steps 1.1–1.3 in Figure 6.9. Then the AAA procedure is performed at the end of this step.

Step A.2. After creating the PDP context, the GGSN serves as a MIP FA to issue the **MIP Agent Advertisement** message to the MS. Upon receipt of the FA advertisement message, the MS performs the MIP registration to the HA via the GGSN (MIP FA). This step is the same as Steps 1.4 and 1.5 in Figure 6.9.

Step A.3. The GGSN (new FA) and the PDSN (old FA) exchange the **MIP Binding Update** and **MIP Binding Acknowledge** message pair to

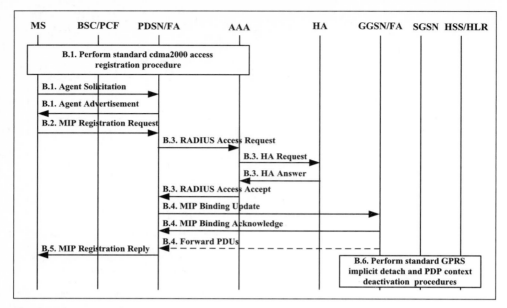

Figure 6.15   Roaming from UMTS to cdma2000

perform data forwarding, which provides smooth handoffs between the old and new FAs [Per 98]. Then the PDSN starts to forward the arriving datagrams to the GGSN. The forwarding operation is used to minimize packet loss during the inter-system roaming procedure.

Step A.4. The GGSN responds to the MS's MIP registration request by sending the **MIP Registration Reply** message. This step is the same as Step 1.6 in Figure 6.9.

Step A.5. Finally, the PDSN performs the standard cdma2000 packet data session clearing operation and de-registration procedure with the BSC/PCF and AAA.

The inter-system roaming message flow from UMTS to cdma2000 is illustrated in Figure 6.15, and the executed steps are described below:

Step B.1. The standard cdma2000 access registration procedure is performed. Then the PDSN and the MS exchange the **Agent Solicitation** and **Agent Advertisement** message pair to perform the standard MIP agent discovery operation. This step is the same as Step 1 in Figure 6.8.

Step B.2. Upon receipt of **Agent Advertisement**, the MS performs the MIP registration to the PDSN as described in Step 2 of Figure 6.8.

Step B.3. The PDSN, the AAA, and the HA perform standard authentication and MIP registration procedures. This step is the same as Step 3 in Figure 6.8.

Step B.4. The PDSN and the GGSN exchange the **MIP Binding Update** and **MIP Binding Acknowledge** message pair as described in Step A.3 of Figure 6.14.

Step B.5. Through **MIP Registration Reply**, the PDSN informs the MS that the MIP registration is successful. This step is the same as Step 4 in Figure 6.8.

Step B.6. When the MRT (see Section 6.2.1) expires, the implicit detach procedure is performed by the SGSN. At this point, the PDP contexts of the MS are also deactivated at the GGSN and the SGSN.

6.5   Concluding Remarks

Based on [Pan 04a], this chapter described mobility and session management mechanisms for UMTS and cdma2000, and compared the design guidelines for these mobile core network technologies. We first introduced network architectures and protocol stacks for UMTS and cdma2000. Then we elaborated on UMTS and cdma2000 mobility management and session management, and discussed the differences between the design guidelines of these two systems. The IP mobility mechanisms implemented in UMTS and cdma2000 were also demonstrated and compared. Based on the UMTS and cdma2000 mobility management mechanisms, we described an architecture and intersystem roaming procedures for UMTS-cdma2000 IP-level interworking. In this approach, a mobile user can roam between the UMTS and cdma2000 systems without losing the ongoing communication session.

In this chapter, we focused on the UMTS and cdma2000 PS service domains. The reader is referred to [Kim 03] for interoperability between the CS service domains of UMTS and cdma2000. For more details about Mobile IP and AAA operations in UMTS, the reader is referred to [Par 02, 3GP00b].

Section 6.4 addressed UMTS and cdma2000 interworking issues. For radio-level interworking, a dual mode terminal that contains both UMTS and cdma2000 radio interfaces and the capability to perform real-time handoff between these two systems is required. For core network interworking, the MM and the PDP contexts have to be migrated between UMTS and cdma2000 networks. These issues are very complex and have not been effectively solved from the business and technical perspectives. These issues warrant further study.

6.6   Questions

1. Compare the UMTS SGSN/GGSN functions with that for PDSN in cdma2000.

2. Describe SDU in cdma2000. Which component in UMTS/GPRS performs the same function? (Hint: See Figure 10.12 in Section 10.2.)

3. What are the differences between the device authentications for UMTS and cdma2000?

4. What are the common guidelines for the radio and the core network resource management design followed by both UMTS and cdma 2000?

5. In cdma2000, both MM and SM protocols are implemented on the same lower-layer protocols. UMTS takes a difference approach. Describe the tradeoffs.

6. Compare the UMTS and the cdma2000 MM/SM message transfer mechanisms between the MS and the core network.

7. Compare the UMTS RLC with the cdma2000 LAC.

8. Does cdma2000 have a similar mechanism as PDP and MM contexts in UMTS?

9. What mechanisms in UMTS and cdma2000 support IP mobility?

10. What mechanisms in UMTS and cdma2000 support packet re-routing (when handoff occurs at the radio network)?

11. Describe the packet header compression mechanisms in UMTS and cdma2000. Which approach is better?

12. Describe the timers used to perform periodic RA location update in UMTS.

13. How does cdma2000 support periodic location update? How does it support the MS detach action?

14. Combine the MM state diagram in Figure 6.5 and the PDP state diagram in Figure 6.6. Is the resulting state diagram the same as that in Figure 6.7 for cdma2000?

15. Compare the authentication procedures conducted in registration for UMTS and cdma2000.

16. Describe the relationship between the SRNC relocation in Chapter 5 and the scenario for changing SGSNs and GGSNs in UMTS Stage 2 (see Figure 6.10).

17. Draw the message flows of the registration and intersystem handoff procedures for IGSN.

18. The HA is the key network node for roaming between cdma2000 and UMTS. Describe the location update procedure for roaming from cdma2000 to UMTS. From the HA's viewpoint, what are the differences between this procedure and that for roaming from a UMTS network to another UMTS network?

19. What are the advantages and disadvantages of MIP compared to GTP?

20. Evaluate the performance of the cdma2000-to-UMTS and the UMTS-to-cdma2000 location update procedures. What are the bottleneck steps in these procedures?

CHAPTER 7

UMTS Charging Protocol

In UMTS, the extension of the *GPRS Tunneling Protocol (GTP)* called *GTP'* is utilized to transfer the *Charging Data Records* or *Call Detail Record (CDRs)* from *GPRS Support Nodes (GSNs)* to *Charging Gateways (CGs)*. Figure 7.1 illustrates a simplified UMTS network to show the relationship between the GSNs (Figure 7.1 (d) and (e)) and the CG (Figure 7.1 (f)). The CG collects the billing and charging information from the GSNs. Several IP-based interfaces are defined among the GSNs, the CGs, and the external PDN. In the *Gn* interface, the GTP [3GP05d] transports user data and control signals among the GSNs. The GGSN connects to the PDN through the Gi interface. In the *Ga* interface, the GTP' protocol is utilized to transfer CDRs from the GSNs to the CGs. When a Mobile Station (MS) is receiving a UMTS PS service, the CDRs are generated based on the charging characteristics (data volume limit, duration limit, and so on) of the subscription information for that service. Each GSN only sends the CDRs to the CG(s) in the same UMTS network. A CG analyzes and possibly consolidates the CDRs from various GSNs, and passes the consolidated data to a billing system. The CG maintains a GSN list. An entry in the list represents a GTP' connection to a GSN. This entry consists of pointers to a CDR database and the sequence numbers of possibly duplicated packets. The CDR database is a nonvolatile

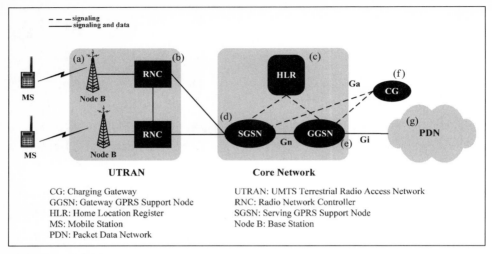

- - - signaling
——— signaling and data

UTRAN

Core Network

CG: Charging Gateway
GGSN: Gateway GPRS Support Node
HLR: Home Location Register
MS: Mobile Station
PDN: Packet Data Network

UTRAN: UMTS Terrestrial Radio Access Network
RNC: Radio Network Controller
SGSN: Serving GPRS Support Node
Node B: Base Station

Figure 7.1 UMTS Network Architecture with Charging Gateway

storage. Data stored in this database is analyzed and consolidated before the CG sends it to the billing system. The CG is associated with a *restart counter* that records the number of restarts performed at the CG. Details of this counter are described in Section 7.1. For redundancy reasons, a CG may also maintain a configurable list of peer CG addresses (e.g., to be able to recommend other CGs to the GSNs).

A GSN maintains a list of CGs in the priority order (typically, ranges from 1 to 100). This CG list can be configured by the *Operations and Maintenance Center (OMC)*. If a GSN unexpectedly loses its connection to the current CG, it may send the CDRs to the next CG in the priority list. An entry in the CG list describes parameters for GTP′ transmission. The entry includes pointers to the buffers containing the unacknowledged CDR packets and the sequence numbers of possibly duplicated packets. The entry also stores the restart counter of the corresponding CG.

After sending a GTP′ request, a GSN may not receive a response from the CG due to network failure, network congestion, or temporary node unavailability. In this case, 3GPP TS 29.060 [3GP05d] defines a mechanism for request retry whereby the GSN will retransmit the message until either a response is received within a timeout period or the number of a retry threshold is reached. In the latter case, the GSN-CG communication link is considered disconnected, and an alarm is sent to the OMC. For a GSN-CG link failure, the OMC may cancel CDR packets in the CG and unacknowledged sequence numbers in the GSN.

7.1 The GTP' Protocol

The GTP' protocol is used for communications between a GSN and a CG, which can be implemented over UDP/IP or TCP/IP. GTP' utilizes some aspects of GTP defined in 3GPP TS 29.060 [3GP05d]. Specifically, *GTP Control Plane (GTP-C)* is partly reused. Figure 7.2 illustrates a GTP' service model. In our design in [Hun05], the GTP' protocol is built on top of UDP/IP. Above the GTP' protocol, a *charging agent* (or CDR sender) is implemented in the GSN and a *charging server* is implemented in the CG. Our GTP' service model follows the *primitive flow model* described in Sections 3.2 and 8.3. In this model, a GSN communicates with a CG through a dialog by invoking GTP' service primitives. A service primitive can be one of four types: Request (REQ), Indication (IND), Response (RSP), Confirm (CNF) and Reject (REJ). A service primitive is initiated by a GTP' service user of the dialog initiator.

In Figure 7.2, the dialog initiator is a GSN and the service user is a charging agent. The charging agent issues a service primitive with type REQ. This service request is sent to the GTP' service provider of the GSN. The service provider sends the request to the dialog responder (the CG in Figure 7.2) by creating a GTP' message. This GTP' message is delivered through lower-layer protocol, i.e., UDP/IP. When the GTP' service provider of the CG receives the request, it invokes the same service primitive with type IND to the charging

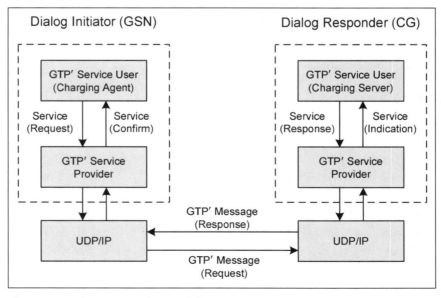

Figure 7.2 The GTP' Service Model

				Bits				
Octets	8	7	6	5	4	3	2	1
1		Version		PT	Spare ' 1 1 1 '			'0/1'
2				Message Type				
3–4				Length				
5–6				Sequence Number				
7–8				FlowLabel				
9				SNDCP N-PDULLC Number				
10				Spare ' 1 1 1 1 1 1 1 1 '				
11				Spare ' 1 1 1 1 1 1 1 1 '				
12				Spare ' 1 1 1 1 1 1 1 1 '				
13–20				TID				

(a) GTP Header (Version 0)

				Bits				
Octets	8	7	6	5	4	3	2	1
1		Version		PT	Spare ' 1 1 1 '			'0/1'
2				Message Type				
3–4				Length				
5–6				Sequence Number				

(b) 6-octet Header

Figure 7.3 GTP′ Header Formats

server (GTP′ service user). The charging server then performs appropriate operations, and invokes the same service primitive with type RSP. This response primitive is a service acknowledgment sent from the CG to the GSN. After the GTP′ service provider of the GSN receives this response, it invokes the same service primitive with type CNF.

If a dialog is initiated by the CG, then the roles of the CG and the GSN are exchanged in Figure 7.2. Based on the preceding GTP′ service model, this section describes the GTP′ message format. As defined in 3GPP TS 32.215 [3GP04j], the GTP′ header may follow the standard 20-octet GTP header format (Figure 7.3 (a)) [3GP02a] or a simplified 6-octet format (Figure 7.3 (b)). The 6-octet GTP′ header is the same as the first 6 octets of the standard GTP header. Octets 7–20 of the GTP header are used to specify a data session between a GSN and the MS. These octets are not needed in GTP'. In Figure 7.3, the first bit of octet 1 is used to indicate the header format. If the value is 1, the 6-octet header is used. If the value is 0, the 20-octet standard GTP header is used. Note that better GTP' performance is expected by using the 6-octet format, because the unused GTP header fields are eliminated. On the other hand, it is easier to support GTP' in an existing GTP environment if the standard GTP header format is used. In Figure 7.3, the *Protocol Type (PT)* and the *Version* fields are used to specify the protocol being used (GTP or GTP' in R99, R4, R5, and so on). For a GTP' message, PT=0. The *Length* field indicates the length of the payload. The *Sequence Number* is used as the transaction identity.

The GTP' message types are listed in Figure 7.4. Three GTP message types are reused in GTP', including **Echo Request, Echo Response,** and **Version Not Supported.** The **Echo Request/Response** message pair is typically used to check if the peer is alive. These path management messages are required if GTP' is supported by UDP. Specifically, the **Echo Request** is sent from a GSN to find out if the peer CG is alive. In 3GPP TS 29.60 [3GP05d], the **Echo Request** is periodically sent for more than 60 seconds on each connection. Whenever a CG receives the **Echo Request,** it replies with the **Echo Response,** which contains the value of its local restart counter. As we previously mentioned, this counter is maintained in both the GSN and the CG to indicate the number of restarts performed at the CG. If the restart counter value received by the GSN is larger than the value previously stored, the GSN assumes that the CG has restarted since the last **Echo Request/Response**

Message Type Value	GTP' Message
1	Echo Request
2	Echo Response
3	Version Not Supported
4	Node Alive Request
5	Node Alive Response
6	Redirection Request
7	Redirection Response
240	Data Record Transfer Request
241	Data Record Transfer Response

Figure 7.4 GTP' Message Types

message pair exchange. In this case, the GSN may retransmit the earlier unacknowledged packets to the CG, rather than wait for expiration of their timers.

The **Node Alive Request/Response** message pair is used to inform that a CG has restarted its service after a service break. For example, the service break may be caused by hardware maintenance. When a CG's service is stopped, the CG sends the **Redirection Request** message to inform a GSN to redirect its CDRs to another CG. This message can also be used to balance the workloads among the CGs.

The **Data Record Transfer Request/Response** message pair is used for CDR delivery. In the **Data Record Transfer Request** message, the header is followed by two *Information Elements (IEs)*. The first IE is a code indicating "Send Data Record Packet". The second IE consists of one or more CDRs. In the **Data Record Transfer Response** message, the header is followed by a cause IE. This IE is a code that indicates how a CDR is processed in the CG (e.g., Request Accepted, No Resource Available, and so on).

7.2 Connection Setup Procedure

Before a GSN can send CDRs to a CG, a GTP′ connection must be established between the charging agent in the GSN and the charging server in the CG. The GTP′ connection setup procedure is described in the following steps (see Figure 7.5):

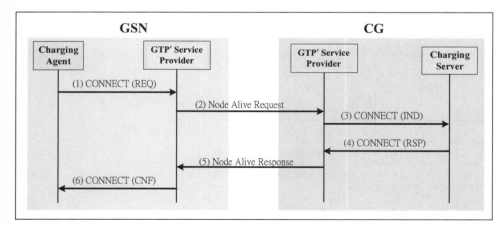

Figure 7.5 GTP′ Connection Setup Message Flow

Step 1. The charging agent instructs the GTP′ service provider to set up a GTP′ connection. This task is performed by issuing the CONNECT (REQ) primitive with the CG address.

Step 2. The service provider generates the **Node Alive Request** message and delivers it to the CG through UDP/IP. The UDP source port number is locally allocated at the GSN. On the CG side, the default UDP destination port number 3386 is reserved for GTP′ [3GP04j]. The CG is allowed to reconfigure this destination port number.

Step 3. The GTP′ service provider of the CG interprets the **Node Alive Request** message and reports this connection setup event to the charging server via the CONNECT (IND) primitive.

Step 4. The charging server creates and sets a new entry (for this new connection) in the GSN list, and responds to the service provider with the CONNECT (RSP) primitive. The charging server is either ready to receive the CDRs or it is not available for this connection. In the latter case, the charging server may include the address of a recommended CG in the CONNECT (RSP) primitive for a further redirection request.

Step 5. Suppose that the CG is available. The GTP′ service provider generates the **Node Alive Response** message, and delivers this message to the GSN.

Step 6. The GTP′ service provider of the GSN receives the **Node Alive Response** message. It interprets the message and reports this acknowledgment event to the charging agent through the CONNECT (CNF) primitive. The charging agent creates and sets the CG entry's status as active in the CG list. At this point, the setup procedure is complete.

7.3 GTP′ CDR Transfer Procedure

The charging agent is responsible for CDR generation in a GSN. The CDRs are encoded, using, for example, the *Abstract Syntax Notation One (ASN.1)* format [3GP04j]. The charging server is responsible for decoding the CDRs, and returns the processing results to the GSN. The CDR transfer procedure is illustrated in Figure 7.6 and is described in the following steps:

Step 1. The charging agent encodes the released CDR. Then it invokes the CDR_TRANSFER (REQ) primitive. This primitive instructs the GTP′ service provider to generate the **Data Record Transfer Request** message.

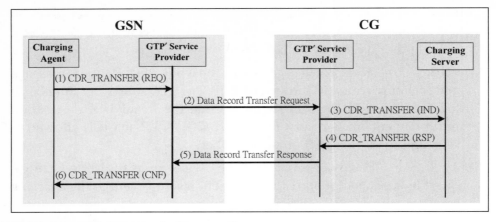

Figure 7.6 GTP' CDR Transfer Message Flow

Step 2. The service provider includes the CDR in the **Data Record Transfer Request** message and sends it to the CG.

Step 3. When the service provider of the CG receives the GTP' message, it issues the CDR_TRANSFER (IND) primitive to inform the charging server that a CDR is received. The charging server decodes the CDR and stores it in the CDR database. This CDR may be consolidated with other CDRs, and is later sent to the billing system.

Steps 4 and 5. The charging server invokes the CDR_TRANSFER (RSP) primitive that requests the GTP' service provider to generate the **Data Record Transfer Response** message. The cause IE value of the message is "Request Accepted". The service provider sends this GTP' message to the GSN.

Step 6. The GTP' service provider of the GSN receives the **Data Record Transfer Response** message and reports this acknowledgment event to the charging agent via the CDR_TRANSFER (CNF) primitive. The charging agent deletes the delivered CDR from its unacknowledged buffer.

7.4 GTP' Failure Detection

This section describes the *path failure detection algorithm* that detects path failure between the GSN and the CG. Figure 7.7 illustrates the data structures utilized to implement the path failure detection algorithm. In a GSN, an entry

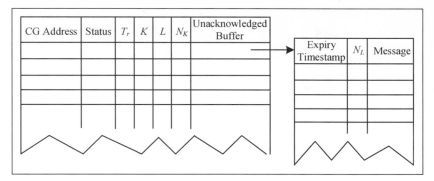

Figure 7.7   Data Structures for Path Failure Detection Algorithm

in the CG list represents a GTP′ connection to a CG. We describe the entry attributes related to the path failure detection algorithm as follows:

- The CG Address attribute identifies the CG connected to the GSN.

- The Status attribute indicates whether the connection is "active" or "inactive".

- *Charging Packet Ack Wait Time T_r* is the maximum elapsed time the GSN is allowed to wait for the acknowledgment of a charging packet; typical allowed values range from 1 millisecond to 65 seconds.

- *Maximum Number of Charging Packet Tries L* is the number of attempts (including the first attempt and the retries) the GSN is allowed to send a charging packet; L typically ranges from 1 to 16. When $L = 1$, there is no retry.

- *Maximum Number of Unsuccessful Deliveries K* is the maximum number of consecutive failed deliveries that are attempted before the GSN considers a connection failure occurs. Note that a delivery is considered failed (or timed out) if it has been attempted for L times without receiving any acknowledgment from the CG.

- The *Unsuccessful Delivery Counter N_K* attribute records the number of consecutive failed delivery attempts.

- The *Unacknowledged Buffer* stores a copy of each GTP′ message that has been sent to the CG but has not been acknowledged. A record in the unacknowledged buffer consists of an *Expiry Time Stamp t_e*, the *Charging Packet Try Counter N_L*, and an unacknowledged GTP′ message. The expiry time stamp t_e is equal to T_r plus the time when the GTP′ message was sent, which represents the expiry of the message.

The counter N_L counts the first attempt and retries that have been performed for this charging packet transmission.

The path failure detection algorithm works as follows:

Step 1. After the connection setup procedure in Section 7.2 is complete, both N_L and N_K are set to 0, and the Status is set to "active". At this point, the GSN can send GTP' messages to the CG.

Step 2. When a GTP' message is sent from the GSN to the CG at time t (see Step 2, Section 7.3), a copy of the message is stored in the unacknowledged buffer, where the expiry time stamp is set to $t_e = t + T_r$.

Step 3. If the GSN has received the acknowledgment from the CG before t_e (see Step 6, Section 7.3), both N_L and N_K are set to 0.

Step 4. If the GSN has not received the acknowledgment from the CG before t_e, N_L is incremented by 1. If $N_L = L$, then the charging packet delivery is considered failed. N_K is incremented by 1.

Step 5. If $N_K = K$, then the GTP' connection is considered failed. The Status is set to "inactive".

When Step 5 of the path failure detection algorithm is encountered, it is assumed that the path between the GSN and the CG is no longer available, and the GSN should be switched to another CG. However, besides link failure, unacknowledged packet transfers may also be caused by temporary network congestion. In this case, it is not desirable to perform CG switching (which is an expensive operation). An alternative to avoid this kind of "false" failure detection is to set large values for parameters T_r, L, and K. However, large parameter values may result in delayed detection of "true" failures. Therefore, it is important to select appropriate parameter values so that true failures can be quickly detected while false failures can be avoided.

7.5 Concluding Remarks

In UMTS, the GTP' protocol is used to deliver the CDRs from GSNs to CGs. This chapter described the UMTS charging protocol. To ensure that the mobile operator receives the charging information, availability for the charging system is essential. One of the most important issues regarding GTP' availability is connection failure detection. In [Hun05], we studied the GTP' connection failure detection mechanism specified in 3GPP TS 29.060 and 3GPP TS 32.215 (see also question 1 in Section 7.6). The output measures

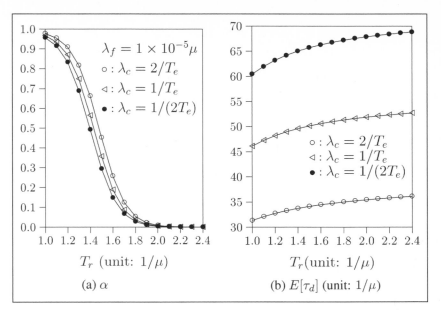

Figure 7.8   Effects of T_r and λ_c ($K = 6, L = 1$)

considered are the false failure detection probability α and the expected elapsed time $E[\tau_d]$ for true failure detection. We proposed an analytic model to investigate how these two output measures are affected by input parameters, including the charging packet ack wait Time T_r, the maximum number L of charging packet tries, and the maximum number K of unsuccessful deliveries. Figures 7.8 and 7.9 show α and $E[\tau_d]$ as functions of T_r, the charging packet arrival rate λ_c, K, and L. In these figures, T_e is the inter-**Echo** message arrival time (a fixed interval) and $1/\mu$ is the expected round-trip delay of the GTP's message pair. We have the following observations:

- In Figure 7.8, when T_r is small, increasing T_r decreases the false failure detection probability significantly. When T_r is sufficiently large, increasing T_r only has insignificant impact on the false failure detection probability. On the other hand, increasing T_r always non-negligibly increases the expected elapsed time for true failure detection.

- The false failure detection probability increases as the charging packet arrival rate λ_c increases. This effect is insignificant when T_r becomes large. On the other hand, Figure 7.9 shows that the effects of λ_c on the expected elapsed time for true failure detection are not the same for different (K, L) setups. In our examples (where $1 \leq L \leq 6$), when λ_c is large, the expected failure detection time is larger for $L = 6$ than for

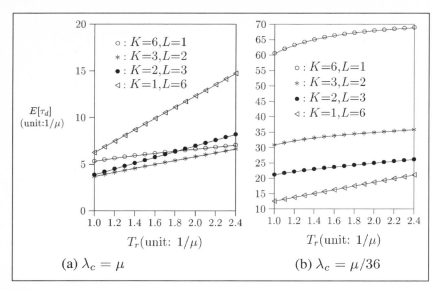

Figure 7.9   Effects of T_r and L on $E[\tau_d]$

$L = 1$. When λ_c is small, the expected time of true failure detection is smaller for $L = 6$ than for $L = 1$. Therefore, the effects of λ_c should be considered when selecting the L value.

7.6   Questions

1. The performance of the GTP′ connection failure detection mechanism can be measured by the probability of false failure detection and the expected true failure detection time. Consider the derivation of α as follows. Let random variable t_f be the lifetime between when the GTP′ connection is established and when a true failure occurs. During this period, undesirable false failures (temporary network congestion) may be detected, and the GSN is unnecessarily switched to another CG. Let α be the probability that the path failure detection algorithm detects a false failure (and therefore the GSN is switched to another CG before a true failure occurs). Suppose that t_f has the density function $f_f(t_f)$. Let the arrivals of charging packets be a Poisson stream with rate λ_c. Note that the charging packets delivered between a GSN and the CG are generated by all users in this GSN. Each CDR stream of an individual user may have an arbitrary distribution, but

the net traffic of all users becomes a Poisson stream [Mit87]. We observed that the charging packets form a Poisson stream when there are more than 20 users. Let the **Echo** message arrivals be a deterministic stream with the fixed interval T_e. For any reasonable setting, an **Echo** message should not be issued before the previous one is acknowledged or timed out. Thus, in CG configuration, we set $T_e \geq LT_r$. Based on the above assumptions, please derive α. (Hint: See the solution in http://liny.csie.nctu.edu.tw/supplementary)

2. In Section 7.5, we stated that the effects of λ_c on the expected elapsed time for true failure detection are not the same for different (K, L) setups. In our examples (where $1 \leq L \leq 6$), when λ_c is large, the expected failure detection time is larger for $L = 6$ than for $L = 1$. When λ_c is small, the expected time of true failure detection is smaller for $L = 6$ than for $L = 1$. Explain why. (Hint: See the discussion in [Hun05].)

3. Use the GTP$'$ protocol to design a prepaid mechanism for the GPRS or UMTS PS services. (Hint: See Section 12.3 for the description of a prepaid mechanism.)

4. GTP$'$ can be implemented in UDP, TCP, or SCTP (described in Chapter 8). Describe the advantages and disadvantages of these approaches.

5. With the **Redirection Request** message, the CGs can be switched. Design a load-balancing algorithm using GTP$'$. What are the criteria for switching the CGs? How does the switching cost affect the switching decision in your algorithm?

Mobile All-IP Network Signaling

Signaling System Number 7 (SS7) [Rus02] provides control and management functions in the telephone network. SS7 consists of supervisory functions, addressing, and call information provisioning. An SS7 channel conveys messages to

- initiate and terminate calls,
- determine the status of some part of the network, and
- control the amount of allowed traffic.

Traditional SS7 signaling is implemented in a *Message Transfer Part (MTP)*-based network, which is utilized in the existing mobile networks including GSM and GPRS. Also, in the UMTS Packet Switched (PS) service domain, the SS7 messages are delivered between the Serving GPRS Support Nodes (SGSNs) and the Home Location Register (HLR). Conversely, in UMTS all-IP architecture, the SS7 signaling will be carried by an IP-based network. The low costs and the efficiencies for carriers to maintain a single, unified telecommunications network guarantee that all telephony services will eventually be delivered over IP [Che05c]. This chapter describes the design and implementation of IP-based network signaling for the mobile all-IP network.

8.1 Signaling System Number 7

Separate from the voice trunk network, SS7 uses an out-of-band network to carry signaling messages (i.e., the signaling messages and user data are delivered in different networks). Figure 8.1 shows a typical SS7 network—the thin lines represent the signaling links and the thick lines represent voice trunks. The SS7 network consists of three distinct components:

Service Switching Point (SSP) is a telephony switch that performs call processing on calls that originate, tandem, or terminate at that node. In UMTS, a Mobile Switching Center (MSC) is an SSP.

Service Control Point (SCP) contains databases for providing enhanced services. An SCP accepts queries from an SSP and returns the requested information to the SSP. In UMTS, the HLR and the VLR are implemented in SCPs.

Signal Transfer Point (STP) is a switch that relays SS7 messages between SSPs and SCPs. Based on the address fields of the SS7 messages, the STPs route the messages to the appropriate outgoing signaling link. To meet the stringent reliability requirements, STPs are provisioned in

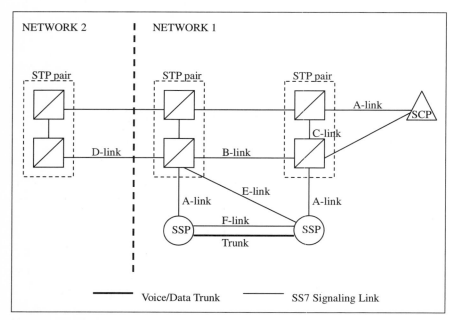

Figure 8.1 SS7 Architecture

mated pairs. Some UMTS number portability solutions are implemented in STPs, as described in Chapter 13. In UMTS, the STP provides *Global Title Translation* (*GTT*), which may be used to route queries from a *Gateway MSC* (*GMSC*) to the HLR. Note that for every call to a Mobile Station (MS), the call is first routed to the MS's GMSC, as described in Section 9.2.1.

In an SS7 network, the trunks (voice circuits) connect SSPs to carry user data/voice information. The signaling links provide direct connection between the following pairs:

- SSPs and STPs
- SCPs and STPs
- STPs and STPs

The SSPs and the SCPs are connected indirectly through the STPs. As shown in Figure 8.1, there are six types of SS7 signaling links:

Access Links (A-links) connect the SSP/STP pairs or the SCP/STP pairs. Each SSP and SCP will have a minimum of one A-link to each STP pair.

Bridge Links (B-links) connect STPs in different pairs. B-links are deployed in a quad arrangement with 3-way path diversity.

Cross Links (C-links) connect mated STPs in a pair.

Diagonal Links (D-links) are the same as B-links except that the connected STPs belong to different SS7 networks, for example, one in the PSTN and the other in UMTS.

Extended Links (E-links) provide extra connectivity between an SSP and the STPs other than its home STP.

Fully-Associated Links (F-links) connect SSPs directly.

In the existing SS7 networks, E-links and F-links typically are not deployed. The signaling links are monitored such that the failure links are automatically detected and the traffic load is shared by the active links.

The SS7 protocol follows the International Organization for Standardization's Open System Interconnect (ISO/OSI) model, which was designed for ease of adding new features. The SS7 protocol layers and the corresponding OSI layers are shown in Figure 8.2. These layers are described as follows:

Message Transfer Part (MTP) [ANS96] consists of three levels corresponding to the OSI physical layer, data link layer, and network layer,

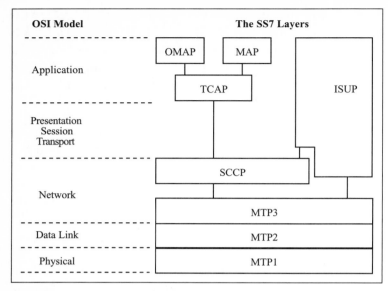

Figure 8.2 SS7 Signaling Protocol Stack

respectively. The MTP level 1 (MTP1) defines the physical, electrical, and functional characteristics of the signaling links connecting SS7 components. The MTP level 2 (MTP2) provides reliable transfer of signaling messages between two directly connected signaling points. The MTP level 3 (MTP3) provides the functions and procedures related to message routing and network management.

Signaling Connection Control Part (SCCP) [ANS01] provides additional functions such as *Global Title Translation (GTT)* to the MTP. The MTP utilizes GTT to transfer non-circuit-related signaling information such as the 1-800 services and the mobile telecommunications service (e.g., registration and cancellation). In UMTS, SCCP is responsible for reliable connection between UTRAN and the Serving GPRS Support Node (SGSN).

Transaction Capabilities Application Part (TCAP) [ANS00b] provides the capability to exchange information between applications using non-circuit-related signaling.

Integrated Services Digital Network User Part (ISUP) [ANS00a] establishes circuit-switched network connections (e.g., for call setup). Pass-along signaling service sends the signaling information to each switching point involved in a call connection.

Operations, Maintenance, and Administration Part (OMAP) is a TCAP application for network management.

Mobile Application Part (MAP) is a TCAP application that supports mobile roaming management. GSM/UMTS MAP [3GP05e] is implemented at this layer, and is elaborated in Section 8.3.

The MTP and the SCCP provide routing services between a mobile telecommunications network and the PSTN. In UMTS, the wireless call setup/release are executed by using the ISUP. The ISUP messages for call setup and release are used in wireless call control throughout this book, and are introduced here:

- When a phone number is dialed, the originating SSP sends the ISUP **Initial Address Message** (IAM) to the terminating SSP, which initiates trunk setup. The originating SSP marks the circuit busy, and the circuit information is carried by the IAM.

- When the IAM arrives at the terminating SSP, the SSP sends an **Address Complete Message** (ACM) to the originating SSP. There are three possibilities:

    - If the called line is idle, the **ACM** indicates that the routing information required to complete the call has been received by the terminating SSP. The message also informs the originating SSP of the called party information, charge indications, and end-to-end protocol requirements. The terminating SSP provides a ring-back tone through the setup trunk to the calling party.

    - If the called line is busy, the terminating SSP sends the busy tone to instruct the caller to hang up the phone. There are several alternatives. The calling party may be asked to leave a voice message if the voice mail service is activated. The called party may still be alerted if the call waiting service is available. The call may also be forwarded to another telephone line if the call-forwarding-on-busy service is enabled.

    - If the called line is not busy, but call setup cannot be completed (for example, the called number is not allocated), then the terminating SSP plays an appropriate tone or announcement.

- When the called party answers the call, the **Answer Message** (ANM) is sent from the terminating SSP to the originating SSP. The message indicates that the call has been answered. At this moment, the call is established (through the trunk path) and the conversation begins.

- Assume that the calling party hangs up the phone, the originating SSP sends the **Release Message** (REL) to indicate that the specified trunk is being released from the connection. The specified trunk is released before the **REL** is sent, but the trunk is not set to idle at the originating SSP until the terminating SSP releases the trunk.

- When the terminating SSP receives the **REL** message from the originating SSP, it replies with the **Release Complete Message** (**RLC**) to confirm that the indicated trunk has been placed in an idle state.

8.2   Stream Control Transmission Protocol (SCTP)

Transmission Control Protocol (TCP) provides reliable data transfer in IP networks. However, TCP is not an ideal protocol for SS7 signaling transport for the following reasons:

- TCP provides strict order of transmission. This feature is not required in SS7 transport, but will cause the *Head of the Line* (*HOL*) blocking problem. HOL blocking occurs when the messages are transferred over a single TCP connection. If a message gets lost, other in-sequence messages must be postponed for delivery until the lost message has been retransmitted.

- The TCP socket does not support *multi-homing*. A host is considered multi-homed if two or more IP addresses can be used as destination addresses to reach that host. The multi-homed host is often equipped with multiple network interfaces, and the IP addresses are allocated and processed in different network interfaces. A TCP application can only bind a single IP address to the TCP connection. When the network interface fails, the connection is lost and probably cannot be reestablished.

- TCP is vulnerable to blind *Denial-of-Service* (*DoS*) attacks such as flooding SYN attacks.

The IETF *Signaling Transport* (*SIGTRAN*) working group addresses issues regarding the transport of packet-based PSTN signaling (i.e., SS7 signaling) over IP networks [Str00]. SIGTRAN defines not only the architecture but also a suite of protocols to carry SS7 messages over IP. This suite of protocols consists of a new transport protocol, the *Stream Control Transmission Protocol* (*SCTP*) and a set of *user adaptation* layers, which provides the same services of the lower layers of the traditional SS7. For example, *MTP3 User Adaptation Layer* (*M3UA*) provides the same MTP3 services to the MTP3-user (i.e., the SCCP or the ISUP) [IET02].

Like TCP, SCTP provides reliable IP connection. Specifically, SCTP employs TCP-friendly congestion control (including slow-start, congestion avoidance, and fast retransmit). Unlike TCP, SCTP provides selective acknowledgments for packet loss recovery, and message-oriented data delivery service. Moreover, SCTP offers new delivery options (ordered or unordered) and features that are particularly desirable for SS7 signaling. An example is *multi-homing*, whereby an SCTP association (i.e., a connection) allows the SCTP endpoints of the association to have multiple IP addresses for reliability enhancement. Another useful feature is *multi-streaming*, which independently delivers among data streams to prevent additional delay caused by the HOL blocking problem. SCTP also features a *four-way handshake* to establish an association, which is resistant to blind DoS attacks and thus improves overall protocol security.

Figure 8.3 illustrates the SCTP packet format. The SCTP packet begins with an SCTP common header. This header includes the following fields:

Source Port Number (Figure 8.3 (1)) is the SCTP sender's port number, which identifies the association to this SCTP packet.

Destination Port Number (Figure 8.3 (2)) is the SCTP port number for the destination.

Verification Tag (Figure 8.3 (3)) is used to verify the sender of the SCTP packet.

Checksum (Figure 8.3 (4)) is used by the Adler32 algorithm [Str00] to maintain the packet's integrity.

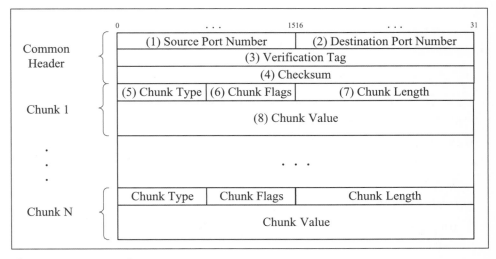

Figure 8.3 SCTP Packet Format

```
ID Value        Chunk Type
--------        ----------------
0               Payload Data (DATA)
1               Initiation (INIT)
2               Initiation Acknowledgement (INIT ACK)
3               Selective Acknowledgement (SACK)
4               Heartbeat Request (HEARTBEAT)
5               Heartbeat Acknowledgement (HEARTBEAT ACK)
6               Abort (ABORT)
7               Shutdown (SHUTDOWN)
8               Shutdown Acknowledgement (SHUTDOWN ACK)
9               Operation Error (ERROR)
10              State Cookie (COOKIE ECHO)
11              Cookie Acknowledgement (COOKIE ACK)
12              Reserved for Explicit Congestion Notification Echo (ECNE)
13              Reserved for Congestion Window Reduced (CWR)
14              Shutdown Complete (SHUTDOWN COMPLETE)
```

Figure 8.4   SCTP Chunk Types

The body part of the SCTP packet consists of one or more *chunks*, which contain either control or data information. Each chunk has a chunk header that includes the following fields:

Chunk Type (Figure 8.3 (5)) identifies the type of information contained in the Chunk Value field. Figure 8.4 lists the defined chunk types. These types determine the usage of the **Chunk Flags** field (Figure 8.3 (6)) and the Chunk Value field.

Chunk Length (Figure 8.3 (7)) indicates the length of the chunk.

Chunk Value (Figure 8.3 (8)) contains the actual information to be transferred.

The DATA chunk format is illustrated in Figure 8.5. The DATA chunk header includes the following fields:

Type for this chunk is 0 (DATA; see Figure 8.4).

Unordered bit indicates whether the DATA chunk uses an ordered or unordered delivery service.

Beginning and **Ending Fragment Bits** are used for the SCTP message fragmentation when the message size is larger than the maximum transmission units allowed in the transport path.

Chunk Length indicates the length of the chunk.

Transmission Sequence Number (TSN) and **Stream Sequence Number** (SSN) (see Figure 8.5 (1) and (3)) provide two separate sequence numbers on every DATA chunk. The TSN is used for per-association

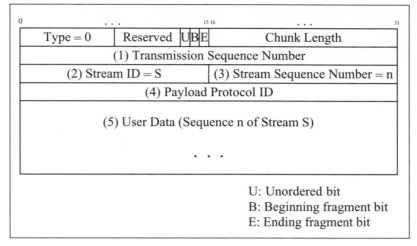

Figure 8.5 DATA Chunk Format

reliability. Each DATA chunk is assigned a TSN that allows the receiving SCTP endpoint to acknowledge the received packets and detect duplicate deliveries. The SSN is used for per-stream ordering. The usage of these two numbers is given in Section 8.7.3.

Payload Protocol Identifier (Payload Id; see Figure 8.5 (4)) is used to identify the type of information being carried in this DATA chunk.

User Data (see Figure 8.5 (5)) contains the user data.

Stream Identifier (Stream Id; see Figure 8.5 (2)) identifies the stream of the User Data.

The SCTP also defines *control chunks* to carry information needed for association functionality such as association establishment, association termination, data acknowledgment, failure detection and recovery, etc.

The SCTP uses a four-way handshake and cookie mechanism to establish an association that prevents blind DoS attacks. After the association is established, the two SCTP endpoints can transfer data by using the DATA chunk. The association termination is a three-way handshake process which can be initiated by either one of the endpoints engaged in the association (to be elaborated in Section 8.7.2).

8.3 UMTS Network Signaling

UMTS utilizes the MAP protocol for network signaling. As shown in Figure 8.6 (a), GSM/UMTS MAP is an application of the SS7 protocol. The MAP

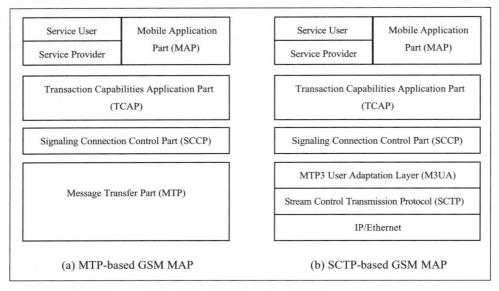

Figure 8.6   MAP Protocol Hierarchy (MTP-based vs SCTP-based)

uses SCCP classes 0 and 1 connectionless services to provide efficient routing with or without maintaining message sequencing. The network entities
may consist of several *Application Service Elements* (*ASEs*). The SCCP addresses these ASEs with subsystem numbers. The subsystem numbers for
MAP ASEs such as HLR, VLR, and MSC are 6, 7, and 8, respectively. For
intra-UMTS network message delivery, the destination address of the message may be a simple *Destination Point Code* (*DPC*) that can be used by
the MTP for direct routing. For inter-UMTS network message delivery, the
originating node does not have enough knowledge to identify the actual
address of the destination. In this case, the SCCP translates the actual
destination address by GTT [ANS01]. Then the SCCP layer invokes the
MTP_TRANSFER primitive to provide the MTP layer message transfer services. The MTP layer generates the MTP message, and the message is sent to
the destination over the MTP-based SS7 network. The MTP-based protocol
layers can be replaced by the SCTP-based protocol hierarchy as shown in
Figure 8.6 (b). When the SCCP layer invokes the MTP_TRANSFER primitive
to the M3UA layer, the M3UA layer generates the appropriate M3UA packet.
Then the packet is sent to the destination through the SCTP/IP/Ethernet
layers.

The UMTS network entities (such as SGSN and HLR) communicate with
each other through MAP dialogues by invoking MAP service primitives. Following the primitive flow model shown in Figure 3.6, a MAP service primitive

can be one of the following types: REQ (Request), IND (Indication), RSP (Response), CNF (Confirm) or REJ (Reject). The service primitive is initiated by a MAP service user of a network entity called the *dialogue initiator*. The service type is REQ. This service request is sent to the MAP service provider of the network entity. The service provider delivers the request to the peer network entity (i.e., the *dialogue responder*) by the TCAP. When the MAP service provider of the peer network entity receives the request, it invokes the same service primitive with type IND to inform the destination MAP service user. In most cases, the information (parameters) of the service with type IND is identical to that with type REQ. The primitive is typically a query, which asks the dialogue responder to perform some operations.

A corresponding service acknowledgment, with or without results, may be sent from the dialogue responder to the dialogue initiator. The same service primitive with type RSP is invoked by the MAP service user of the dialogue responder. After the MAP service provider of the dialogue initiator receives this response, it invokes the same service primitive with type CNF. The parameters of the CNF and the RSP services are identical in most cases, except that the CNF service may include an extra provider error parameter to indicate a protocol error.

A MAP dialogue consists of several MAP services to perform a task. The services are either specific or common. The specific services include the following:

- Mobility services
- Operation and maintenance services
- Call-handling services
- Supplementary services
- SMS management services

An example of mobility services is the MAP_SEND_AUTHENTICATION_INFO primitive, described later. The common MAP services are used to establish and clear MAP dialogue between peer MAP service users. They invoke functions supported by the TCAP and report abnormal situations. Some common MAP services defined in [3GP05e] are described as follows:

MAP_OPEN is used to establish a MAP dialogue. This service is confirmed by the service provider; that is, MAP_OPEN has REQ/IND and RSP/CNF types.

MAP_CLOSE is used to clear a MAP dialogue. This service is not confirmed by the service provider; that is, the service primitive only has the REQ/IND types, but not the RSP/CNF types.

MAP_DELIMITER is used to explicitly request the TCAP to transfer the MAP protocol data units to the peer entities. This service does not have any parameters and is not confirmed by the service provider.

8.4 UMTS MAP Software Architecture

The same MAP layer implementation can be supported by both the MTP-based and the SCTP-based approaches. Although the lower-layer implementations for these two approaches are very different, the modular design of the MAP enables quick porting to different implementations of the TCAP layer. In this section, we describe a modular MAP design and two SS7 implementations.

8.4.1 The MAP Layer Architecture

This section describes a MAP layer architecture based on the MAP version 1.4 software developed by Trillium Digital Systems, Inc. [Tri]. Figure 8.7

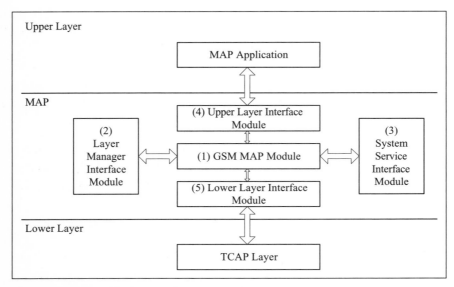

Figure 8.7 MAP Software Architecture

illustrates the software architecture of the MAP layer, which includes the following modules:

MAP Module (Figure 8.7 (1)) implements standard MAP service primitives based on the 3GPP TS 09.02 [3GP05e]. An example of specific primitives is MAP_ SEND_ AUTHENTICATION_ INFO, described in Section 8.6. Examples of the common primitives are MAP_OPEN, MAP_CLOSE, and MAP_ DELIMITER, described in Section 8.3. The session management software described in Chapter 3 is implemented in this module.

Layer Manager Interface Module (Figure 8.7 (2)) provides functions to control and monitor the status of the MAP layer. The functions include layer resources configuration, layer resources activation/deactivation, layer information tracing, and layer status indication.

System Service Interface Module (Figure 8.7 (3)) provides functions required for the MAP layer buffer management, timer management, date and time management, resource checking, and initialization. All functions related to the Operating System (OS) are also included in this module.

Upper Layer Interface Module (Figure 8.7 (4)) provides a function-based interface to interact with the upper layer (i.e., the application program) through the function calls.

Lower Layer Interface Module (Figure 8.7 (5)) provides an interface to communicate with the lower layer (i.e., TCAP).

The modular design of the MAP layer implementation provides flexibility and portability, which enables quick porting of the MAP software to different TCAP implementations. We only need to modify the Lower Layer Interface Module for the specific TCAP implementations. We can also port the MAP software to various OSs by modifying the System Service Interface module.

8.4.2 An MTP-based SS7 Implementation

This section illustrates an SS7 implementation based on the Performance Technologies, Inc., 372 series SS7 card. The SS7 card is a host-independent SS7 controller board embedded with full SS7 functionality based on MTP [ADC00]. The SS7 card can be plugged into a host (i.e., a computer) with the Peripheral Component Interconnect (PCI) or Compact Peripheral

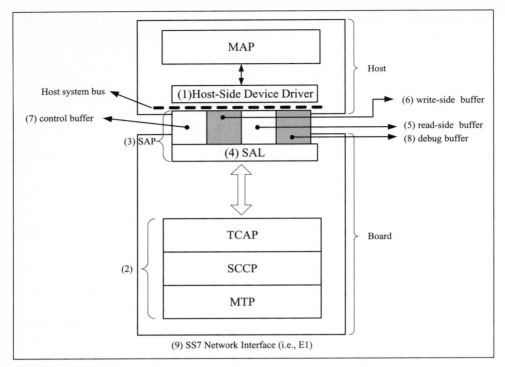

Figure 8.8   SS7 Card-based Architecture (an MTP-based Approach)

Component Interconnect (CPCI) interface. Figure 8.8 illustrates the MTP-based SS7 architecture consisting of a host and the SS7 card. The MAP layer is implemented in the host. The configuration and protocol interfaces for the SS7 card are implemented in the Lower Layer Interface Module of the MAP layer at the host (Figure 8.7 (5)). The MAP layer interacts with the TCAP layer in the SS7 card through the device driver (Figure 8.8 (1)). All SS7 protocol layers (i.e., TCAP, SCCP, and MTP; see Figure 8.8 (2)) necessary for communication with the SS7 network entity reside on the SS7 card. Each protocol layer is implemented as a software component on the SS7 card.

The lower-layer software components (i.e., TCAP, SCCP, and MTP) are installed in the SS7 card by a host-resident download program each time the SS7 card is reset. The host device driver communicates with the SS7 card through the Service Access Point (SAP; see Figure 8.8 (3)) on the SS7 card. The SAP consists of the *Service Access Layer* (SAL; see Figure 8.8 (4)) software, and four buffers implemented in a shared memory region on the SS7 card. The SAL manages the license key and the shared memory, and distributes messages to the appropriate layers and buffers. The read-side buffer (Figure 8.8 (5)) and write-side buffer (Figure 8.8 (6)) are used to exchange messages between the host and the SS7 card. The control buffer (Figure 8.8 (7)) is used by the SAL to manage the buffers and to keep track of information, such as read-side and write-side buffer pointers and semaphores. The debug buffer (Figure 8.8 (8)) is used to transfer diagnostic information from the SS7 card to the host.

The SS7 card also includes an SS7 physical layer interface (i.e., E1; Figure 8.8 (9)) and a Programmable Read-Only Memory (PROM; not shown in Figure 8.8) that is used to download the SAL software component from the host. To initialize the SS7 card, the PROM software is activated with a hardware reset, and remains active until the first software component (i.e., the SAL) is downloaded. After the SAL is downloaded, it takes control from the SS7 card PROM and disables the PROM. Then the MTP2, the MTP3, the SCCP, and the TCAP software components can be downloaded into the SS7 card. At this point, the host can invoke the TCAP services through the device driver. Readers are referred to [ADC00] for more details. The SS7 software components are implemented as modules similarly to the approach described in the next subsection. The details are omitted.

8.4.3 An SCTP-based SS7 Implementation

The SCTP protocol is implemented on top of the Internet Protocol (IP). Figure 8.9 illustrates the SCTP-based protocol stack. A *Stack Entity* (represented by a dashed rectangle) is a set of modules that implement the functionalities defined in the corresponding specifications (e.g., RFC 2960 [Str00]). The software architecture includes the following modules:

OS Interface Module provides OS-independent functions to be invoked by the stack entities. These functions are mapped to actual function calls provided by the underlying OS. The functions include buffer management, timer management, and so on.

System Management Interface Module in each layer provides functions to control and monitor the specific stack entity. Examples of the functions include system initialization, statistic collection, and layer information tracing.

User Interface and Transport Layer Interface Modules provide function-based interfaces to interact with upper and lower layers through the function calls. For example, the functions include

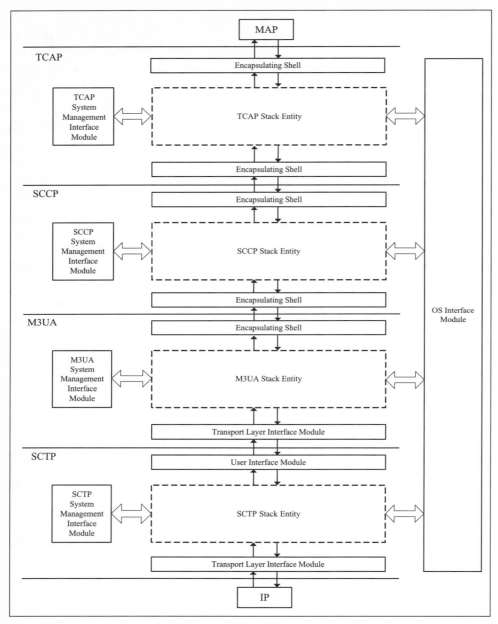

Figure 8.9   Software Architecture of an SCTP-based Approach

transmission service invoking and incoming data indication. These modules are used for communication between the M3UA and the SCTP.

Encapsulating Shell provides a message-based interface between upper and lower layers. The message-based interface is implemented by socket to allow communications between two different processes (e.g., communication between the TCAP and the SCCP).

The TCAP, the SCCP, and the M3UA/SCTP are implemented as individual processes. These processes communicate with each other by the message-based interface (i.e., socket). With this process architecture, we can independently deploy each layer. This modular design of the process structure is also easy to debug, maintain, analyze, and port to various underlying OSs.

8.5 TCAP and SCCP Based on M3UA

Both SCTP and M3UA stack entities include the modules shown in Figure 8.10, and are described as follows:

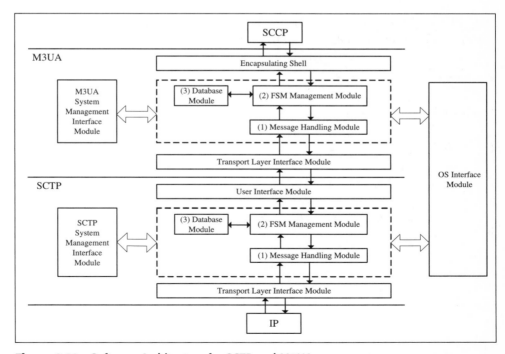

Figure 8.10 Software Architecture for SCTP and M3UA

Message Handling Module (Figure 8.10 (1)) provides functions for message parsing (e.g., message parameter checking) and message building (e.g., message header generation).

Finite State Machine (FSM) Management Module (Figure 8.10 (2)) implements the functionalities and maintains the FSMs defined in the specific protocol recommendations. At the SCTP layer, the *association state machine* is implemented [Str00]. At the M3UA layer, the *adjacent signaling point state machine* is implemented [IET02]. These state machines implement the SCTP and M3UA functions described in Section 8.7.

Database Module (Figure 8.10 (3)) maintains information required in the specific protocol layer. For example, in the SCTP layer, the Database Module maintains the association status, including the number of streams, IP addresses of the multi-homed SCTP endpoint, and so on.

The remainder of this section describes the SCCP and the TCAP stack entities built on top of the M3UA stack entity.

8.5.1 SCCP Stack Entity

The SCCP stack entity is designed based on SCCP functionality [ANS01], and includes the modules shown in Figure 8.11:

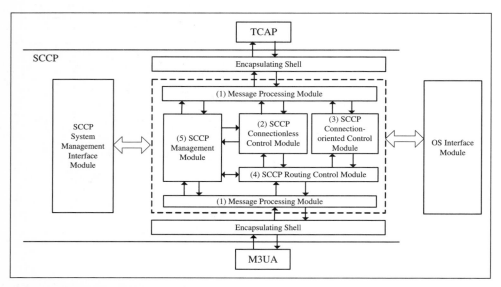

Figure 8.11 SCCP Software Architecture

Message Processing Module (Figure 8.11 (1)) provides needed functions for message parsing, message building, and message dispatch (e.g., connectionless service request or connection-oriented service request from the TCAP).

SCCP Connectionless Control Module (SCLC) (Figure 8.11 (2)) and **SCCP Connection-oriented Control Module (SCOC)** (Figure 8.11 (3)) handle message transfer in the corresponding services.

SCCP Routing Control Module (SCRC; Figure 8.11 (4)) determines the routes of the outgoing messages and dispatches the incoming messages to the SCLC or the SCOC.

SCCP Management Module (SCMG; Figure 8.11 (5)) provides the capabilities to handle congestion or failure at the SCCP layer.

8.5.2 TCAP Stack Entity

The TCAP stack entity consists of two sublayers: *Component Sublayer* (TC) and *Transaction Sublayer* (TR). Figure 8.12 illustrates the relationship between the TCAP sublayers and among the upper and lower stack entities. Figure 8.13 shows details of the modules implemented for the TCAP sublayers:

Component Sublayer

This layer is responsible for component handling and dialogue handling. This sublayer includes the *Component Coordinator Module* (Figure 8.13 (1)), the

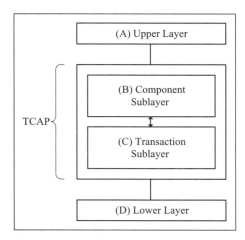

Figure 8.12 Abstract TCAP Software Architecture

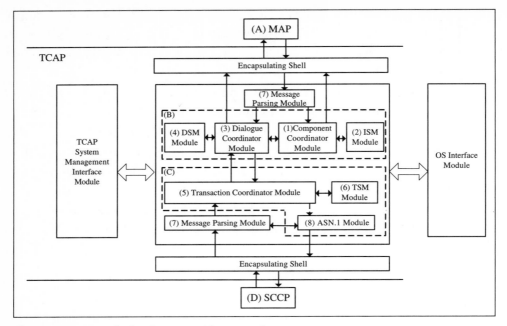

Figure 8.13   Detailed Software Architecture of TCAP

Invoke State Machine Module (ISM; see Figure 8.13 (2)), the *Dialogue Coordinator Module* (Figure 8.13 (3)), and the *Dialogue State Machine Module* (DSM; see Figure 8.13 (4)). In the implementation, the *Message Parsing Module* (Figure 8.13 (7)) and the *Encapsulating Shell Module* are used as interfaces between layers. They are not defined in the specifications, but are required in implementation. The Message Parsing Module parses and dispatches the message, and the Encapsulating Shell Module provides socket communications between two layers.

An operation is performed at the remote end. Invocation of an operation is identified by an invoke Id so that simultaneous invocations can be distinguished. Four operation classes are defined:

Class 1: Both success and failure are reported.

Class 2: Only failure is reported.

Class 3: Only success is reported.

Class 4: Neither success nor failure is reported.

Different types of state machines are defined for different classes of operations (referred to [ANS00b] for the details). The ISM Module (Figure 8.13 (2))

Table 8.1 Primitives for component handling

NAME	TYPE	DESCRIPTION
TC_INVOKE	REQ / IND	Invocation of an operation
TC_RESULT_L	REQ / IND	Returned result
TC_RESULT_NL	REQ / IND	Result of non-final part
TC_U_ERROR	REQ / IND	Indication of a failed operation
TC_L_CANCEL	IND	Operation termination due to timeout
TC_U_CANCEL	REQ	Operation termination due to TC-user decision
TC_L_REJECT	IND	Detection of invalid component
TC_R_REJECT	IND	Component rejection from the remote side
TC_U_REJECT	REQ / IND	Rejection of a component by the TC-user
TC_TIMER_RESET	REQ	Reset of an operation invocation timer

maintains the state machine for each invoked operation. A Finite State Machine (FSM) is associated with each of the four operation classes. A reply corresponds to one operation, and may be a returned result indicating success, a returned error indicating operation failure, or a reject indicating inability to perform the operation. Table 8.1 lists the primitives for component handling in the component sublayer. The Request (REQ) type indicates the primitive issued from the upper layer to the component sublayer, and the Indication (IND) type indicates the primitive issued from the component sublayer to the upper layer.

Dialogue handling is supported by the *Dialogue Coordinator Module* (Figure 8.13 (3)) to exchange TCAP components within a dialogue. The component sublayer provides two kinds of dialogues: *unstructured dialogue* and *structured dialogue*. The unstructured dialogue is used to deliver components that do not require replies (i.e., the class 4 operations). An unstructured dialogue is also called a *unidirectional dialogue*, whereby the components are grouped in one single unidirectional message. Conversely, in a structured dialogue, the TC-users (i.e., the MAP layers) indicate the beginning, continuation, and end of a dialogue. The MAP layer may exchange successive messages that contain components within this dialogue. Multiple structured dialogues can run simultaneously, and each of them is identified by a unique dialogue Id. The DSM Module (Figure 8.13 (4)) maintains the state machine of each structured dialogue [ANS00b]. Table 8.2 lists the primitives related to dialogue handling. In this table, the primitives TC_UNI, TC_BEGIN, TC_CONTINUE, and TC_END cause invoked component(s) to be delivered

Table 8.2   Primitives for dialogue handling

NAME	TYPE	DESCRIPTION
TC_UNI	REQ / IND	Request an unstructured dialogue
TC_BEGIN	REQ / IND	Start a dialogue
TC_CONTINUE	REQ / IND	Continue a dialogue
TC_END	REQ / IND	End a dialogue
TC_U_ABORT	REQ / IND	Terminate a dialogue abruptly
TC_P_ABORT	IND	Terminate a dialogue from the service provider
TC_NOTICE	IND	Indicate that the service provider (e.g., SCCP) cannot provide the requested service

to the remote end. When one of those primitives is issued to the component sublayer, the Dialogue Coordinator Module will query the Component Coordinator Module to obtain the TCAP components invoked by the TCAP-user, and then issue the corresponding primitive with the User Data Field (i.e., the TCAP components) to the transaction sublayer.

Transaction Sublayer

This layer provides primitives to deal with the TR-user message exchanges. This sublayer includes the *Transaction Coordinator Module* (Figure 8.13 (5)), the *Transaction State Machine Module* (TSM; see Figure 8.13 (6)), and the *Abstract Syntax Notation One (ASN.1) Module* (Figure 8.13 (8)). The Transaction Coordinator Module (Figure 8.13 (5)) deals with transaction handling primitives and manages message exchange within a transaction. Multiple transactions can be run simultaneously, and each of them is

Table 8.3   Primitives for the transaction sublayer

NAME	TYPE	DESCRIPTION
TR_UNI	REQ / IND	Delivery of TR-user Information without establishing an explicit association
TR_BEGIN	REQ / IND	Start of a transaction
TR_CONTINUE	REQ / IND	TR-user message exchange in a transaction
TR_END	REQ / IND	End of a transaction
TR_U_ABORT	REQ / IND	Abrupt termination of a transaction
TR_P_ABORT	IND	Termination of a transaction due to abnormal situations
TR_NOTICE	IND	Indication that the requested service cannot be executed

Table 8.4 Mapping between TC dialogue handling primitives and TR transaction handling primitives

TC PRIMITIVE	TR PRIMITIVE
TC_UNI	TR_UNI
TC_BEGIN	TR_BEGIN
TC_CONTINUE	TR_CONTINUE
TC_END	TR_END
TC_U_ABORT	TR_U_ABORT
TC_P_ABORT	TR_P_ABORT
TC_NOTICE	TR_NOTICE

identified by a unique transaction Id. The state machine of each transaction is maintained by the TSM [ANS00a]. The ASN.1 Module provides ASN.1-related encoding/decoding functions for encoding/decoding the TCAP message format [ITU97]. Table 8.3 shows the primitives provided by the transaction sublayer. There is a one-to-one relationship between a dialogue and a transaction (see Table 8.4). When the transaction sublayer is used to support a dialogue, dialogue handling primitives of the component sublayer are mapped to transaction handling primitives of the transaction sublayer with the similar generic names.

As an example, when the MAP layer issues the TC_BEGIN REQ primitive to the component sublayer, the component sublayer groups the previous invoked TCAP components (e.g., the component conveying SEND_AUTHENTICATION_INFO operation) of the same dialogue, and issues the TR_BEGIN REQ primitive with the User Data Field (i.e., the grouped TCAP components) to the transaction sublayer. Then the transaction sublayer generates the appropriate TCAP message (including the components) and delivers this message to the remote end.

8.6 MAP Message Delivery

This section illustrates MAP message delivery for the SCTP-based approach. Consider a scenario in which the SGSN sends the **MAP Send Authentication Info** to the HLR (see Section 9.1). Figure 8.14 shows the interaction among the layers for this example. Note that in the implementation, the Message Parsing Module, the Message Processing Module, and the Encapsulating

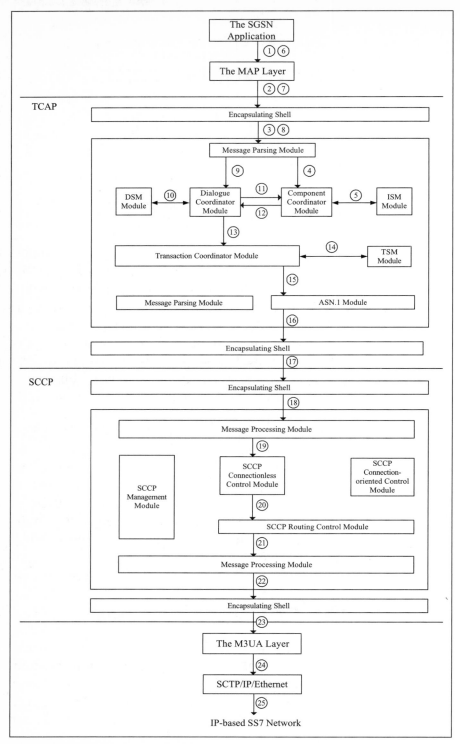

Figure 8.14 An Example of SCTP-based MAP Message Delivery

Shell Module are used as interfaces between layers. We omit the descriptions of these modules in the following steps:

Step 1. The SGSN application initiates a MAP dialogue by invoking the MAP_OPEN REQ primitive of the MAP layer. Then the SGSN application invokes the MAP_SEND_AUTHENTICATION_INFO REQ primitive of the MAP layer to request remote operation (at the HLR).

Steps 2–4. Based on the request primitives from the SGSN application, the MAP layer invokes the TC_INVOKE REQ primitive of the TCAP layer to set the operation code (i.e., SEND_AUTHENTICATION_INFO) and parameters of the TCAP component.

Step 5. The Component Coordinator Module creates the TCAP component for invocation of the operation (i.e., SEND_AUTHENTICATION_INFO), and the corresponding state machine of the operation is maintained by the ISM Module.

Step 6. The SGSN application invokes the MAP_DELIMITER REQ primitive of the MAP layer to explicitly request the TCAP layer for MAP message transfer.

Steps 7–9. The MAP layer invokes the TC_BEGIN request primitive of the TCAP layer to trigger TCAP component transmission.

Steps 10–13. The Dialogue Coordinator Module in the TCAP layer starts a dialogue, where the state machine of the dialogue is maintained by the DSM Module. The Dialogue Coordinator Module then queries the Component Coordinator Module to obtain the invoked TCAP component (i.e., the component conveys the MAP_SEND_AUTHENTICATION_INFO operation), and issues the TR_BEGIN request primitive with parameter "User Data" (i.e., the TCAP component) to the Transaction Coordinator Module.

Steps 14–19. The Transaction Coordinator Module (TCM) starts a transaction, whereby the state machine of the transaction is maintained by the TSM Module. The TCM utilizes the ASN.1 Module to encode the TCAP message. The resulting message is **MAP Send Authentication Info** (described in Section 9.1). The TCM then invokes the SCCP_N_UNITDATA REQ primitive of the SCCP layer to use the SCCP Connectionless Service for delivering the message to the remote end.

Steps 20–23. The SCCP Connectionless Control Module handles the requested connectionless service by determining the route of the message, and invokes the MTP_TRANSFER REQ primitive of the M3UA layer to use the MTP message transfer services. Details of M3UA messages are given in Section 8.7.

Steps 24–25. The M3UA layer checks whether the adjacent signaling point in the route is active. If so, the M3UA layer generates the M3UA packet. This packet is sent to the destination through the SCTP/IP/Ethernet layers and the IP-based SS7 network.

8.7   SCTP and MTP Approaches

This section describes and compares MTP-based and SCTP-based implementations. We first elaborate on the message format. Then we discuss issues on connection setup and data transmission.

8.7.1   Message Format

When SS7 signaling messages are transferred over the MTP-based SS7 network, the MTP2 and the MTP1 layers provide reliable transfer of messages between two adjacent signaling points. The *Signal Unit (SU)* used by the MTP2 is shown in Figure 8.15. There are three SU types:

Message Signal Unit (MSU) is used to carry signaling information of the upper layer protocol (i.e., the MTP3) in the *Signaling Information Field (SIF)*, which is limited to 272 octets.

Link Status Signal Unit (LSSU) is used to notify the adjacent signaling point of the link's status.

Fill-in Signal Unit (FISU) provides signaling link failure detection, and is transmitted when no MSUs or LSSUs are transmitted.

In the MSU, the *Service Indicator Octet (SIO)* consists of the following:

- *Service Indicator (SI)*
- *Network Indicator (NI)*
- Two spare bits

The SI is used for message distribution to determine the MTP-user part (i.e., the SCCP or the ISUP). The NI is used for discrimination between international and national messages. For national messages, the two spare bits may be used to indicate *Message Priority (MP)* for the optional procedure in national applications. The *Length Indicator (LI)* field is used by the MTP2

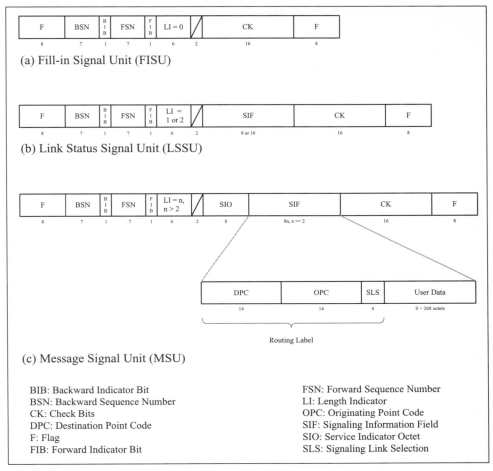

Figure 8.15 MTP Signal Unit Formats

to determine the SU type, and indicates the number of octets. The values of the LI can be any of the following:

- 0 for FISU
- 1 or 2 for LSSU
- 3 or more for MSU

Descriptions of other fields are omitted and can be found in [ANS96]. The MTP3 provides the functions and procedures related to message routing and network management. The MTP3 uses Routing Label of the MSU to (in SIF; see Figure 8.15 (c)) determine how to route the messages. The Routing Label consists of the following:

- *Destination Point Code (DPC)*
- *Originating Point Code (OPC)*
- *Signaling Link Selection (SLS)*

The DPC and the OPC fields are used to indicate the signaling point addresses. The SLS field indicates the signaling link between the destination point and originating point to which the message refers. The User Data field (i.e., the SCCP or the ISUP message) is up to 268 octets in length. Figure 8.15 (c) illustrates the MTP message format. The message header is about 11 bytes in length.

Like the MTP-based approach, SCTP-based implementation provides the same set of primitives and services to the MTP3-User (i.e., the SCCP or the ISUP). Figure 8.16 illustrates the M3UA packet format for transferring signaling data of the upper layer (i.e., the SCCP or the ISUP). The packet includes a common message header and a mandatory parameter "Protocol Data". The "Protocol Data" parameter contains information about the MTP message, including the SIO and the Routing Label (see Figure 8.15 (c)). The fields OPC, DPC, SI, NI, MP, and SLS are similar to those in the MTP message. The User Protocol Data field contains the MTP-User information (i.e., the SCCP or the ISUP message), and is similar to the User Data field in the MTP message (see Figure 8.15 (c)). The M3UA layer does not impose the 272-octet SIF length limit as specified by the MTP2. Larger information blocks can be accommodated directly by the M3UA/SCTP. However, the maximum 272-octet block size must be enforced when the signaling point interworks with

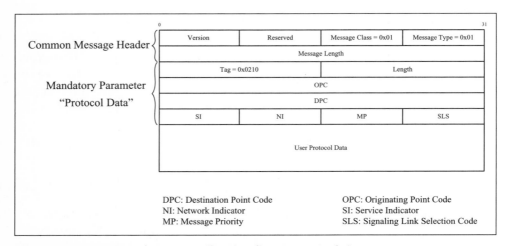

Figure 8.16 M3UA Packet Format (for Signaling Data Transfer)

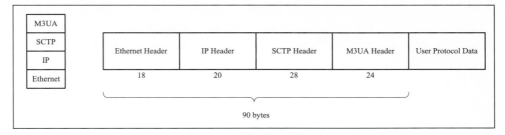

Figure 8.17 Message Headers of M3UA and Lower Layers

an MTP-based SS7 network that does not support larger information block transfer. Figure 8.17 illustrates the packet format, including M3UA, SCTP, IP, and Ethernet for SS7 signaling transport over an IP-based SS7 network. The message headers are about 90 bytes. The message header overhead for the M3UA and the lower layers is about 8 times greater than that for the MTP.

8.7.2 Connection Setup

In the MTP-based SS7, a signaling data link is a bidirectional transmission path for signaling, which consists of standard E1 64 Kbps channels connected to the SS7 card. A signaling point is typically equipped with multiple SS7 network interfaces (i.e., E1 interfaces) to an adjacent signaling point so that multiple signaling links can be used to carry the SS7 messages with enhanced availability. In this approach, the MTP2 provides reliable transfer within a signaling link, which uses FISUs (see Figure 8.15 (a)) as a keep-alive signal to detect whether the signaling link is available for carrying messages.

The MTP3 provides functions and procedures related to message routing and network management. MTP3 defines a link set that consists of the signaling links connected to the same adjacent signaling point. In addition to the concept of link set, MTP3 also defines a route as a collection of the link sets used to reach a particular destination. A link set can belong to more than one route. MTP3 uses the MSU Routing Label (see Figure 8.15 (c)) to determine how to route the messages. The MSU Routing Label consists of OPC, DPC, and SLS. If the DPC of the received MSU is the local signaling point, then the message is processed by the MTP3. If the DPC identifies another signaling point, MTP3 selects an appropriate route for message transfer according to the information stored in its routing table. Selection of the particular signaling link is determined by the SLS field of the MSU. If the MTP-user messages are transferred in sequence, the SLS field should be coded with the same value for the messages. If the MTP-user messages do not need in-sequence delivery service, the messages may be assigned to any SLS or to a default SLS such as 0000 to allow load sharing for message delivery. The MTP3 will route the MTP-user messages through appropriate signaling links based on the load of these signaling links. If a signaling link fails in that path or a signaling point becomes congested, the messages can be alternatively rerouted by the MTP3 network management.

For each signaling point in an MTP-based SS7 network, the signaling links, the link sets, the routes, and the routing tables are configured in advance. When the MTP3-user (i.e., the SCCP) uses the MTP_TRANSFER service primitive for message transfer, the MTP3 does not need to conduct additional connection setup.

In the SCTP-based SS7 network, M3UA/SCTP/IP provide the same MTP functions to the M3UA-user (i.e., the SCCP or the ISUP). In this approach, Internet protocol provides functions for packet routing in the IP network. The SCTP provides reliable transfer, with two features that are particularly desirable for SS7 signaling:

- *Multi-streaming*
- *Multi-homing*

An SCTP stream is a unidirectional logical channel established from one SCTP endpoint to another SCTP endpoint, and can be considered a signaling link in the MTP-based approach. The SCTP stream is identified by the stream Id field in the DATA chunk (see Figure 8.5). The SSN field in the DATA chunk is used to preserve the data order within a stream. Each stream independently delivers messages so that the HOL blocking problem in TCP can be avoided. While one stream is blocked and is waiting for the next in-sequence message, message deliveries of other streams are not affected. The M3UA may use the SLS value to select the SCTP stream. The messages that need to be transferred in sequence are assigned to the same SLS value.

The SCTP also supports a multi-homed endpoint that has two or more IP addresses. The IP addresses are typically assigned to different network interfaces of the endpoint. The multi-homed SCTP endpoint specifies available IP addresses during the SCTP association establishment (to be elaborated later), and selects a primary path (i.e., a primary destination address). To ensure reachability, an endpoint sends the HEARTBEAT chunk to its peer endpoint to probe a particular destination IP address defined in the present association. This mechanism is equivalent to the keep-alive signal (i.e., FISU) in the MTP-based network. Each endpoint sends the DATA chunks through the primary path for normal transmission. Retransmitted DATA chunks use

an alternate path. Continuous failures of the primary path resu
sion to transmit all chunks using the alternate destination unt
destination becomes available again. Note that the SCTP doe
load sharing of the multi-homed endpoint by simultaneously
tiple paths. The M3UA provides the MTP3 functions to the M
the SCCP). The M3UA also provides management of SCTP a
address mapping from SS7 point codes to IP addresses. Because the MTP3
routing is based on OPC, DPC, and SLS, these parameters are used to determine the IP addresses of SCTP endpoints and the specific stream of the
association between the endpoints.

Like the MTP approach, in an IP-based SS7 network, the IP routing tables
are configured in advance. To support multi-streaming and multi-homing
features, SCTP needs to use a four-way handshake procedure to exchange
information and allocate resources to establish connection (i.e., SCTP association) between the peer multi-streaming (multi-homing) endpoint. The
M3UA-user (i.e., the SCCP) invokes MTP_TRANSFER service of the M3UA
to transfer messages after the SCTP association is established.

A FSM for SCTP association establishment is implemented in each of the
endpoints. Figure 8.18 illustrates a simplified state transition diagram for
association establishment in which the error conditions and timeout events
are omitted. The details can be found in [Str00]. The events that drive the
FSM include

- the primitives invoked by the SCTP-user (i.e., ASSOCIATE and
  ABORT),
- reception of the SCTP chunks (i.e., INIT, INIT ACK, COOKIE ECHO,
  COOKIE ACK, and ABORT), and
- expiry of the timers.

The chunk described in Section 8.2 is the basic structure to carry information
in the SCTP packet.

Suppose that SCTP endpoint A attempts to set up an association with
SCTP endpoint Z. The SCTP association establishment process consists of
the following steps (see Figure 8.19).

Step 1. Initially, the FSMs of both endpoints are in the **CLOSED** state.
The SCTP-user in endpoint A invokes the ASSOCIATE primitive to initiate an SCTP association. A *Transmission Control Block* (*TCB*) is created to contain all status and operational information for the endpoint
to maintain and manage this association. Then endpoint A sends an
INIT chunk to endpoint Z. In the INIT chunk, endpoint A specifies the

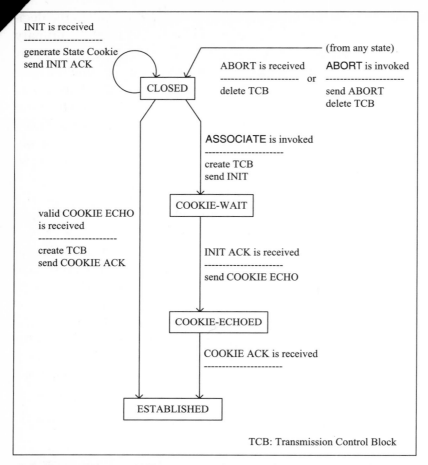

Figure 8.18 State Transition Diagram for SCTP Association Establishment

number of the streams that it can support in this association. Two or more IP addresses can be included in this INIT chunk if endpoint A is multi-homed. After sending the SCTP packet, endpoint A moves to the **COOKIE-WAIT** state.

Step 2. Upon receipt of the INIT chunk, endpoint Z immediately responds with an INIT ACK chunk. In this chunk, endpoint Z specifies the number of streams that it can support in this association, and multiple IP addresses are included if endpoint Z is multi-homed. This chunk also includes a State Cookie [Str00] that is generated by endpoint Z to contain all information necessary to establish the association. The State Cookie is private and only useful to the generator (i.e., endpoint Z) for TCB creation later. The action of sending the INIT ACK chunk does not

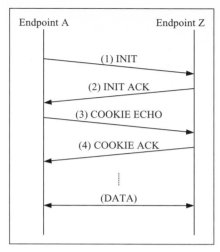

Figure 8.19 SCTP Association Establishment by Four-way Handshake

enable endpoint Z to allocate any resources or keep any states for this new association. Endpoint Z remains in the **CLOSED** state. In this way, DoS attacks (such as flooding SYN attacks presented in TCP) can be avoided.

Step 3. Upon receipt of the INIT ACK chunk, endpoint A sends the COOKIE ECHO chunk that includes the same State Cookie received in the INIT ACK chunk. The received State Cookie is private and only useful to the sender of the INIT ACK chunk (i.e., endpoint Z). Endpoint A does not use this State Cookie and must not modify it. At the end of this step, endpoint A moves to the **COOKIE-ECHOED** state.

Step 4. Upon receipt of the COOKIE ECHO chunk, endpoint Z authenticates the received State Cookie to ensure that the State Cookie was previously generated by endpoint Z. If the authentication succeeds, endpoint Z uses this State Cookie to create a TCB for this association. Endpoint Z replies with a COOKIE ACK chunk, and its FSM moves to the **ESTABLISHED** state. Upon receipt of the COOKIE ACK chunk from endpoint Z, endpoint A moves to the **ESTABLISHED** state, and the association is established.

8.7.3 Data Transmission

In the MTP-based approach, the MTP2 provides reliable in-sequence transfer for a signaling link directly connecting two signaling points. Two error correction methods are defined in MTP: *basic error correction* and *preventive cyclic retransmission*. Preventive cyclic retransmission is used whenever satellite transmission is required for signaling links, and is not elaborated on here. Basic error correction operates independently in two transmission directions and ensures error-free, in-sequence MSU transfer (see Figure 8.15 (c)) over the signaling link. Basic error correction utilizes positive/negative acknowledgment and retransmission mechanisms for MSU delivery. Several sequence numbers and indication bits are included in these messages. The *Forward Sequence Number* (*FSN*) and *Forward Indicator Bit* (*FIB*) of MSUs (see Figure 8.15 (c)) in one direction are associated with the *Backward Sequence Number* (*BSN*) and *Backward Indicator Bit* (*BIB*) of SUs in the other direction. FSN is the sequence number of an MSU. FIB is a one-bit field of MSU to indicate retransmission start. If an MSU has an FIB value different from the previous MSU, it represents the first retransmitted MSU. BSN is used to acknowledge the last correctly received MSU. BIB is a one-bit field of SU with the following usage:

- If the BIB value of an SU is equal to the BIB value of the previous SU, it positively acknowledges the MSU whose FSN is equal to the BSN of the SU.

- If an SU has a BIB value different from the previous SU, it represents a negative acknowledgment that results in retransmission of the corresponding MSU.

In normal operation, the FIB included in the transmitted MSU is equal to the BIB value of the received SU to indicate that the sender has received the last SU sent by the receiver.

Suppose that signaling point A sends MSUs to signaling point Z. When signaling point Z receives an MSU, it first checks whether the FSN of the received MSU exceeds the FSN of the last correctly received MSU by 1.

- If the received MSU exceeds the FSN of the last correctly received MSU by 1, in-sequence delivery is guaranteed, and signaling point Z checks the FIB of the received MSU.

- If the FIB of the received MSU is equal to the BIB of the last delivered SU, the MSU is forwarded to the MTP3. If the received MSU is not in the FSN sequence, and the FIB of the received MSU is equal to the BIB of the last sent SU, signaling point Z sends a negative acknowledgment (i.e., the SU whose BIB value differs from the BIB value of the last SU) to request signaling point A for retransmission.

■ Upon receipt of the negative acknowledgment from signaling point Z, signaling point A checks the BIB of the received SU. If the BIB of the received SU is not equal to the FIB of the last MSU, all unreceived MSUs are transmitted in sequence starting with the MSU whose FSN exceeds the FSN of the most recently acknowledged MSU by 1.

Signaling point A can send new MSUs only when the last unreceived MSU has been transmitted. At the beginning of a retransmission, the FIB value of the first retransmitted MSU is inverted. Therefore, this FIB value equals the BIB value of the last received SU (i.e., the negative acknowledgment). The new FIB bit is maintained in subsequently transmitted MSUs until a new retransmission is started. Thus, under normal operations, the FIB included in the transmitted MSU is equal to the BIB value of the received SU.

Figure 8.20 shows an example of basic error correction for MSUs sent from signaling point A to signaling point Z. In this example, before the transmission of MSU(FIB=0, FSN=14), the BIB value of the last received

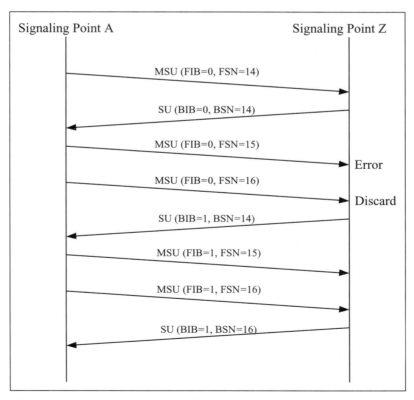

Figure 8.20 An MTP Message Delivery Scenario with Basic Error Correction

SU in signaling point A is 0, and signaling point Z has received MSUs with FSNs up to 13. In this scenario, signaling point A sends MSU (FIB=0, FSN=14). Signaling point Z positively acknowledges MSUs with FSNs up to 14 by replying SU(BIB=0, BSN=14). Signaling point A continues to send MSU(FIB=0, FSN=15) and MSU(FIB=0, FSN=16). Signaling point Z receives MSU(FIB=0, FSN=15) with error and discards MSU(FIB=0, FSN=15). Later, MSU(FIB=0, FSN=16) is received by signaling point Z. This MSU will fail the "sequence test" because its FSN exceeds the FSN of the last received MSU (i.e., MSU(FIB=0, FSN=14)) by 2. Signaling point Z discards MSU(FIB=0, FSN=16) and negatively acknowledges MSUs with FSNs up to 14 by replying SU(BIB=1, BSN=14). Note that the BIB of SU(BIB=1, BSN=14) is not equal to the BIB of the previous SU (i.e., SU(BIB=0, BSN=14)). Upon receipt of the negative acknowledgment SU(BIB=1, BSN=14), signaling point A immediately starts retransmission of MSU(FIB=1, FSN=15) and MSU(FIB=1, FSN=16). Note that the FIB of MSU(FIB=1, FSN=15) is not equal to the FIB of the previous MSU (i.e., MSU(FIB=0, FSN=16)), which indicates the start of retransmission. Upon receipt of MSU(FIB=1, FSN=15) and MSU(FIB=1, FSN=16), signaling point Z positively acknowledges MSUs with FSNs up to 16 by replying SU(BIB=1, BSN=16).

In the IP-based SS7 network, an SCTP stream is considered a signaling link in the MTP-based SS7 network. Like the MTP, the SCTP also provides reliable in-sequence transfer within a stream. As described in Section 8.2, an SCTP packet (see Figure 8.3) is composed of an SCTP common header and one or more SCTP chunks that contain control or data information. The SCTP-user (i.e., the M3UA) message passed to the SCTP layer for transmission will be carried in the DATA chunk (see Figure 8.5). The in-sequence delivery service is achieved by utilizing the stream Id (see Figure 8.5 (2)) and the SSN (see Figure 8.5 (3)) in the DATA chunk. The SCTP layer forwards the received messages (i.e., the DATA chunks) associated with the specific stream Id to the SCTP-user (i.e., the M3UA) in the SSN order. Every stream is associated with an SSN. Unlike the MTP2, the DATA chunks that are not received in the SSN sequence are stored in the *receive buffer* allocated at the SCTP layer. The receive buffer stores all received DATA chunks that have not yet been forwarded to the SCTP-user (e.g., if the DATA chunks are not received in the SSN sequence or the SCTP-user is busy). A DATA chunk in the receive buffer is not released until it has been forwarded to the SCTP-user. The SCTP uses *selective acknowledgment* [Str00] to ensure reliable transfer. In this approach, each DATA chunk is assigned a sequence number TSN (see Figure 8.5 (1)) that is used to acknowledge the received DATA chunks and to detect duplicate deliveries. TSN is used within an association that may contain two or more streams for either ordered delivery services or unordered delivery services, and is independent of any SSN assigned at the stream level. Note that SSN cannot be used for the purpose of TSN because SSN is only associated with a particular stream.

The receiver uses the SACK (Selective Acknowledgment) chunk to acknowledge the received DATA chunks, to inform the sender of the gaps found in a received TSN sequence, and to indicate the available receive buffer size of the receiver. (A "gap" represents not-yet-received messages of consecutive TSNs.) The sender may retransmit the DATA chunks based on the SACK chunk sent from the receiver. Figure 8.21 illustrates the SACK chunk format, which includes the following fields:

Type value is 3 for SACK.

Chunk Length indicates the size of the chunk.

Cumulative TSN Ack (Figure 8.21 (1)) contains the TSN of the last in-sequence DATA chunk received before a gap.

Advertised Receiver Window Credit (a_rwnd) (Figure 8.21 (2)) indicates the available receive buffer size for the sender of this SACK chunk.

Number of Gap Ack Blocks (Figure 8.21 (3)) indicates the number of Gap Ack Blocks included in this SACK chunk.

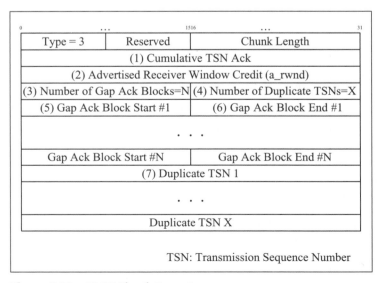

Figure 8.21 SACK Chunk Format

Number of Duplicate TSNs (Figure 8.21 (4)) indicates the number of duplicate TSNs received by the endpoint.

Gap Ack Block contains Gap Ack Block Start offset (Figure 8.21 (5)) and Gap Ack Block End offset (Figure 8.21 (6)), which indicate the range of TSNs received.

Optional Fields include Gap Ack Block and Duplicate TSN. In order to calculate the actual TSNs, these offsets (i.e., Gap Ack Block Start and Gap Ack Block End) are added to the Cumulative TSN Ack (Figure 8.21 (1)). Duplicate TSN (Figure 8.21 (7)) indicates that the chunk of the TSN has been received more than once. Other usage of Duplicate TSN has not yet been defined in the SCTP specification.

Figure 8.22 shows an example of the field values for the SACK chunk after a sequence of DATA chunk deliveries. In this example, endpoint A sends 11 DATA chunks to endpoint Z in an SCTP stream. The size of each DATA chunk is 200 bytes. Each SCTP packet contains one DATA chunk, and the receive buffer size of endpoint Z is 5,600 bytes. Suppose that endpoint Z has received the DATA chunks with TSNs 1–4, 7, and 8. Since endpoint Z has not received DATA chunks with TSNs 5 and 6, the DATA chunks with TSNs 7 and 8 cannot be forwarded to the SCTP-user and are stored in the receive buffer. At this point, the available receive buffer size becomes 5,200 bytes.

When the SCTP packet containing the DATA chunk with TSN 11 is received, endpoint Z detects that the DATA chunks with TSNs 5, 6, 9, and 10 have not been received, and the DATA chunk with TSN 11 is stored in the receive buffer. At this point, the available receive buffer size becomes 5,000 bytes.

After processing the SCTP packet containing the DATA chunk with TSN 11, endpoint Z immediately sends a SACK chunk to endpoint A as the acknowledgment. The format of the SACK chunk is illustrated in Figure 8.22. Upon receipt of the SACK chunk from endpoint Z, endpoint A will retransmit the DATA chunks with TSNs 5, 6, 9, and 10.

0	1516	31
Type = 3	Reserved	Chunk Length = 24
Cumulative TSN Ack = 4		
a_rwnd = 5000		
Number of Gap Ack Blocks = 2	Number of Duplicate TSNs = 0	
Gap Ack Block Start #1 = 3	Gap Ack Block End #1 = 4	
Gap Ack Block Start #2 = 7	Gap Ack Block End #2 = 7	

Figure 8.22 An Example of the Field Values for SACK Chunk

8.8 Concluding Remarks

SS7 signaling is used to provide control and management functions in the mobile telecommunications network. This chapter presented the implementations for SS7 signaling transport. We described the MTP-based approach and the SCTP-based approach. While the MTP-based approach is used in GSM, GPRS, and UMTS R99, the SCTP-based approach is used in the UMTS all-IP network. We compared these two approaches from three perspectives: message format, connection setup, and data transmission. We have used the **Send Authentication Info** procedure defined in 3GPP TS 23.060 [3GP05q] as an example to illustrate the performance of the SCTP-based and the MTP-based MAP approaches. In our experiment (not shown in this book), SS7 signaling message transport over an IP-based SS7 network has better performance than that for an MTP-based SS7 network. As for further reading, the reader is referred to Chapter 5 in [Lin 01b] for the details of MTP-based SS7 signaling.

8.9 Questions

1. Describe the SS7 network components and the types of links connecting the network components.

2. In most commercial SS7, E links and F links are not deployed. Why? (Hint: A commercial switch product typically builds in both the STP and the SSP functions.)

3. Describe the MTP-based protocol stack and the SCTP-based protocol stack.

4. Section 8.2 claims that TCP is not appropriate for supporting IP-based signaling. Is UDP appropriate for IP-based signaling implementation? Why or why not?

5. Describe TSN and SSN. Can you use one sequence number to serve for the purposes for both TSN and SSN?

6. Describe the four types of service primitives in the primitive flow model.

7. Can you port the SS7 implementation in Section 8.4.3 to the SS7 card in Section 8.4.2?

8. Based on the example in Section 8.6, show how the MAP message **MAP Send Authentication Info Ack** is delivered in SCTP-based MAP.

9. Compare the M3UA packet format with the MTP3 message format.

10. How does MTP2 detect the availability of a connection? How does SCTP support the same function?

11. Which features in MTP correspond to the multi-streaming and multi-homing mechanisms in SCTP?

12. Describe the usage of State Cookie in SCTP.

13. How are DoS attacks avoided in SCTP?

14. Compare the in-sequence delivery mechanisms in MTP and SCTP.

## CHAPTER 9

UMTS Security and Availability Issues

This chapter describes some issues related to UMTS security and availability. In Section 9.1, we introduce authentication signaling for UMTS and discuss the network traffic generated by authentication. In Section 9.2, we investigate potential fraudulent usage in UMTS and show how to detect such illegal activities. In Section 9.3, we describe how a mobile user could be eavesdropped, and how this situation can be avoided. In Section 9.4, we elaborate on failure restoration for HLR.

9.1 Authentication Signaling for UMTS

In UMTS, authentication functions are utilized to identify and authenticate a Mobile Station (MS) or User Equipment (UE), and to validate the service request type to ensure that the user is authorized to use particular network services. The authenticating parties are the *Authentication Center* (*AuC*) in the home network and the MS. To describe UMTS authentication, a simplified UMTS architecture is illustrated in Figure 9.1. In this architecture, packet data services of an MS are provided by the SGSN connecting to the UTRAN that covers the MS. For this discussion, we refer to the area covered by an SGSN as the *SGSN service area*. The SGSN connects the MS to the external

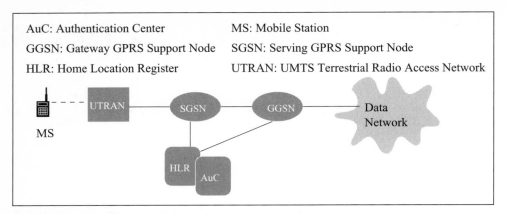

Figure 9.1   A Simplified UMTS Architecture

data network through the GGSN. Furthermore, the SGSN communicates with the HLR and the AuC to receive subscriber data (user profile) and authentication information of an MS. The HLR maintains the current location (identified by an SGSN number) of an MS. The AuC is used in security data management for authentication of subscribers. The AuC may be co-located with the HLR.

As described in Section 2.2, an SGSN service area is partitioned into several *Routing Areas* (*RAs*). When the MS moves from one RA to another, a location update is performed, which informs the SGSN of the MS's current location.

A crossing between two RAs within an SGSN area requires an *intra-SGSN* location update, while a crossing between two RAs of different SGSN areas requires an *inter-SGSN* location update.

In UMTS, the authentication function identifies and authenticates an MS (i.e., a mobile user), and validates the service request type to ensure that the user is authorized to use particular network services. Authentication is performed for every location update (either inter-SGSN or intra-SGSN), call origination, and (possibly) call termination. UMTS supports mutual authentication, i.e., authentication of the MS by the network and authentication of the network by the MS. The procedure also establishes a new UMTS *Cipher Key* (*CK*) and *Integrity Key* (*IK*) agreement between the SGSN and the MS. In UMTS authentication, authenticating parties are the HLR/AuC in the home network and the *User Service Identity Module* (*USIM*) in the user's MS. Two major authentication procedures are described as follows:

Distribution of Authentication Vector. This procedure distributes *Authentication Vectors* (*AVs*) from the HLR/AuC to the SGSN. An AV contains temporary authentication data, which enables an SGSN to engage in UMTS *Authentication and Key Agreement* with a particular user. This procedure assumes the following:

- The HLR/AuC trusts that the SGSN will handle authentication information securely.

- The communication path between the SGSN and the HLR/AuC is adequately secure.

- The user (MS/USIM) trusts the HLR/AuC.

Authentication and Key Establishment. GSM only provides one-way authentication. In UMTS, mutual authentication is achieved by sharing a secret key between the USIM and the AuC [3GP04g]. This procedure follows a challenge/response protocol identical to the GSM subscriber authentication and key establishment protocol combined with a sequence-number-based, one-pass protocol for network authentication derived from ISO/IEC 9798-4 [ISO99].

Based on the above description, the remainder of this section describes the UMTS authentication procedure and then analyzes the network traffic due to UMTS authentication.

9.1.1 UMTS Authentication Procedure

Signaling flows for the Distribution of Authentication Vector procedure and the Authentication and Key Establishment procedure are described in the following steps (see Figure 9.2):

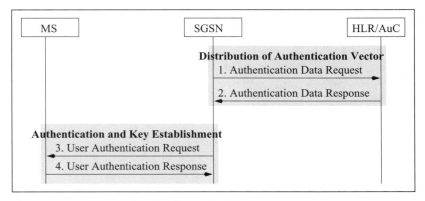

Figure 9.2 UMTS Authentication Procedure

Step 1. When an MS moves into a new SGSN area, the SGSN does not have previously stored authentication information (i.e., the AVs). The SGSN invokes the distribution of authentication vector procedure by sending the **Authentication Data Request** message to the HLR/AuC. This message includes the IMSI that uniquely identifies the MS.

Step 2. Upon receipt of a request from the SGSN, the IMSI is used to identify the HLR/AuC record of the MS, and the HLR/AuC sends an ordered array of K AVs (generated based on the MS record) to the SGSN through the **Authentication Data Response** message. An AV consists of a random number (RAND), an expected response (XRES), a cipher key (CK), an integrity key (IK), and an authentication token (AUTN). Each AV is good for one authentication and key agreement between the SGSN and the USIM.

The HLR/AuC may pre-compute the required AVs and then retrieve these AVs from the HLR database later or compute them on demand.

The next two steps authenticate the user and establish a new pair of cipher and integrity keys between the SGSN and the USIM:

Step 3. When the SGSN initiates an authentication and key agreement, it selects the next unused authentication vector from the ordered AV array and sends the parameters RAND and AUTN (from the selected authentication vector) to the USIM through the **User Authentication Request** message.

Step 4. The USIM checks whether the AUTN can be accepted and if so, produces a response, RES, which is sent back to the SGSN through the **User Authentication Response** message. The SGSN compares the received RES with the XRES. If they match, then the authentication and key agreement exchange is successfully completed. Note that in this mutual authentication procedure, the AUTN is used by the USIM to authenticate the network, and the RES/XRES is used by the network to authenticate the USIM.

In this step, the USIM also computes the CK and the IK using the received AUTN. During data delivery, the CK and the IK are utilized to perform ciphering and integrity functions in the MS. On the network side, the SGSN retrieves the CK and the IK from the AV and passes them to the UTRAN for data ciphering and integrity.

Note that the message names in the above descriptions are based on [3GP04g]. In [3GP05c], the actual SS7 messages sent in Steps 1 and 2 are **MAP Send Authentication Info** and **Send Authentication Info Ack**.

In [3GP05f], the messages exchanged in Steps 3 and 4 are **Authentication Request** and **Authentication Response**.

9.1.2 Network Traffic Due to UMTS Authentication

3GPP TS 33.102 [3GP04g] describes a procedure to distribute authentication data from a previously visited SGSN to the newly visited SGSN. This procedure assumes that the links between SGSNs are adequately secure. In reality, this may not be true, especially when SGSNs are owned by different service providers or even different countries. Thus, it is likely that when the MS moves from the old SGSN area to a new SGSN area, the authentication data stored in the old SGSN is not sent to the new SGSN. Instead, the new SGSN obtains a new AV array from the HLR/AuC through an **Authentication Data Request & Response** (ADR) message pair exchange (see Steps 1 and 2 in Figure 9.2).

Consider the timing diagram in Figure 9.3. Suppose that an MS enters a new SGSN area at time $\tau_{1,1}$. The MS sends a registration message to the SGSN. Since the SGSN does not have authentication information, the distribution of authentication vector procedure is executed through an ADR, where the number of AVs obtained from the HLR/AuC is K. After the SGSN has obtained the AV array from the HLR/AuC, mutual authentication is performed between the SGSN and the MS/USIM through a **User Authentication Request & Response** (UAR) message pair exchange (see Steps 3 and 4 in Figure 9.2) by using the first AV. After $\tau_{1,1}$, the second *authentication event* (a call request or an inter-SGSN RA update) occurs at time $\tau_{1,2}$. The MS/USIM

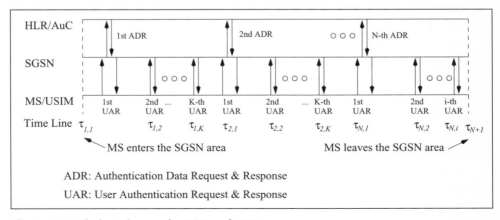

Figure 9.3 Timing Diagram for ADR and UAR

initiates the second UAR, and the SGSN uses the second AV in the array for mutual authentication. At time $\tau_{1,K}$, the last AV in the array is used in the UAR for the Kth authentication event. After $\tau_{1,K}$, the next authentication event occurs at $\tau_{2,1}$. The SGSN realizes that no AV is available, and issues the second ADR to obtain the next AV array from the HLR/AuC. Then a UAR is performed. For the next incoming authentication events, ADRs and UARs are executed accordingly as described above. At time τ_{N+1}, the MS leaves the SGSN area. The last authentication event before τ_{N+1} occurs at $\tau_{N,i}$ (where $1 \leq i \leq K$), which utilizes the ith AV in the AV array. Thus, during the period $[\tau_{1,1}, \tau_{N+1}]$ $(N-1)K + i$ UARs and N ADRs are performed, and at time τ_{N+1}, $K - i$ AVs are left unused in the SGSN.

Note that the ADR operation is expensive (especially when SGSN and HLR/AuC are located in different countries). Therefore, one may increase the AV array size K to reduce the number of ADRs performed when an MS is in an SGSN area. Conversely, with a large K, the AVs may occupy too much network bandwidth for every transmission from the AuC to the SGSN. Thus, it is desirable to select an appropriate K value to minimize the authentication network signaling cost. In [Lin03a], we proposed an analytic model to investigate the impact of K on network signaling traffic (see also question 1 in Section 9.6).

9.2   Fraudulent Usage in UMTS

Mobile telecommunications services have become very popular. As the penetration rate of mobile services significantly increases, fraudulent usage has become a serious issue. On the mobile subscriber side, modern mobile services such as GSM, GPRS, and UMTS are secured by the Subscriber Identity Module (SIM). Without the SIM card, an MS cannot access mobile services (except for emergency calls). The SIM can be a smart card (the size of a credit card), a smaller sized *plug-in SIM*, or a smart card that can be performed, which contains a plug-in SIM that can be removed. A SIM contains subscriber-related information, including the user's identity, authentication key, and encryption key. The subscriber identity and authentication key are used to authorize mobile service access, and to avoid fraudulent access of a cloned MS. Through interaction between the SIM and the AuC, GSM provides one-way authentication. In UMTS, mutual authentication is achieved by sharing a secret key between the SIM and the AuC [3GP04g]. This procedure follows a challenge/response protocol identical to the GSM subscriber authentication and key establishment protocol described in Section 9.1. The above authentication procedures assume that the SIM is securely kept by the corresponding legal mobile user. Unfortunately, we have seen cloned SIMs that have the same identities and authentication keys as the legal SIMs. These cloned SIMs will result in fraudulent usage as well as misrouting of incoming calls to the illegal users. In this section, we investigate how fraudulent usage may occur, when fraudulent usage may be detected by the existing UMTS/GSM mechanisms, and how to reduce the possibility of fraudulent usage.

9.2.1 Circuit-Switched Registration and Call Termination

In GSM/UMTS, authentication is typically performed in the following events:

- Registration (MS attach or location update)
- Call origination (i.e., outgoing call)
- Call termination (i.e., incoming call)

An illegal user gains access to the mobile network through registration and then enjoys "free" services through call originations. Such fraudulent usage may be detected by the GSM/UMTS network when the legal user performs the next registration or call origination.

We briefly describe the circuit-switched domain registration and call setup procedures as follows (the packet-switched domain procedures are given in Chapter 2): Mobility management of mobile services involves two types of mobility databases: HLR and VLR. When a user subscribes to the services of a mobile operator, a record is created in the operator's HLR (information maintained in a HLR record is described in Section 9.4). The service areas of mobile networks are partitioned into several *Location Areas* (LAs). One or more LAs are associated with a VLR, as described in Chapter 2. When a mobile user visits an LA, a temporary record for the mobile user is created in the corresponding VLR. The VLR typically co-locates with a *Mobile Switching Center* (*MSC*) that is responsible for call setup of MSs. The VLR stores subscription information (replicated from the HLR) for the visiting subscribers so that the visited system can provide services. The VLR record also maintains the location information (i.e., the address of the LA where the MS resides). In other words, the VLR is the location register that provides the location information of a visiting mobile user at the visited network for handling calls to or from this mobile user. To track the location of an MS, the MS automatically reports its location (to the visited VLR and then to the

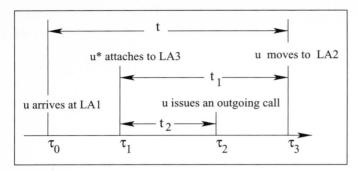

Figure 9.4 Timing Diagram for Movements of u and u^*

HLR) when it moves to a new location. This procedure is called *registration*. Before the user accesses the VLR/HLR, authentication is performed between the user and the AuC, as described in Section 9.1. To deliver a call to an MS, the network retrieves the location information stored in the HLR and the VLR, and the network sets up a trunk based on this location information. Like registration, authentication is required before call setup is performed.

The timing diagram in Figure 9.4 illustrates how a legal user u enters a location area LA_1 at time τ_0, and moves to another location area LA_2 at time τ_3. Following the standard GSM/UMTS mobility management procedure, u informs the network that it has moved into LA_1 at τ_0, and that it has left LA_1 and moved into LA_2 at τ_3.

Normal Registration

Figure 9.5 illustrates the registration procedure performed at τ_0. In this figure, the MS is first authenticated by the AuC. (Note that through network signaling optimization [3GP04g, Lin03a], the MS may be authenticated by the VLR instead of the AuC; see Section 9.1.) Then the MS registers to VLR1 and the HLR.

Step 1. Authentication is exercised between u and the AuC (through VLR1 and the HLR). If authentication is successful, Steps 2–5 are executed.

Step 2. The MS u initiates the registration procedure to VLR1 through the MSC. Specifically, the MSC sends the **MAP Update Location Area** message to VLR1. This message includes the following:

- Address of the MSC
- Temporary Mobile Subscriber Identity (TMSI) of the MS

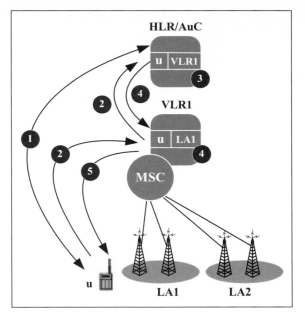

Figure 9.5 Circuit-Switched Registration of a Legal User u at Time τ_0

- Target Location Area Identification (LAI); in this example, the LAI is LA_1
- Other related information listed in [3GP05e] and [ETS93a]

VLR1 forwards the registration request **MAP Update Location** to the HLR. This message includes the following:

- IMSI of u
- Address of the MSC
- Address of VLR1
- Other related information listed in [3GP05e]

Note that at this step, VLR1 may use u's TMSI to query the VLR previously visited by u, and obtain u's IMSI from this VLR (see Steps 3 and 4 in Section 11.2.1).

Step 3. By using the received IMSI, the HLR retrieves the u's record. The MSC number is updated. For inter-VLR movement, the VLR address is also modified.

Step 4. The HLR informs VLR1 that the registration is successful by replying with the message **MAP Update Location Ack**. This message includes the subscriber data of u. VLR1 updates the record of u.

Step 5. VLR1 informs u that the registration operation is successful, and assigns a new TMSI to u.

After registration, the HLR record of u indicates that the MS is in VLR1, and the record in VLR1 indicates that the MS is in LA_1. For call termination to u (an incoming call), routing information must be obtained from the serving VLR. Details of the basic call termination procedure are given as follows:

Normal Call Termination

Figure 9.6 illustrates the message flow for circuit-switched call termination, described here:

Step 1. When the Mobile Station ISDN Number (MSISDN) of u is dialed by the caller (e.g., a PSTN user), the call is first routed to u's *Gateway MSC* (*GMSC*) by an ISUP **IAM** message (see Section 8.1 for the details of ISUP messages).

Step 2. To obtain the routing information, the GMSC interrogates the HLR by sending **MAP Send Routing Information** to the HLR. The message consists of the MSISDN of the MS and other related information (see [Lin01b]).

Step 3. The HLR sends **MAP Provide Roaming Number** to the VLR (i.e., VLR1) to obtain the *Mobile Station Roaming Number* (*MSRN*). The message consists of the IMSI, the MSC number, and other related information. The MSRN provides the address of the target MSC, and is used to set up the voice trunk.

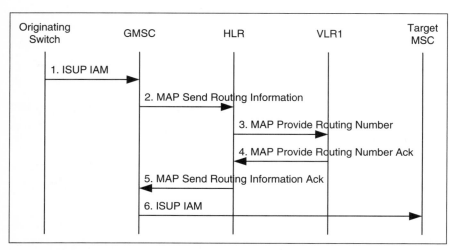

Figure 9.6 Circuit-Switched Call Termination Message Flow (Simplified Version)

Steps 4 and 5. The VLR creates the MSRN by using the MSC number stored in the VLR record of u. This roaming number is sent back to the GMSC through the HLR.

Step 6. The MSRN provides the address of the target MSC where u resides. An ISUP IAM message is directed from the GMSC to the target MSC to set up the voice trunk.

9.2.2 Fraudulent Registration and Call Setup

The fraudulent usage problem due to cloned SIM is described as follows. Suppose that the SIM card of a legal user u has been cloned by an illegal user u^*. Suppose that u^* attempts to illegally access the mobile network at time τ_1 in the interval $[\tau_0, \tau_3]$ in Figure 9.4. This illegal user u^* either issues an attach or a normal location area update action at τ_1. If u^* and u are in the same location area LA_1, then this abnormal action will be detected; that is, the network notices that the same MS registers to the same location area twice. The network may suspend the services for u until the fraudulent usage issue is cleared. If u^* registers in a different location area LA_3 (controlled by VLR2 in our example), then the network assumes that u has moved from LA_1 to LA_3. Figure 9.7 illustrates the registration of u^* at τ_1.

The fraudulent registration procedure is exactly the same as the normal registration procedure:

Step 1. Authentication is exercised between u^* and the AuC (through VLR2 and the HLR). Since u^* has the same SIM card as u, authentication is successfully performed, and Steps 2–5 are executed.

Step 2. VLR2 forwards the registration request from u^* to the HLR.

Step 3. The HLR updates the HLR record of u.

Step 4. The HLR informs VLR2 that the registration is successful. VLR2 updates the record of u.

Step 5. VLR2 informs u^* that the registration operation is successful.

Step 6. By exchanging the MAP Cancel Location and Ack message pair, the HLR performs location cancellation to delete the VLR record of u in VLR1.

After the registration, the HLR record indicates that u is in VLR2, and the VLR2 record indicates that u is in LA_3. The VLR1 record for u is deleted

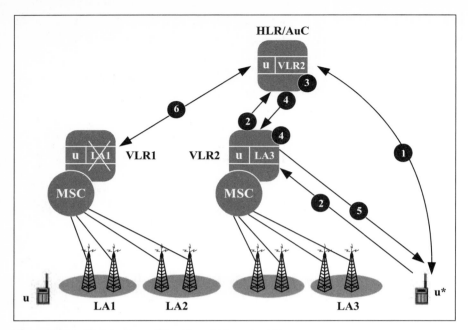

Figure 9.7 Registration of the Illegal User u^* at Time τ_1

through the location cancellation procedure. At this point, u^* gains mobile service access, and fraudulent usage may occur. Such fraudulent usage may last until either there is an outgoing call originated by u at time τ_2 (in Figure 9.4) or u moves to another location area LA_3 and issues location update at time τ_3. That is, one of the following two cases may occur:

Case I

Suppose that after τ_1, the first outgoing call for u occurs before u moves to LA_2 (i.e., $\tau_2 < \tau_3$). Then, at τ_2, u issues a call origination request to VLR1, as illustrated in Figure 9.8 with the following steps:

Step 1. u issues a call origination request to VLR1.

Step 2. VLR1 attempts to initiate the authentication procedure, but cannot locate u's VLR record (because the record has been erased at time τ_1).

Step 3. VLR1 rejects the call origination request.

Step 4. VLR1 informs the HLR that fraudulent usage may occur. The HLR takes necessary actions to resolve the issue.

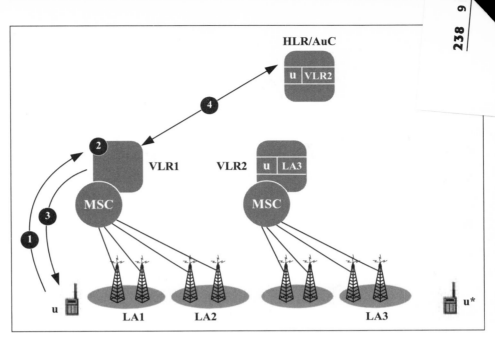

Figure 9.8 Outgoing Call (Call Origination) of *u* at Time τ_2

Since the *u*'s record at VLR1 has been erased or modified (in Figure 9.8, the record has been erased), the network will detect that *u* has registered with VLR2 but is issuing an outgoing call from VLR1. At this point, the mobile service of *u* will be suspended until the network has resolved the fraudulent usage issue.

Case II

If $\tau_3 < \tau_2$, then after τ_1, the first event of *u* is a registration request due to movement from LA_1 to LA_2 (assume that both location areas are covered by VLR1). Similar to the situation in Case I, when *u* issues the registration request to VLR1, VLR1 cannot identify the record for *u*, and the potential fraudulent usage situation is detected.

In Figure 9.4, let $t_1 = \tau_3 - \tau_1$ and $t_2 = \tau_2 - \tau_1$. Then $T = \min(t_1, t_2)$ is the potential fraudulent usage period. During this period, u^* can illegally originate calls. Also, all incoming calls to *u* will be directed to u^* during T; i.e., these calls are *misrouted*, as illustrated in Figure 9.9 with the following steps:

Step 1. The calling party dials *u*'s phone number. The call is connected to *u*'s GMSC.

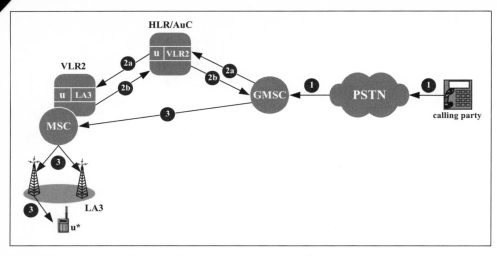

Figure 9.9 An Incoming Call (Call Termination) to u in Period $[\tau_1, \tau_3]$

Step 2. The GMSC queries the HLR and then VLR2, and determines that u is connected to the MSC of VLR2.

Step 3. The call is directed to u^* through the MSC of VLR2. This call is misrouted.

The calling parties of these misrouted calls may mistakenly think that the mobile operator accidentally routed the calls to wrong places.

In [Lin 02b], potential fraudulent usage is investigated by two output measures: the expected potential fraudulent usage period and the probability that there are misrouted calls during the potential fraudulent usage period (see also question 2 in Section 9.6). The study indicated the following:

- If the legal users exhibit irregular movement patterns (i.e., the LA residence times have large variance), then fraudulent usage is more likely to occur.

- If the legal users often make outgoing phone calls, then the network has better opportunity to detect the fraudulent usage.

- Incoming calls to the legal user are misrouted during the potential fraudulent usage period. The probability of misrouted calls increases as the movement pattern of a legal user becomes more irregular.

- *Periodic LA Update (PLAU)* can effectively speed up the fraudulent usage detection, and reduce the probability of misrouted calls. Therefore, by exercising PLAU, we can enhance security at the cost of increasing the signaling traffic.

As a final remark, note how the HLR/AuC can detect a potential abnormal event due to fraudulent access of an illegal user. When the HLR/AuC receives a report of potential fraudulent usage from the VLR, several approaches can assist to determine whether the reported case is actually due to fraudulent usage. For example, in the *rare event approach* [Lin97b], the MS is requested to include statistics of some rare events (such as the time stamp of last registration, the number of calls made in the last location area, and so on) when it issues location update or call origination requests. The network then compares these statistics received from the MS with that obtained from network *Operations and Maintenance Center* (*OMC*) to ensure that the request is issued from a legal user.

9.3 Eavesdropping a Mobile User

By modifying the call control software of an MS, the mobile user of the MS could be eavesdropped when he/she is not in phone conversation. Figure 9.10 illustrates the normal call flow for MS call termination. According to the descriptions of ISUP messages in Section 8.1 and mobile call termination in Section 9.2.1, the following steps are executed:

Step 1. The IAM message is sent from the originating switch to the GMSC of the called MS. The GMSC obtains the MSRN of the target MSC and forwards the IAM message to the target MSC. The target BSS pages the MS.

Step 2. If the called MS is in the radio's coverage, it sends the page response signal to the target BSS. The target MSC then returns the ACM message to the originating switch.

Step 3. A ringing tone is sent to the called MS and a ringback tone is sent to the calling party.

Step 4. When the called party picks up the handset, the MS sends the answer signal to the BSS, and the target MSC sends the ANM message to the originating switch. The ringing and the ringback tones are removed and the conversation starts.

To eavesdrop through an MS, the MS is modified to accommodate the eavesdropping procedure illustrated in Figure 9.11:

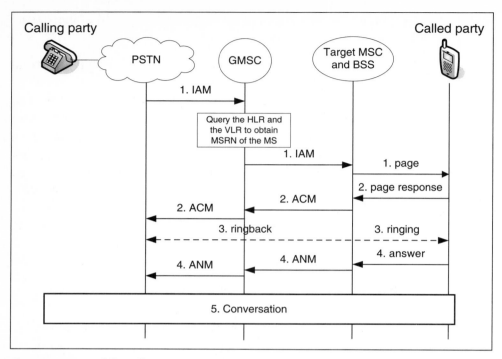

Figure 9.10   Mobile Call Termination Message Flow (with Radio Interaction)

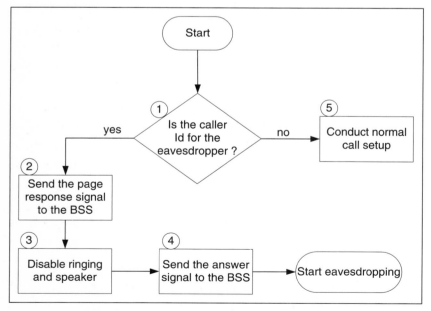

Figure 9.11   Mobile Call Termination Message Flow (with Radio Interaction)

- When the MS receives a page signal from the BSS, it obtains the caller identification (caller Id) from the page signal. It checks whether the caller Id is the phone number of the eavesdropper (Figure 9.11 (1)). If not, the normal call setup procedure is performed (Figure 9.11 (5)).

- If the caller Id matches, the eavesdropping procedure is triggered and Steps (2)–(4) in Figure 9.11 are executed.

- At Figure 9.11 (2), the MS responds to the page as the normal call setup procedure.

- At Figure 9.11 (3), the MS disables the ringing and the speaker (therefore, the eavesdropped user will not be alerted).

- At Figure 9.11 (4), the MS sends the answer signal to the BSS (without having the called user pick up the handset).

At this point, the call is connected, and the eavesdropper can hear all sounds from the eavesdropped MS. Since the speaker of the eavesdropped MS is disabled, any noise generated by the eavesdropper will not be detected by the eavesdropped user.

To eavesdrop an MS, only the call setup software at the MS needs to be changed. The PSTN and the mobile network are not modified. If the calling party is not the eavesdropper, the call is set up as a normal call. Therefore, the eavesdropped user can receive calls from other calling parties normally.

When an incoming call arrives during an eavesdropping session, the eavesdropping session is immediately terminated and the newly arrived call is connected. Therefore, the eavesdropped user can receive other normal calls.

Note that the eavesdropped user will detect abnormal call records if the billing method is called-party-pay as exercised in, e.g., the U.S. If the billing method is calling-party-pay as exercised in, e.g., Taiwan, the eavesdropped calls will not be shown in the telephone bill. If the eavesdropped user roams to other countries, the eavesdropped calls are marked and shown as international calls in the telephone bill.

Based on the above discussion, the eavesdropping scenario works as follows:

1. The eavesdropper purchases the eavesdropped handset. The handset is initialized with the eavesdropper's phone number as the caller Id that will trigger the eavesdropping procedure illustrated in Figure 9.11.

2. The eavesdropper gives this handset to the eavesdropped user perhaps as a birthday gift. In the countries exercising called-party-pay, the
eavesdropper also covers the billing of this handset (i.e., the telephone
bill will not go to the eavesdropped user).

To avoid eavesdropping, it is always a good idea for a mobile user to purchase
his/her own handset, and to investigate/pay his/her own telephone bill.

9.4   HLR Failure Restoration

This section studies failure restoration of mobility databases for
GPRS/UMTS. In these networks, the HLR is a database used for mobile
user information management. All permanent subscriber data is stored in
this database. An HLR record consists of three types of information:

Mobile Station (MS) Information, such as the MSISDN and the IMSI
that identify the MS

Service Information (user profile or subscriber data), such as service
subscription (call waiting, call forwarding, voice mailbox, and so on),
service restrictions, and supplementary services

Location Information, such as the address of the SGSN on which the
MS resides

The location information in the HLR is updated whenever the MS moves to a
new SGSN. To access the MS, the HLR is queried to identify the current SGSN
location of the MS. Note that both the MS and the service information items
are only occasionally updated. Conversely, an MS may move frequently and
the location information is often modified.

If the HLR fails, one will not be able to access the MSs. To guarantee service availability to the MSs, database recovery is required after an
HLR failure. In UMTS/GPRS, the HLR recovery procedure works as follows [3GP05q]: The HLR database is periodically checkpointed into a backup
storage. After an HLR failure, the database is restored by reloading the
backup information. There are several approaches to checkpoint the HLR
database. In the *all-record checkpoint* approach, all HLR records are saved
into the backup at the same time [Haa98, ETS93b, Lin95]. The checkpoint
overhead for this approach is very high and is typically performed at mid-
night when the HLR activities are infrequent. Alternatively, checkpoint can
be exercised for individual mobile users, which is referred to as *per-user*

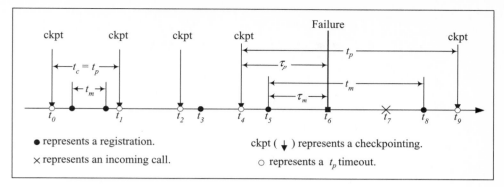

Figure 9.12 Timing Diagram for Algorithm I

checkpoint [Cao02, Lin96a, Lin03c]. We describe two per-user checkpoint algorithms:

Algorithm I is the same as all-record checkpoint except that the checkpoint frequencies for individual MSs may be different. For every MS, we define a timeout period t_p. In Figure 9.12, the t_p timeouts occur at time t_0, t_1, t_2, t_4, and t_9. When this timer expires, checkpoint is performed to save the HLR record of the MS. Therefore, the checkpoint interval t_c is equal to the timeout period t_p. After a failure (see t_6 in Figure 9.12), the HLR record in the backup database is restored to the HLR. The backup copy is *obsolete* if the HLR record is updated between the last checkpoint and when the failure occurs (i.e., a registration occurs in $[t_4, t_6]$ in Figure 9.12). After the HLR record is restored, one of the following two events may occur next:

- The record may be updated again if the MS issues a registration (i.e., $t_8 < t_7$ in Figure 9.12)

- The record may be accessed due to an incoming call to the MS (i.e., $t_7 < t_8$ in Figure 9.12).

   After a failure, if the backup record is obsolete and the next event to the MS is an incoming call ($t_7 < t_8$ in Figure 9.12), then the call is lost. Conversely, if the next event is a registration, then the location information of the HLR record is modified and the record is up-to-date again.

Algorithm II is designed around the following intuition: If registration activities are very frequent (i.e., a registration always occurs before the t_p timer expires), then Algorithm II behaves exactly the same as Algorithm I. On the other hand, if no registration has occurred before the

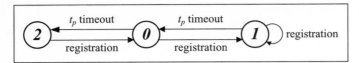

Figure 9.13   State Diagram for Algorithm II

t_p timer expires, then there is no need to checkpoint the record (because the backup copy is still valid). In this case, checkpoint is performed when the next registration occurs.

In Algorithm II, a three-state Finite State Machine (FSM) is implemented for an HLR record. The state diagram for the FSM is shown in Figure 9.13.

- Initially, the FSM is in **state 0**, and the t_p timer starts to decrement. If a registration event occurs before the t_p timer expires, the FSM moves to **state 1**. If a t_p timeout event occurs at **state 1**, the FSM moves back to **state 0**, the HLR record is checkpointed into the backup, and the t_p timer is restarted.

- If the timeout event occurs at **state 0**, then the FSM moves to **state 2**, and the t_p timer is stopped.

- If a registration event occurs at **state 2**, the FSM moves to **state 0**, a checkpoint is performed, and the t_p timer is restarted.

Consider the timing diagram in Figure 9.14. At time t_0, the FSM is at **state 0** (when a registration occurs). At time t_1, the next registration occurs, and the FSM moves from **state 0** to **state 1** (where $t_m = t_1 - t_0$

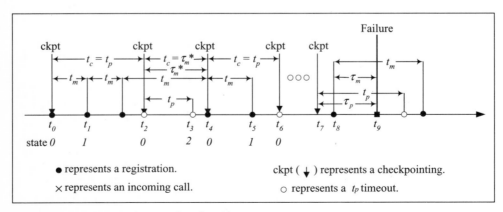

Figure 9.14   Timing Diagram for Algorithm II

is the inter-registration interval). At time t_2, the t_p timer expires, and the FSM moves from **state 1** to **state 0** (where $t_c = t_p = t_2 - t_0$). At time t_3, the t_p timer expires again, and the FSM moves from **state 0** to **state 2**. At time t_4, a registration occurs. The FSM moves from **state 2** to **state 0**, and $t_c = \tau_m^* = t_4 - t_2$ (where τ_m^* is the *excess life* or *residual time* of t_m).

Two output measures are considered to investigate the checkpoint performance.

- The expected checkpoint interval: The larger the checkpoint interval, the lower the checkpoint overhead. That is, the checkpoint cost is proportional to the checkpoint frequency.

- The *obsolete probability*, the probability that the HLR record in the backup is obsolete when a failure occurs: The smaller the obsolete probability, the better the checkpoint performance.

In [Lin05a], an analytic model was developed to compare Algorithms I and II in terms of the checkpoint cost and the obsolete probability. The study indicates that Algorithm II can save more than 50% of the checkpoint cost over Algorithm I. For the obsolete probability, Algorithm II demonstrates 20%–55% improvement over Algorithm I.

As a final remark, failure restoration for a SGSN (or a VLR in the circuit-switched service domain) is very different from the HLR failure restoration described in this section. No checkpoint is performed for a SGSN because all MS records in the SGSN are temporary, and it is useless to store these temporary records into backup. Details of SGSN failure restoration can be found in [Fan00a, Cha98, Lin95] and Chapter 11 in [Lin 01b].

9.5 Concluding Remarks

This chapter described security and availability issues for UMTS. We first described UMTS mutual authentication that can be effectively conducted by sending multiple authentication vectors (AVs) from the AuC to the SGSN, and then performing the authentication procedure locally between the SGSN and the MS. The authentication procedure also generates the CK (for ciphering) and the IK (for integrity). More details are given in Sections 6.5 and 6.6 in [3GP04g]. An important issue is determining the number of AVs that should be sent to the SGSN. We have developed a dynamic AV size selection algorithm that is a ROC patent (Patent Number 0942024950). The reader is referred to [Lin03a] for the details.

In Section 9.2, we described how potential fraudulent usage can be detected by the UMTS registration and call setup procedures. In [Lin 02b], we observed that the movement pattern of a legal user significantly affects the detection time of potential fraudulent usage. In Section 9.3, we showed how an MS could be used to eavesdrop on a mobile user when he/she is not in phone conversation.

In Section 9.4, we described checkpoint approaches to support backup HLR records in a nonvolatile database. We also showed how checkpoint can be utilized for HLR failure restoration.

9.6   Questions

1. Authentication signaling for UMTS (see Section 9.1) can be modeled as follows: Let N be the total number of ADRs performed when the MS resides in an SGSN area. For each ADR, the number of AVs obtained from the HLR/AuC is K. Suppose that the aggregate incoming/outgoing call and registration arrivals form a Poisson process with rate λ. As mentioned in Section 9.1, for every incoming/outgoing call and registration, a UAR is performed. For a specific period τ, let $\Theta(n, K, \tau)$ be the probability that there are n ADRs to the HLR/AuC. Note that n ADRs are performed if there are $(n-1)K + k$ UARs in the period τ (where $1 \le k \le K$). According to the probability function of the Poisson distribution, we have

$$\Theta(n, K, \tau) = \sum_{k=1}^{K} \left\{ \frac{(\lambda\tau)^{(n-1)K+k}}{[(n-1)K + k]!} \right\} e^{-\lambda\tau} \tag{9.1}$$

Let t be the period that an MS resides in an SGSN service area. That is, $t = \tau_{N+1} - \tau_{1,1}$ in Figure 9.3. Suppose that t has a general distribution with the density function $f(t)$, the mean $1/\mu$, and the Laplace Transform $f^*(s) = \int_{t=0}^{\infty} f(t)e^{-st}dt$. Let $P(n, K)$ be the probability that there are n ADRs during the MS's residence in the SGSN area. Then

$$P(n, K) = \int_{t=0}^{\infty} \Theta(n, K, t)f(t)dt$$

$$= \sum_{k=1}^{K} \int_{t=0}^{\infty} \left\{ \frac{(\lambda t)^{(n-1)K+k}}{[(n-1)K + k]!} \right\} e^{-\lambda t} f(t)dt$$

$$= \sum_{k=1}^{K} \left\{ \frac{\lambda^{(n-1)K+k}}{[(n-1)K+k]!} \right\} \int_{t=0}^{\infty} t^{(n-1)K+k} f(t) e^{-\lambda t} dt \qquad (9.2)$$

$$= \sum_{k=1}^{K} \left\{ \frac{\lambda^{(n-1)K+k}}{[(n-1)K+k]!} \right\} (-1)^{(n-1)K+k} \left[\frac{d^{(n-1)K+k} f^*(s)}{ds^{(n-1)K+k}} \right] \Bigg|_{s=\lambda} \qquad (9.3)$$

where (9.3) is derived from (9.2) using Rule P.1.1.9 in [Wat81]. Let $E[N]$ be the expected number of ADRs when the MS resides in an SGSN service area. Then

$$E[N] = \sum_{n=1}^{\infty} n P(n, K) \qquad (9.4)$$

Derive $P(n, K)$ and $E[N]$ based on three SGSN residence time distributions: Exponential, Gamma, and Hyper-Erlang. (Hint: See the solution in http://liny.csie.nctu.edu.tw/supplementary.)

2. Derive the expected potential fraudulent usage period $E[T]$ and the probability $E[N = n]$ of n misrouted calls in Section 9.2. (Hint: See the solution in http://liny.csie.nctu.edu.tw/supplementary.)

(a) Consider Figure 9.4. Suppose that the residence time $t = \tau_3 - \tau_0$ of a mobile user u in an LA is a random variable with the distribution function $F(t)$, the density function $f(t)$, the Laplace Transform $f^*(s)$, and the mean value $1/\mu$. Suppose that the illegal user u^* issues a location area update (or attach) message at time τ_1 during the interval $[\tau_0, \tau_3]$. Then τ_1 is a random observer of the period t. The excess life $t_1 = \tau_3 - \tau_1$ has the distribution function $R(t_1)$, density function $r(t_1)$, and Laplace Transform $r^*(s)$, where from the excess life theorem [Ros96],

$$r(t_1) = \mu[1 - F(t_1)]$$

and

$$r^*(s) = \int_{t_1=0}^{\infty} r(t_1) e^{-st_1} dt_1 = \left(\frac{\mu}{s} \right) [1 - f^*(s)] \qquad (9.5)$$

Suppose that the outgoing calls originated by u are a Poisson stream with arrival rate λ. After τ_1, the first outgoing call of u occurs at time τ_2. Then from the excess life theorem and the memoryless property of the exponential distribution, $t_2 = \tau_2 - \tau_1$ is also exponentially distributed with mean $1/\lambda$. The potential fraudulent usage period is $T = \min(t_1, t_2)$. Let $h(t)$ be the density function of T. Then

$$h(t) = \int_{t_2=t}^{\infty} r(t)\lambda e^{-\lambda t_2} dt_2 + \int_{t_1=t}^{\infty} r(t_1)\lambda e^{-\lambda t} dt_1 \qquad (9.6)$$

The expected value $E[T]$ of the potential fraudulent usage period is

$$\begin{aligned}
E[T] &= E[\min(t_1, t_2)] \\
&= E[t_1 | t_1 = \min(t_1, t_2)] \Pr[t_1 = \min(t_1, t_2)] \\
&\quad + E[t_2 | t_2 = \min(t_1, t_2)] \Pr[t_2 = \min(t_1, t_2)] \qquad (9.7)
\end{aligned}$$

If $f(t)$ is a Gamma density function with mean $1/\mu$ and variance v, then

$$f^*(s) = (1 + \mu v s)^{-\frac{1}{\mu^2 v}} \qquad (9.8)$$

Based on (9.5)–(9.8), show that

$$E[T] = \frac{1}{\lambda} - \left(\frac{\mu}{\lambda^2}\right)\left[1 - (1 + \lambda \mu v)^{-\frac{1}{\mu^2 v}}\right]$$

(Note that the Gamma distribution is often used to model user movement in mobile networks; see [Lin01a, Fan00b, Lin03a] and the references therein. It has been shown that the distribution of any positive random variable can be approximated by a mixture of Gamma distributions; see Lemma 3.9 in [Kel79]. One may also measure the LA residence times of an MS in a real mobile network, and the measured data can be approximated by a Gamma distribution [Joh70a, Joh70b] as the input to the analytic model.)

(b) Let random variable N be the number of misrouted calls during the potentially fraudulent usage period. The probability distribution of N is derived as follows: Assume that the incoming calls to u are a Poisson stream with rate γ. Then N has a Poisson probability mass function $g(n, t)$ in the potentially fraudulent usage period t, where

$$g(n, t) = \left[\frac{(\gamma t)^n}{n!}\right] e^{-\gamma t} \qquad (9.9)$$

Let $\Pr[N = n]$ be the probability that there are n misrouted calls during the potentially fraudulent usage period. From (9.6) and (9.9), we have

$$\Pr[N = n] = \int_{t=0}^{\infty} g(n, t)h(t)dt$$

$$= \int_{t=0}^{\infty} \left[\frac{(\gamma t)^n}{n!} \right] e^{-\gamma t} \int_{t_2=t}^{\infty} r(t)\lambda e^{-\lambda t_2} dt_2 dt$$

$$+ \int_{t=0}^{\infty} \left[\frac{(\gamma t)^n}{n!} \right] e^{-\gamma t} \int_{t_1=t}^{\infty} r(t_1)\lambda e^{-\lambda t} dt_1 dt \quad (9.10)$$

Based on (9.10), show that

$$\Pr[N = 0] = \frac{\lambda}{\lambda + \gamma} + \left[\frac{\gamma\mu}{(\lambda + \gamma)^2} \right] [1 - f^*(\lambda + \gamma)]$$

$$\Pr[N = 1] = \frac{\lambda\gamma}{(\lambda + \gamma)^2} + \left[\frac{\gamma\mu(\gamma - \lambda)}{(\lambda + \gamma)^3} \right] [1 - f^*(\lambda + \gamma)]$$

$$+ \left[\frac{\gamma^2\mu}{(\lambda + \gamma)^2} \right] \left[\frac{df^*(s)}{ds} \Big|_{s=\lambda+\gamma} \right]$$

If $f(t)$ is an exponential density function, show that

$$\Pr[N = n] = \frac{(\lambda + \mu)\gamma^n}{(\lambda + \mu + \gamma)^{n+1}} \quad (9.11)$$

Equation (9.11) says that if the outgoing call arrival rate λ and the mobility rate μ are much higher than the incoming call arrival rate γ, then it is unlikely that misrouted calls will occur.

3. Per-user checkpointing in Section 9.4 can be modeled as follows. (Hint: See the solution in http://liny.csie.nctu.edu.tw/supplementary.)

(a) Modeling of Algorithm I. Consider Exponential timeout period t_p with the density function

$$f_p(t_p) = \lambda e^{-\lambda t_p}$$

Since the checkpoint interval is $t_c = t_p$ in Algorithm I, the expected checkpoint interval is

$$E_I[t_c] = E[t_p] = \frac{1}{\lambda} \quad (9.12)$$

After a failure, the HLR record is restored from the backup. This backup copy is obsolete if the record in the HLR has been modified since the last checkpoint. In Figure 9.12, a failure occurs at time t_6, which is a random observer of the inter-checkpoint interval $[t_4, t_9]$ and the inter-registration interval $[t_5, t_8]$. In this figure, the interval $t_m - \tau_m$ is the residual time of t_m, and τ_m is the *reverse residual time*. Similarly, τ_p is the reverse residual time of t_p. Consider a random variable t with the probability density function $f(t)$, the distribution function $F(t) = \int_{y=0}^{t} f(y)dy$, the expected value $E[t]$, and the Laplace Transform $f^*(s) = \int_{t=0}^{\infty} f(t)e^{-st}dt$. Let τ be the residual time of t with the density function $r(\tau)$, the probability distribution function $R(\tau)$, and the Laplace transform $r^*(s)$. From (9.5)

$$r(\tau) = \frac{1 - F(\tau)}{E[t]} \quad \text{and} \quad r^*(s) = \frac{1 - f^*(s)}{E[t]s} \qquad (9.13)$$

Note that the reverse residual time distribution is the same as the residual time distribution [Kel79], and (9.13) also holds for the reverse residual time. From (9.13), the density function $r_p(t)$ is the same as $f_p(t)$ for the Exponential distribution. That is,

$$r_p(t) = f_p(t) = \lambda e^{-\lambda t}$$

Let α_I be the probability that the HLR backup is obsolete when a failure occurs in Algorithm I. In Figure 9.12, the backup copy is obsolete if $t_4 < t_5 < t_6$. Let $\tau_c = \tau_p$. Then

$$\alpha_I = \Pr[\tau_c > \tau_m] \qquad (9.14)$$

Use (9.12) and (9.13) to derive (9.14).

Another output measure of interest is the number of lost calls after a failure. In Figure 9.12, consider the period between t_6 (when the failure occurs) and t_8 (when the next registration occurs). Let K be the number of incoming calls arriving during (t_6, t_8). If the HLR backup record is obsolete at time t_6, then these K calls cannot be delivered to the MS. These incoming calls are either lost or forwarded to another phone number (or voice mailbox). Suppose that the incoming calls to the MS form a Poisson process with arrival rate γ. Compute $\Pr[K = k]$ or the probability that $K = k$ in the period $\tau_m^* = t_m - \tau_m = t_8 - t_6$. (Note that from the renewal theory, $\Pr[K = k]$ is determined by the inter-incoming call arrival time distribution, the inter-registration interval distribution, and when the failure occurs. In other words, $\Pr[K = k]$ is independent of the checkpoint algorithms, and therefore the derived $\Pr[K = k]$ equation should hold for both Algorithms I and II.)

Modeling of Algorithm II. The expected checkpoint interval $E_{II}[t_c]$ and the probability α_{II} of obsolete HLR backup record for Algorithm

II are derived as follows. Consider the timing diagram in Figure 9.14. In Algorithm II, if the t_p timer is restarted due to the t_p timeout event (i.e., a transition from **state 1** to **state 0**; see t_2 in Figure 9.14), then the next checkpoint interval is $t_c = \max(\tau_m{}^*, t_p)$. Conversely, if the t_p timer is restarted due to a registration event (i.e., a transition from **state 2** to **state 0**; see t_4 in Figure 9.14), then the next checkpoint interval is $t_c = \max(t_m, t_p)$. The two random variables $\max(\tau_m{}^*, t_p)$ and $\max(t_m, t_p)$ are not the same in general. To distinguish the above two cases, **state 0** is split into **state 01** and **state 02**. If a checkpoint occurs due to a registration event, then the FSM moves from **state 2** to **state 02**. If a checkpoint occurs due to a t_p timeout event, then the FSM moves from **state 1** to **state 01**. Figure 9.15 redraws the state diagram in Figure 9.13 with these two new states. Let π_x be the probability that the FSM is in **state x**. Then with probabilities

$$p_1 = \frac{\pi_{01}}{\pi_{01} + \pi_{02}} \quad \text{and} \quad p_2 = \frac{\pi_{02}}{\pi_{01} + \pi_{02}} \tag{9.15}$$

the random variable t_c can be expressed as

$$t_c = p_1 \max(\tau_m{}^*, t_p) + p_2 \max(t_m, t_p) \tag{9.16}$$

In Figure 9.15, it is clear that the transition probability from **state 1** to **state 01** is 1. Similarly, the transition probability from **state 2** to **state 02** is 1. Let the transition probabilities from **state 02** to **state 1** and **state 2** be p_a and p_b, respectively. Similarly, let the transition

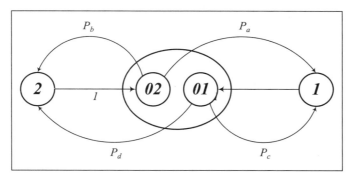

Figure 9.15 Modified State Diagram for Algorithm II

probabilities from **state 01** to **state 1** and **state 2** be p_c and p_d, respectively. These transition probabilities are derived as

$$
\left.
\begin{aligned}
p_a &= \Pr[t_p > t_m] = \int_{t_m=0}^{\infty} f_m(t_m) \int_{t_p=t_m}^{\infty} \lambda e^{-\lambda t_p} dt_p dt_m = f_m^*(\lambda) \\
p_b &= 1 - p_a = 1 - f_m^*(\lambda) \\
p_c &= \Pr[t_p > \tau_m] = \int_{\tau_m=0}^{\infty} r_m(\tau_m) \int_{t_p=\tau_m}^{\infty} \lambda e^{-\lambda t_p} dt_p d\tau_m = r_m^*(\lambda) \\
p_d &= 1 - p_c = 1 - r_m^*(\lambda)
\end{aligned}
\right\}
\tag{9.17}
$$

From Figure 9.15, the limiting probabilities π_x are expressed as

$$
\left.
\begin{aligned}
1 &= \pi_{01} + \pi_{02} + \pi_1 + \pi_2 \\
\pi_1 &= \pi_{01} \\
\pi_2 &= \pi_{02} \\
\pi_1 &= p_c \pi_{01} + p_a \pi_{02} \\
\pi_2 &= p_d \pi_{01} + p_b \pi_{02}
\end{aligned}
\right\}
\tag{9.18}
$$

Substituting (9.17) into (9.18), we obtain

$$
\pi_1 = \pi_{01} = \frac{f_m^*(\lambda)}{2\left[1 + f_m^*(\lambda) - r_m^*(\lambda)\right]}
$$

and

$$
\pi_2 = \pi_{02} = \frac{1 - r_m^*(\lambda)}{2\left[1 + f_m^*(\lambda) - r_m^*(\lambda)\right]}
$$

With the above equations, (9.15) is solved to yield

$$
p_1 = \frac{f_m^*(\lambda)}{1 + f_m^*(\lambda) - r_m^*(\lambda)} \quad \text{and} \quad p_2 = 1 - p_1 = \frac{1 - r_m^*(\lambda)}{1 + f_m^*(\lambda) - r_m^*(\lambda)}
\tag{9.19}
$$

Compute $E_{II}[t_c]$ and α_{II} by using (9.16) and (9.19).

4. Design a fraudulent detection algorithm using rare events. What are the costs for managing the rare event information?

5. How can you eavesdrop a mobile user when he/she is in phone conversation?

6. Describe the information maintained in the HLR. What information in the HLR is significantly affected by HLR failure?

7. Design a per-user HLR checkpoint algorithm that is different from Algorithms I and II.

8. Does it make sense to utilize the checkpoint technique for VLR/SGSN failure restoration?

## CHAPTER 10

VoIP for the Non-All-IP Mobile Networks

With the explosive growth of the Internet subscriber population, supporting Internet telephony services, also known as *Voice over IP* (*VoIP*), is a promising trend in the telecommunication business, and the momentum is clearly in favor of VoIP. Industry analysts project that the number of residential VoIP subscribers in the U.S. will rise to about 12 million by 2009, and total U.S. revenue for business and residential VoIP products and services will be nearly $21 billion US dollars [Che05c].

VoIP services provide real-time and low-cost voice communications over the IP network. Thus, incorporating VoIP services into the existing telecommunication systems is essential. The European Telecommunications Standards Institute's *Telecommunications and Internet Protocol Harmonization over a Network* (*TIPHON*) [ETS98a] specifies mechanisms to integrate IP telephony systems with *Switched Circuit Networks* (*SCN*; e.g., GSM, PSTN, and ISDN). Specifically, TIPHON defines several scenarios to illustrate different ways for integrating IP with mobile networks. Systems such as *iGSM* (see Chapter 16 [Lin 01b]) and *GSM on the Net* [Gra98] have been proposed or implemented based on the TIPHON scenarios. These systems utilize the ITU-T H.323 standard [ITU03] as the VoIP protocol.

This chapter describes how VoIP signaling protocols are utilized in the non-all-IP networks, specifically second-generation (2G) mobile networks (i.e., GSM and GPRS). Two VoIP signaling protocols are considered. *Media gateway control protocol* (*MGCP*) [And03] was proposed by Telcordia Technologies (formerly Bellcore) in 1998. Based on the concept of gateway decomposition, MGCP assumes a call control architecture in which the call control "intelligence" is provided by call agents outside the telephony gateways. Specifically, MGCP standardizes the interfaces between the telephony gateways and call agents. H.323, conversely, implements call control and call signaling conversion in the H.323 gateway. Section 10.1 introduces a system called *GSM-IP* that offers VoIP services through GSM using MGCP. Section 10.2 describes *vGPRS*, a GPRS-based VoIP solution that utilizes H.323.

10.1   GSM-IP: VoIP Service for GSM

This section describes GSM-IP, a system that integrates GSM and MGCP-based VoIP networks. Our results can be easily modified to accommodate other protocols such as the IETF *Media Gateway Control* (*MEGACO*) protocol [Gro03]. Figure 10.1 illustrates an MGCP-based VoIP network. The network entities are elaborated as follows:

Media Gateway (MG) is a telephony gateway that provides conversion between the audio signals carried on the SCN and data packets carried

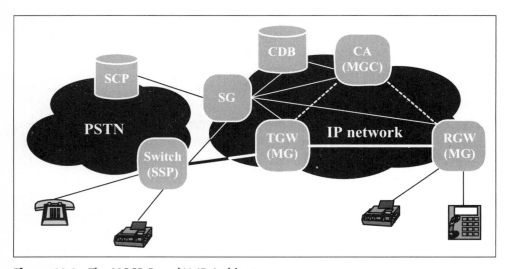

Figure 10.1   The MGCP-Based VoIP Architecture

over the IP network. There are several MG types. Two of them are used in GSM-IP: *Residential Gateway* (*RGW*) and *Trunking Gateway* (*TGW*). The residential gateway provides a traditional analog interface, which connects existing analog telephones and fax machines to the VoIP network. The trunking gateway interfaces between the PSTN and a VoIP network. A TGW is typically a tandem switch connecting to a switch in the PSTN via the T1 or E1 trunks.

Signaling Gateway (SG) interworks the MGCP elements with the SS7 signaling network in the PSTN. This gateway performs conversion between the SS7 signaling protocols such as SS7 *Transaction Capabilities Application Part* (*TCAP*) and *Integrated Services Digital Network User Part* (*ISUP*) in the PSTN and the IETF *Signaling Transport* (*SIGTRAN*) protocol in the IP network. Details of SIGTRAN-based TCAP and ISUP are given in Chapter 8. The SG also maintains the mapping function between the SS7 and the IP addresses. All MGCP elements communicating with the SS7 *Signaling Endpoints* (*SEPs*) follow the IETF SIGTRAN protocol shown in Figures 10.2 and 10.3. In these figures, an SEP can be a Service Switching Point (SSP), a Signal Transfer Point (STP), or a Service Control Point (SCP). The *SCN Signaling Adaptation* (*SSA*)

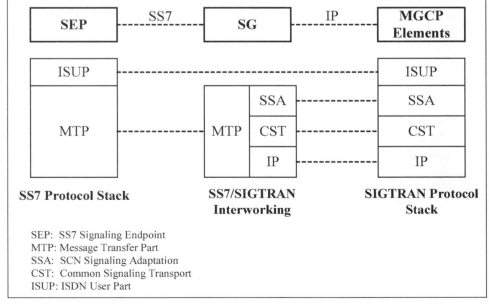

Figure 10.2 ISUP-SIGTRAN Protocol Stacks

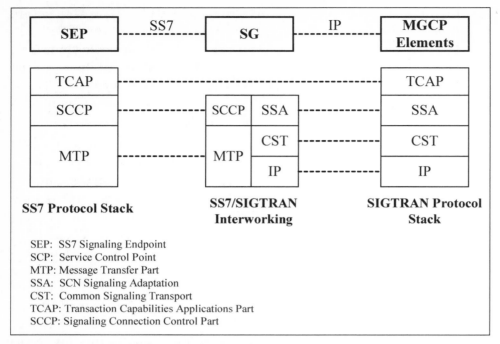

Figure 10.3   TCAP-SIGTRAN Protocol Stacks

layer supports specific primitives (e.g., address translation) required by a particular SCN signaling protocol such as ISDN or SS7. In Chapter 8, the SSA for SS7 in this section is MTP3 User Adaptation Layer (M3UA). The *Common Signaling Transport* (*CST*) layer supports a common set of reliable functions for signaling transport. An example of CST is the *Stream Control Transmission Protocol* (*SCTP*). The layer below CST utilizes the Internet Protocol (IP). Figure 10.2 illustrates the SIGTRAN-ISUP protocol stacks and Figure 10.3 illustrates the SIGTRAN-TCAP protocol stacks [Lou04, Ong99].

Media Gateway Controller (MGC) or Call Agent (CA) is responsible for call setup and release of the media channels in an MG. By utilizing the signaling protocol translation function in an SG, an MGC can handle the SS7 ISUP signaling for call setup between the IP network and PSTN. By exchanging the SS7 TCAP messages, the MGC can also interact with the SCP over the SS7 network to provide *Intelligent Network* (*IN*) services.

Common Database (CDB) serves as an IP SCP, directory server, and authentication center to perform authorization and routing functions.

10.1.1 MGCP Connection Model and the GSM-IP Architecture

Based on the concept of endpoints and connections, the *MGCP connection model* establishes end-to-end voice paths. An endpoint at an MG serves as an interface of the data source or sink. The endpoint either terminates a trunk from a PSTN switch or terminates a connection from Customer Premises Equipment (CPE; e.g., telephone set, key telephone system, or private branch exchange). A connection is an association between two or more endpoints for transmitting audio, video, or data. Based on the MGCP connection model, an MGC instructs an MG to create/delete a connection according to the events (e.g., on-hook or off-hook) occurring at an endpoint. The MGC also requests the MG to generate certain signals (e.g., dial tone) to an endpoint upon the occurrence of specific events. Events and signals are grouped into *packages* that are supported by a particular type of endpoint. In order to integrate the GSM architecture with the MGCP-based VoIP system, an additional GSM-IP package is introduced. All events and signals defined in the GSM-IP package are the GSM A-bis messages. As shown in Figure 10.4, A-bis is the GSM interface between the Base Transceiver Station (BTS) and the Base Station Controller (BSC). Table 10.1 lists some of the events and signals in the GSM-IP package. These events/signals will be elaborated in Section 10.1.2.

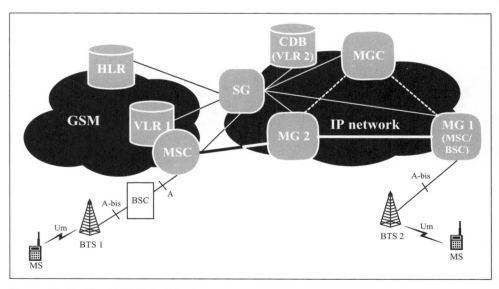

Figure 10.4 The GSM-IP Architecture

Table 10.1 Events and signals defined in the GSM-IP package (an incomplete list)

EVENT	SIGNAL
Setup	Setup
Alerting	Alerting
Connect	Connect
Disconnect	Disconnect
Paging Response	Paging
Handoff Request	Handoff Command
Handoff Complete	

Nine commands are defined in MGCP. Five of them are used in GSM-IP, which are described as follows:

NotificationRequest (RQNT) is sent from an MGC to an MG for requesting notification upon the occurrence of specific events in an endpoint. An MGC uses this command to instruct an MG to generate certain signals to an endpoint when specific events occur.

Notify (NTFY) is sent from an MG to an MGC in response to a **Notification Request** command.

CreateConnection (CRCX) is used to create a connection between two endpoints located in different MGs. The MGC uses this command to instruct each of the two MGs to select a specific IP address and UDP port for its endpoint. The IP address and the UDP port are used to establish an end-to-end connection between the two endpoints.

ModifyConnection (MDCX) is sent from an MGC to an MG for changing the parameters (e.g., audio packet encoding method) associated with a previously established connection.

DeleteConnection (DLCX) is used by an MGC to delete an existing connection in an MG.

All MGCP commands are acknowledged. An acknowledgment message carries a return code indicating the result of the command execution. Examples of the code include the following:

- 100 (the command is currently being executed)
- 200 (the command was executed normally)

- 510 (the command could not be executed because a protocol error was detected)

Figure 10.4 illustrates the GSM-IP architecture. In this architecture, the GSM network is the same as that illustrated in Figure 2.1. Several MGCP elements are modified to perform additional functions for integrating the IP network with the GSM network.

In Figure 10.4, MG1 serves as a Mobile Switching Center (MSC) and BSC. The voice path is connected from the GSM to IP networks via a tandem gateway MG2. A GSM BTS connects to MG1 via the standard GSM A-bis interface. A CDB is used to implement a GSM VLR for GSM-IP subscribers who visit the IP network. Every CDB is assigned an ISDN number that can be recognized by the HLR in the GSM network. The CDB is responsible for performing GSM roaming management procedures based on GSM Mobile Application Part (MAP). With the SIGTRAN (SCTP-based) protocol provided by the SG, the CDB communicates with the HLR and VLR in the GSM network as well as MG1 in the IP network.

10.1.2 GSM-IP Message Flows

This section describes the message flows for registration, call origination, call delivery, call release, and inter-system handoff procedures for the GSM-IP services. In this section, a message with the "ISUP" prefix represents an SS7 ISUP message. The "Um", "A", and "A-bis" prefixes denote the GSM air interface, the A interface, and the A-bis interface, respectively. A message with the "ST" prefix represents a SIGTRAN message that is equivalent to a GSM MAP/TCAP or SS7 ISUP message with the same name. For example, the **ST Update Location Area** message is the SIGTRAN version of the **MAP Update Location Area** message, and the **ST IAM** message is the SIGTRAN version of the **ISUP IAM** message.

Registration

If a GSM-IP user moves around Location Areas (LAs) within a GSM network, the standard GSM registration procedure described in Section 9.2.1 is exercised. When a GSM-IP user moves from a GSM network to an IP network, the registration message flow is executed, as shown in Figure 10.5:

Step 1. An MS moves from the coverage area of BTS1 to that of BTS2, where BTS1 is in the GSM network and BTS2 is in the IP network. The MS sends the **Um Location Update Request** message to BTS2. BTS2

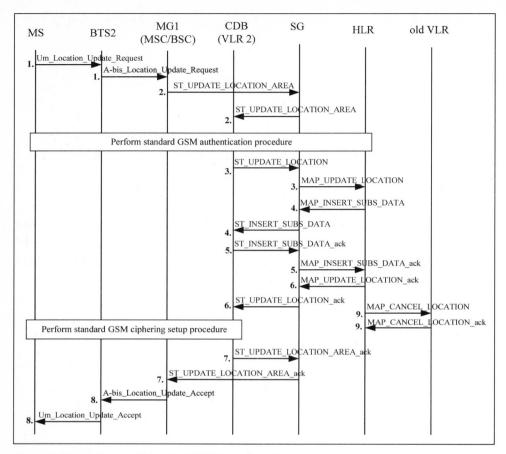

Figure 10.5   Message Flow for GSM-IP Registration

forwards this message to MG1 (MSC/BSC) through the GSM A-bis interface. We assume that the MS provides the International Mobile Subscriber Identity (IMSI) for registration. Our message flow can be easily extended for the case when the Temporary Mobile Subscriber Identity (TMSI) is used for registration [Lin 01b].

Step 2. After receiving the **A-bis Location Update Request** message from BTS2, MG1 sends an **ST Update Location Area** message to the SG. With the address mapping function, the SG identifies the IP address of the destination (i.e., VLR2 in the IP network), and forwards the message to VLR2.

Step 3. VLR2 recognizes that the MS is a new visitor, and creates a VLR record for the MS. After performing the standard GSM authentication procedure, VLR2 sends an **ST Update Location Area** message to the SG. The address mapping indicates that the destination of this message is in the SS7 network. The SG decapsulates the SIGTRAN message by removing the IP, SCTP (CST), and M3UA (SSA) headers, and generates an SS7 message by encapsulating the remaining information in the SIG-TRAN message with the *Signaling Connection Control Part* (*SCCP*) and *Message Transfer Part* (*MTP*) headers. The resulting message **MAP Update Location** is sent to the HLR in the SS7 network.

Steps 4 and 5. The HLR searches the location record of the MS based on the received IMSI, and updates this location record by modifying the MSC and VLR numbers. The HLR sends the *user profile* or *subscriber data* back to VLR2 with the message **MAP Insert Subs Data** through the SG. The user profile indicates the services subscribed by the GSM-IP user (e.g., call waiting). VLR2 records the user profile of the MS and sends an acknowledgment to the HLR via the SG.

Step 6. Upon receipt of the acknowledgment from VLR2, the HLR completes the registration task and sends the **MAP Update Location Ack** message to the SG. The SG translates the **MAP Update Location Ack** message to the **ST Update Location Ack** and forwards this SIGTRAN message to VLR2.

Step 7. VLR2 sets up the standard GSM ciphering with the MS. Then it sends the **ST Update Location Area Ack** message to MG1 through the SG.

Step 8. MG1 informs the MS that the location update request is accepted by the HLR. At this point, the registration procedure is completed.

Step 9. The HLR exchanges the **MAP Cancel Location** message pair with the old VLR to delete the obsolete record of the MS.

Call Origination

Suppose that a GSM-IP subscriber is in the coverage area of a BTS connecting to MG1 in the IP network. When this GSM-IP subscriber originates a call to a PSTN subscriber, the call path illustrated in Figure 10.6 results. In this example, MG2 is a tandem gateway that connects MG1 (the MSC) to the destination switch in the PSTN. For simplicity, we assume that there is a direct link between the destination switch and MG2. The message flow (see Figure 10.7; note that the MGC in this figure is not shown in Figure 10.6) is described in the following steps:

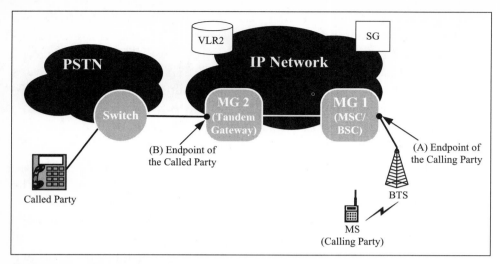

Figure 10.6   PSTN-IP Call Path (Call Origination)

Step 0. At system setup or before call origination, the MGC and every MG that serves as an MSC exchange messages for initializing the status of the endpoints in these MGs. In our example, the MGC sends the **RQNT** command to MG1. This command instructs MG1 to detect the Setup event in the endpoint. Note that all events and signals described in this message flow are defined in the GSM-IP package shown in Table 10.1. Upon receipt of **RQNT**, MG1 acknowledges this command, and is ready to receive the Setup event.

Step 1. To originate a call, the standard GSM traffic channel assignment, authentication, and ciphering setup procedures are performed to set up the radio link between the MS and the BTS [ETS 04, 3GP05e]. Then the digits dialed by the GSM-IP subscriber are sent to the BTS in the **Um Setup** message. The BTS forwards the message to MG1 through the GSM A-bis interface.

Steps 2 and 3. When detecting the Setup event in the MG1 endpoint of the calling party (see Figure 10.6 (A)), MG1 sends the message **ST Send Info For Outgoing Call** to VLR2 through the SG. This message invokes VLR2 to determine whether the service requested by the calling party is legal, e.g., if the calling party is allowed to make international calls or premium rate calls. With the address mapping function, the SG identifies the IP address of the destination (i.e., VLR2 in the IP network), and forwards the SIGTRAN message to VLR2. Upon receipt of this message,

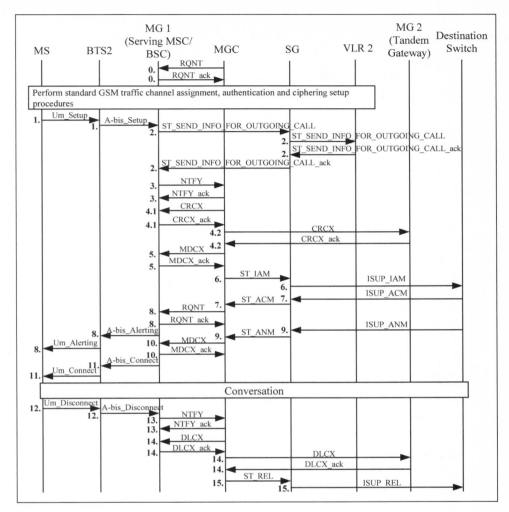

Figure 10.7   Message Flow for GSM-IP Call Origination

VLR2 performs the authorization check and returns the result to MG1 via the SG. After authorization is successfully performed, MG1 sends the received digits to the MGC by issuing the **NTFY** command.

Step 4. Based on the dialed digits, the MGC starts the voice path setup procedure as follows:

Step 4.1. The MGC first sends the **CRCX** command to the originating MG (MG1). Upon receipt of the **CRCX** command, MG1 selects a specific IP address and a UDP port for the endpoint of the calling party.

The **CRCX** acknowledgment is sent from MG1 to the MGC. This acknowledgment carries the return code of the execution result as well as the connection information of the calling party.

Step 4.2. The MGC also sends the **CRCX** command to the terminating MG (MG2) along with the IP address and the UDP port of the calling party endpoint. Upon receipt of the **CRCX** command, MG2 reserves the outgoing trunk connected to the destination switch and selects a specific IP address and the UDP port for the called party endpoint (see Figure 10.6 (B)). The **CRCX** acknowledgment returns the connection information of the called party.

At this point, the called party has not picked up the phone, and the connection information of the called party is not known by MG1. Thus, the connection mode in MG1 is "receive only." Conversely, since the IP address and the UDP port of the calling party have been sent to MG2 in Step 4.2, the connection mode in MG2 is "send/receive."

Step 5. After receiving the connection information of the called party in the **CRCX** acknowledgment from MG2, the MGC sends the information to MG1 by using the **MDCX** command. Upon receipt of **MDCX**, MG1 obtains the IP address and the UDP port of the called party endpoint.

Step 6. The MGC sends the **ST IAM** (Initial Address Message) to the SG to initiate trunk setup between MG2 and the switch of the called party. After address mapping, the SG realizes that the destination of this message is in the SS7 network. The SG decapsulates the SIGTRAN message by removing the IP, SCTP, and M3UA headers, and generates an SS7 message by encapsulating the remaining information in the SIGTRAN message with the MTP header. Then the resulting **ISUP IAM** is sent to the destination switch in the PSTN.

Step 7. The destination switch sends the **ISUP ACM** (Address Complete Message) to the SG. This message indicates that the routing information required to complete the call has been received by the destination switch. The SG translates the **ISUP ACM** message to a SIGTRAN message **ST ACM** and forwards this message to the MGC.

Step 8. After receiving the **ST ACM** message, the MGC instructs MG1 to generate an Alerting signal to the endpoint of the calling party by issuing **RQNT**. MG1 sends an **A-bis Alerting** message to the BTS. The message is then forwarded to the MS through the air interface Um. The **A-bis Alerting** message triggers the ringback tone at the MS.

Step 9. When the called party picks up the phone, the destination switch in the PSTN sends the **ISUP ANM** (Answer Message) to the SG. Then the message is forwarded from the SG to the MGC.

Step 10. The MGC instructs MG1 to change the connection mode of the calling party to "send/receive" by issuing the **MDCX** command. At this point, the voice path between MG1 and MG2 is established. In order to inform the MS that the called party has answered the call, the **MDCX** command requests MG1 to generate a Connect signal to the calling party endpoint. **MDCX** also asks MG1 to detect the Disconnect event, i.e., to detect whether the calling party hangs up the phone after the conversation begins.

Step 11. After receiving the **MDCX** command, MG1 sends the **A-bis Connect** message to the BTS. The BTS forwards the message to the MS through the air interface Um.

At this point, the call is established and the conversation begins. When the conversation is over, the call release procedure is executed with the following steps:

Step 12. Assume that the calling party (GSM-IP subscriber) hangs up the phone first. The MS sends the **Um Disconnect** message to the BTS. The BTS then forwards this message to MG1 through the A-bis interface.

Step 13. The **A-bis Disconnect** message results in a Disconnect event in the endpoint of the calling party at MG1. MG1 informs the MGC of the Disconnect event with the **NTFY** command.

Step 14. The MGC sends the **DLCX** command to both MG1 and MG2 to terminate the connection for the current call. Upon receipt of the **DLCX** command, both MG1 and MG2 release the resources used in the call, and then acknowledge the **DLCX** command.

Step 15. The MGC also sends the **ST REL** (Release) message to the SG to indicate that the voice path for the current call is released. The SG translates this message to the **ISUP REL** message and forwards it to the destination switch in the PSTN.

At this point, the call is disconnected and the MGC instructs MG1 to detect the subsequent Setup event in the endpoint by sending a new **RQNT** command (i.e., Step 0 for the next cycle).

Call Delivery

Suppose that a GSM-IP subscriber is in the coverage area of a BTS connecting to MG1 in the IP network. When a PSTN subscriber makes a call to the GSM-IP subscriber, the call path illustrated in Figure 10.8 results. In this example, the originating switch in the PSTN sets up the call path to the GMSC of the GSM-IP subscriber. Then the GMSC is connected to MG1 (the serving MSC) via the tandem gateway MG2. For simplicity, we assume that there is a direct link between the GMSC and MG2. The message flow (see Figure 10.9) is described in the following steps:

Step 1. When the Mobile Station ISDN Number (MSISDN) of the MS is dialed by a PSTN subscriber, the call is routed to a GMSC by the **ISUP IAM** message.

Step 2. The GMSC sends the **MAP Send Routing Information** message to the HLR to retrieve the routing information. The HLR record indicates that the VLR visited by the MS is VLR2 in the IP network. To obtain the Mobile Station Roaming Number (MSRN; the address of the MSC serving the MS), the HLR sends the **MAP Provide Roaming Number** message to the SG. The SG translates the message to the **ST Provide Roaming Number** and forwards it to VLR2.

Step 3. VLR2 creates the MSRN by using the MSC number stored in the VLR record of the MS. This roaming number is sent back to the GMSC through the SG and the HLR.

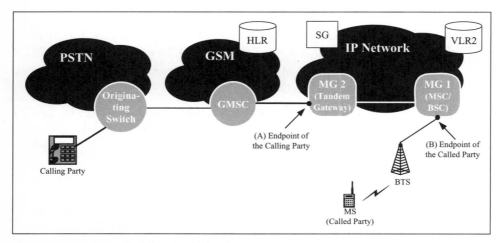

Figure 10.8   PSTN-IP Call Path (Call Delivery)

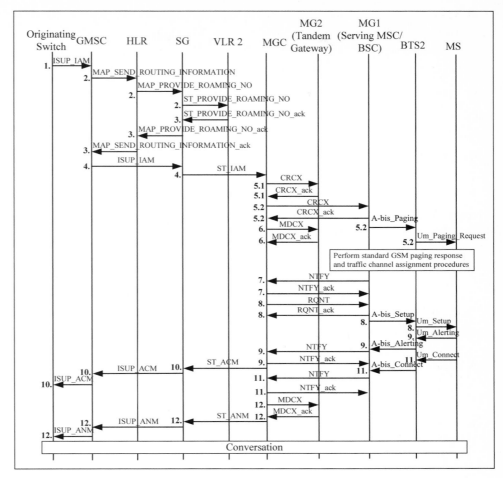

Figure 10.9 Message Flow for GSM-IP Call Delivery

Step 4. The MSRN indicates that the MSC of the called MS is in the IP network. The GMSC sends the **ISUP IAM** message to the SG to initiate signaling for trunk setup. The SG translates this message to the **ST IAM** and forwards it to the MGC.

Step 5. Based on the calling and called party addresses in the **ST IAM** message, the MGC starts the voice path setup procedure between MG2 and MG1.

Step 5.1. The MGC first sends the **CRCX** command to MG2 as described in Step 4.1 in Figure 10.7, which reserves the incoming trunk (i.e., the endpoint of the calling party in MG2) connected to the GMSC.

Step 5.2. The MGC also sends the **CRCX** command to MG1 along with the IP address and the UDP port of the calling party. This step is similar to Step 4.2 in Figure 10.7, except that this **CRCX** command piggybacks a notification request to perform two tasks:

- MG1 is asked to generate a Paging signal to the called party endpoint, which results in an **A-bis Paging** message sent to the BTS. This message is forwarded to the MS through the air interface Um.

- The Paging signal generated by MG1 triggers the detection of a Paging Response event in the called party endpoint.

MG1 waits for the arrival of the Paging Response event.

Step 6. The MGC sends the connection information of the called party to MG2 by issuing the **MDCX** command as described in Step 5 of Figure 10.7.

Step 7. Upon receipt of the **A-bis Paging Response** message from the MS, MG1 performs the standard GSM traffic channel assignment. MG1 also sends the **NTFY** command to the MGC. This command indicates the occurrence of a Paging Response event.

Step 8. When the MGC receives the Paging Response event notification, it sends the **RQNT** command to MG1 to inform the MS that a call will be set up. With the same **RQNT** command, the MGC also requests MG1 to detect the Alerting and Connect events of the called party endpoint. After receiving the **RQNT** command, MG1 sends the **A-bis Setup** message to the BTS. The BTS forwards it to the MS through the air interface Um. Then MG1 is ready to detect the Alerting and Connect events in the called party endpoint.

Step 9. The MS rings and sends the **A-bis Alerting** message to the BTS. The BTS forwards this message to MG1, which results in an Alerting event in MG1. When MG1 detects the Alerting event, it informs the MGC by using the **NTFY** command.

Step 10. The MGC informs the originating switch of the alerting situation in the called party by sending the **ST ACM** message to the SG. The SG translates the **ST ACM** message to the **ISUP ACM** message and forwards this message to the originating switch through the GMSC.

Step 11. The called party answers the call and the MS sends a **Um Connect** message to the BTS. The BTS forwards this message to MG1, which results in a Connect event in MG1. Then MG1 informs the MGC of the Connect event occurrence by issuing the **NTFY** command.

Step 12. Through the **MDCX** command, the MGC instructs MG2 to change the connection mode of the calling party to "send/receive." The MGC also sends the **ST ANM** message to the originating switch to indicate that the called party (MS) has answered the call.

At this point, the call is established and the conversation begins.

Inter-System Handoff

If a mobile user in conversation moves from the coverage area of a BTS to the coverage area of another BTS, the radio link to the old BTS is disconnected and a radio link in the new BTS must be set up to continue the conversation. This process is called *handoff* (see Chapters 2, 3, and 6 in [Lin 01b]). If the handoff occurs between two BTSs connected to the same MSC, then the procedure typically involves the air, the A-bis, and the A interfaces. The wireline transport network is not affected. This type of handoff is called *intra-system handoff*. Conversely, if the two BTSs involved in the handoff are connected to different MSCs, then the handoff is referred to as *inter-system handoff*.

When a GSM-IP subscriber moves from the coverage area of BTS1 in the GSM network to that of BTS2 in the IP network during a conversation, the call path before and after the inter-system handoff illustrated in Figure 10.10 results. In this example, MG2 is a tandem gateway that connects MG1 (the target MSC) in the IP network to MSC1 (the serving MSC) in the GSM

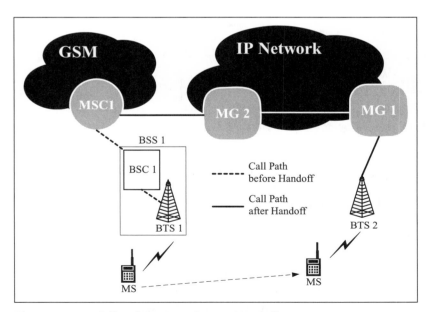

Figure 10.10 Call Path for Inter-System Handoff

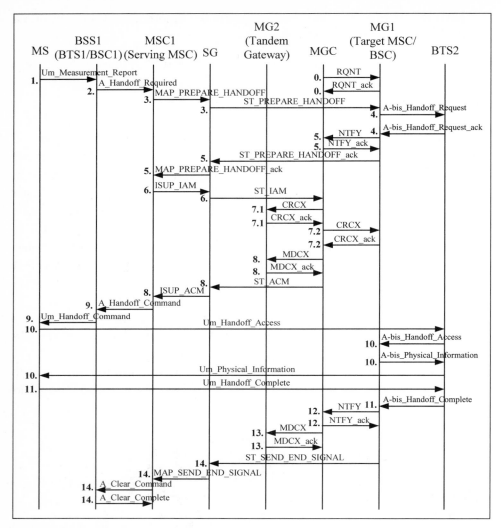

Figure 10.11 Message Flow for GSM-IP Handoff

network. After the handoff, the MS connects to MSC1 through MG1 and MG2. The handoff message flow is illustrated in Figure 10.11. To simplify our discussion, we use the term *Base Station System 1* (*BSS1*) to represent BTS1 and BSC1, and omit the details of the A-bis messages exchanged between BTS1 and BSC1:

Step 0. At system setup or before handoff, the MGC and every MG that serves as an MSC exchange messages for initializing the status of the endpoints in these MGs. In our example, the MGC sends the **RQNT**

command to MG1. This command instructs MG1 to detect the Handoff Request event in the endpoint.

Step 1. The MS periodically monitors the signal quality of the radio link and reports the signal strength to BSS1 via the **Um Measurement Report** message.

Step 2. Based on the measurement result sent from the MS, BSS1 recognizes that handoff is required. BSS1 requests MSC1 to perform the handoff procedure by issuing the **A Handoff Required** message. The **A Handoff Required** message contains a list of the target BTSs that are qualified to serve the MS.

Step 3. After receiving the **A Handoff Required** message, MSC1 selects a target BTS (BTS2) for handoff. Since BTS2 is controlled by MG1 in the IP network, MSC1 sends the **MAP Prepare Handoff** message to MG1 via the SG. This message contains any information needed by MG1 to allocate a new radio channel for the MS.

Step 4. MG1 sends the **A-bis Handoff Request** message to BTS2. This message instructs BTS2 to allocate a traffic channel for the MS. After performing the radio channel allocation, BTS2 sends an acknowledgment to MG1, which results in a Handoff Request event in MG1.

Step 5. When the Handoff Request event is detected, MG1 informs the MGC by invoking the **NTFY** command. MG1 also returns the **ST Prepare Handoff Ack** message to MSC1 via the SG. This message indicates completion of new radio channel allocation.

Step 6. Upon receipt of the **MAP Prepare Handoff Ack** message, MSC1 sends the **ISUP IAM** message to the MGC to initiate signaling for trunk setup between MSC1 and MG1.

Step 7. After receiving the **ST IAM** message, the MGC starts the voice path setup procedure between MG2 and MG1.

Step 7.1. The MGC first sends the **CRCX** command to MG2 as described in Step 4.1 in Figure 10.7, which reserves the incoming trunk (i.e., the endpoint in MG2) connected to MSC1.

Step 7.2. The MGC also sends the **CRCX** command to MG1 (the target MSC) along with the IP address and the UDP port of the endpoint in MG2. This step is similar to Step 4.2 in Figure 10.7, except that this **CRCX** command piggybacks a notification request, which asks MG1 to detect the Handoff Complete event, i.e., to detect whether the MS completes the handoff task.

Step 8. The MGC sends the **MDCX** command to MG2. This command carries the connection information of the MG1 endpoint. The MGC also sends the **ST ACM** message to MSC1 via the SG. This message indicates that the routing information required to set up the voice path has been received.

Step 9. After receiving the **ISUP ACM** message, MSC1 sends the **A Handoff Command** message to the MS through BSS1. This message instructs the MS to perform the handoff operation. This message includes the new radio channel identification supplied by BTS2.

Step 10. The MS tunes to the new radio channel and sends the **Um Hand-off Access** message to BTS2 on the new radio channel. BTS2 forwards the message to MG1 through the GSM A-bis interface. Upon receipt of the **A-bis Handoff Access** message, MG1 sends the physical channel information to the MS through BTS2.

Step 11. After obtaining the physical channel information from BTS2, the MS sends the **Um Handoff Complete** message to BTS2. The message indicates that the handoff is successful. BTS2 then forwards it to MG1 through the GSM A-bis interface, which results in a Handoff Complete event.

Step 12. With the **NTFY** command, MG1 informs the MGC of the Handoff Complete event.

Step 13. The MGC instructs MG2 to change the connection mode of the MG1 endpoint to "send/receive" by issuing the **MDCX** command.

At this point, the handoff procedure is complete and the MS is switched to the new call path.

Step 14. Finally, MG1 sends the **ST Send End Signal** message to MSC1 through the SG. This message indicates that the new radio path of the MS has been connected. Upon receipt of the **MAP Send End Signal** message, MSC1 instructs BSS1 to clear the radio resource with the **A Clear Command** message. Then BSS1 releases the radio channel of the MS and returns the **A Clear Complete** message to MSC1.

10.2   vGPRS: VoIP Service for GPRS

This section describes vGPRS, a VoIP mechanism for GPRS. The vGPRS approach is implemented using standard H.323, GPRS, and GSM protocols.

Thus, existing GPRS and H.323 network elements are not modified. The vGPRS approach provides VoIP service to standard GSM and GPRS MSs. In vGPRS, a new network node called *VoIP MSC* (*VMSC*) is introduced to replace MSC. The VMSC is a router-based softswitch and its cost is anticipated to be cheaper than an MSC. Figure 10.12 (a) shows the interfaces between the VMSC and other GSM/GPRS network nodes. The GSM signaling interfaces of the VMSC are exactly the same as that of an MSC. That is, it communicates with the BSC, the VLR, and the PSTN through the A interface, the B interface (the interface between MSC and VLR), and SS7 ISUP, respectively. Unlike an MSC, the VMSC communicates with Serving GPRS Support Node

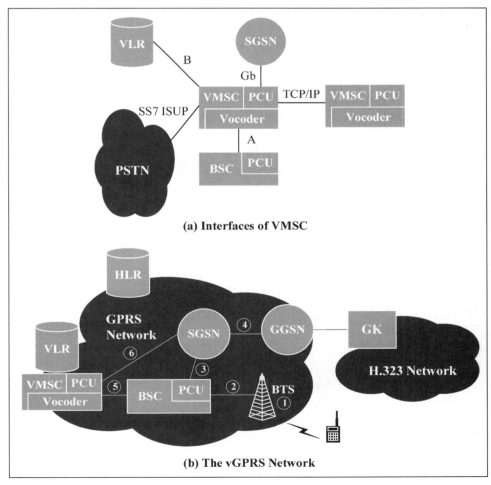

Figure 10.12   The VMSC Interfaces and the VGPRS Network Architecture

(SGSN) through the GPRS Gb interface. In other words, while an MSC delivers voice traffic through circuit-switched trunks, the VMSC delivers VoIP packets through the GPRS network. In Figure 10.12 (b), an SGSN receives and transmits packets between the MSs and their counterparts in the packet data network. To connect to an SGSN, a *Packet Control Unit* (*PCU*) is implemented in the BSC. The BSC forwards circuit-switched calls to the MSC, and packet-switched data (through the PCU) to the SGSN. A BSC can only connect to one SGSN. The data path of a GPRS MS is $(1)\leftrightarrow(2)\leftrightarrow(3)\leftrightarrow(4)$. The voice path is $(1)\leftrightarrow(2)\leftrightarrow(5)\leftrightarrow(6)\leftrightarrow(4)$. In this voice path, $(1)\leftrightarrow(2)\leftrightarrow(5)$ is circuit-switched as in GSM. Thus, both standard GSM MSs and GPRS MSs can set up calls as if they were in the standard GSM/GPRS network.

At the VMSC, the voice information is translated into GPRS packets through the vocoder and the PCU. Then the packets are delivered to the GPRS network through the SGSN. See path $(6)\leftrightarrow(4)$. An IP address is associated with every MS attached to the VMSC. In GPRS, the IP address can be created statically or dynamically for a GPRS MS. Conversely, a standard GSM MS cannot be assigned an IP address through GPRS. Thus, in vGPRS the creation of the IP address is performed by the VMSC through the standard GPRS Packet Data Protocol (PDP) context activation procedure. The VMSC maintains an MS table. This table stores the MS mobility management (MM) and PDP contexts such as TMSI, IMSI, and the QoS profile requested. These contexts are the same as those stored in a GPRS MS (see Chapter 2). In vGPRS, the H.323 protocol is used to support voice applications. The VMSC executes the H.323 protocol just like an H.323 terminal. To support inter-system handoff, VMSCs communicate with each other using TCP/IP (see Section 10.2.4). Clearly, the vGPRS can also be implemented by using *Session Initiation Protocol* (*SIP*; see Chapter 12).

A *gatekeeper* (*GK*) in the H.323 network (see Figure 10.12 (b)) performs standard H.323 gatekeeper functions (such as address translation). Figure 10.13 illustrates the connection between an H.323 terminal and a GSM MS. In this figure, the TCP/IP protocols are exercised in links (1), (2), and (8). The *GPRS Tunneling Protocol* (*GTP*) is exercised in link (3), and the GPRS Gb protocol [ETS97b] is exercised in link (4). The standard GSM protocols are exercised in links (5), (6), (7), (9), and (10). The H.323 protocol is implemented on top of TCP/IP, and is exercised between the VMSC and the H.323 nodes in the IP network. The H.323 packets are encapsulated and delivered in the GPRS network through the GTP. The vGPRS procedures utilize the *Registration, Admission, and Status* (*RAS*) and Q.931 messages as defined in the H.323 and H.225 protocols [ITU03, ITU98]. These procedures also involve the Um, A, A-bis, and MAP interfaces/protocols.

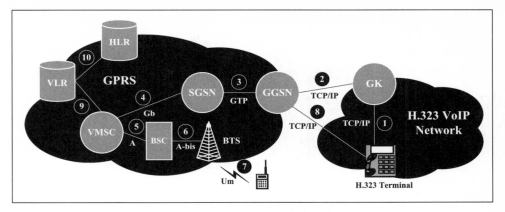

Figure 10.13   Connection between an H.323 Terminal and a GSM MS

10.2.1 Registration

In GSM, a registration procedure is performed when an MS is turned on or when the MS moves to a new location area. Without loss of generality, this section describes the registration procedure by assuming that registration is performed when the MS is turned on. The registration procedure for MS movement is similar and is briefly elaborated at the end of this section. The GSM registration messages (Steps 1.1, 1.2, and 1.3 below) are delivered through the path (7)↔ (6)↔ (5)↔ (9)↔ (10) in Figure 10.13. The H.323 messages (Steps 1.4, 1.5, and 1.6 below) are delivered through the path (7)↔ (6)↔ (5)↔ (4)↔ (3)↔ (2). The message flow illustrated in Figure 10.14 is described in the following steps:

Step 1.1. An MS is turned on. To join in the network, the MS performs registration following the standard GSM location update procedure described in Section 9.2.1. The MS sends the **Um Location Update Request** message to the BTS. The BTS forwards this request to the BSC using the **A-bis Location Update** message. The BSC then sends the **A Location Update** message to VMSC through the A interface. The VMSC issues the **MAP Update Location Area** message to the VLR. The standard GSM authentication procedure is exercised between the MS and the HLR to authenticate the MS. The details are given in Section 9.1.

Step 1.2. The VLR sends the **MAP Update Location** message to the HLR, and obtains the user profile of the MS (the profile indicates, e.g., whether the MS is allowed to make international calls) from the HLR through the **MAP Insert Subs Data** messages exchanged. The VLR then sets up the standard GSM ciphering with the MS. Then it sends the **MAP**

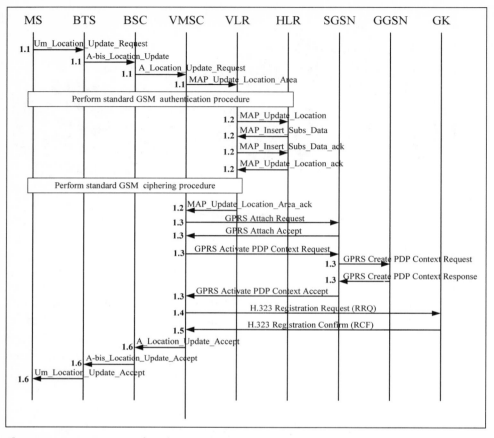

Figure 10.14   Message Flow for vGPRS Registration

Update Location Area Ack message to the VMSC to indicate that the registration is successful.

Step 1.3. The VMSC performs GPRS attach to the SGSN by exchanging the GPRS **Attach Request** and **Accept** message pair. Following the standard GPRS PDP context activation procedure, the VMSC activates a new PDP context just like a GPRS MS does. In the activation procedure, the IMSI of the MS is used by the GGSN to retrieve the HLR record to obtain information such as IP address (we assume that the IP address is allocated dynamically). The PDP context record for the MS is created in the GGSN. The record includes parameters such as IMSI, IP address, negotiated QoS profile, SGSN address, and so on. At this point, an IP session is established so that the VMSC can communicate with the gatekeeper in the external H.323 network.

Step 1.4. The VMSC initiates the endpoint registration to inform the gatekeeper of its transport address and alias address (i.e., MSISDN) through the **RAS Registration Request (RRQ)** message.

Step 1.5. The gatekeeper creates an entry for the MS in the address translation table, which stores the (IP address, MSISDN) pair. Then it confirms the registration by sending the **RAS Registration Confirm (RCF)** message to the VMSC. The VMSC then creates the MS MM and PDP contexts for the MS and stores these contexts in its MS table.

Step 1.6. The VMSC informs the MS that the location update request is accepted by the HLR, and the registration procedure is completed.

For MS movement, the PDP context of the old SGSN is moved to the new SGSN. This is achieved by using the combined inter-SGSN RA/LA update procedure (see Chapter 2), which is also described in the inter-system handoff procedure in Section 10.2.4. Specifically, the location update procedure for MS movement will be a combination of the standard GSM location update and Steps 5.3–5.6 in Section 10.2.4. The details are omitted.

10.2.2 MS Call Origination

This section describes MS call origination in which the MS is the calling party. The called party can be another MS in the same GPRS network or an H.323 terminal in the H.323 VoIP network. The called party can also be a traditional telephone set in the PSTN, which is connected indirectly to the GPRS network through the H.323 network. Without loss of generality, we assume that the called party is an H.323 terminal user. The message flow (see Figure 10.15) is described in the following steps:

Step 2.1. To originate a call, the standard GSM traffic channel assignment, authentication, and ciphering setup procedures are performed to set up the radio link between the MS and the BTS [ETS 04, 3GP05e]. Then the digits dialed by the MS are sent to the BTS in the **Um Setup** message. The BTS forwards the message to the VMSC through the GSM A-bis and A interfaces.

Step 2.2. The VMSC sends the **MAP Send Info For Outgoing Call** message to the VLR, which asks the VLR to check whether the service requested by the calling party is legal, e.g., whether the calling party is allowed to make this call. If authorization is successfully performed, the VMSC checks the PDP context record of the MS and identifies the routing path

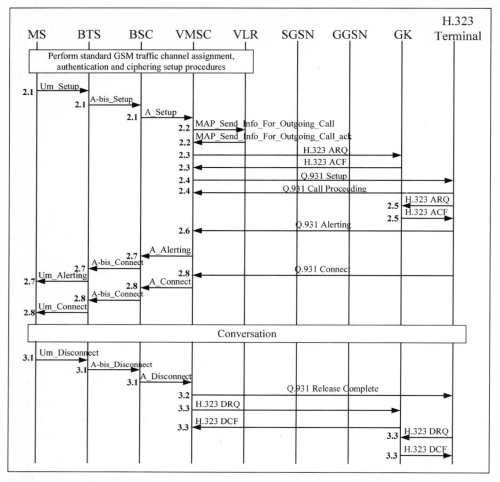

Figure 10.15 The Message Flows for MS Call Origination and Call Release

to the GGSN based on the GPRS *Tunnel Endpoint Identifier* (*TEID*) and the GGSN number.

Step 2.3. Based on the dialed digits, the VMSC starts the voice path setup procedure as follows. Through the **RAS Admission Request** (**ARQ**) and **Admission Confirm** (**ACF**) message pair exchange, the gatekeeper provides the VMSC the destination's call signaling channel transport address. The message path is (4)↔ (3)↔ (2) in Figure 10.13.

Step 2.4. The VMSC sends the **Q.931 Setup** message to the destination through the GGSN. The **Q.931 Call Proceeding** message is sent back to the VMSC to indicate that enough routing information has been received and it does not expect to receive more routing information from the VMSC. The message path is (4) ↔ (3) ↔ (8) in Figure 10.13.

Step 2.5. The H.323 terminal exchanges the **RAS ARQ** and **ACF** message pair with the gatekeeper. It is possible that the **RAS Admission Reject (ARJ)** message is received by the terminal and the call is released. The message path is (1) in Figure 10.13. If the call admission is accepted, the following steps are executed.

Step 2.6. A ringing tone is generated at the H.323 terminal to alert the called party. The **Q.931 Alerting** message is sent to the VMSC.

Step 2.7. The VMSC sends the **A Alerting** message to the BSC. The BSC sends the **A-bis Alerting** message to the BTS. The message is then forwarded to the MS through the **Um Alerting** message. This message triggers the ringback tone at the MS.

Step 2.8. When the called party answers the phone, the H.323 terminal generates the **Q.931 Connect** message to the VMSC. The VMSC sends the **A Connect** message to the BSC. The BSC sends the **A-bis Connect** message to the BTS. The BTS forwards the message to the MS through the air interface Um.

At this point, the call is established and the conversation begins. When the conversation is over, the call release procedure illustrated in Steps 3.1–3.3 of Figure 10.15 results, and is described as follows (we assume that the calling party, the GSM user, hangs up the phone first):

Step 3.1. The MS sends the **Um Disconnect** message to the BTS. The BTS then forwards this message to the VMSC through the BSC.

Step 3.2. The VMSC sends the **Q.931 Release Complete** message to the H.323 terminal to release the call.

Step 3.3. Both the VMSC and the H.323 terminal inform the gatekeeper of call completion by exchanging the **RAS Disengage Request (DRQ)** and **Confirm (DCF)** message pair. At the end of the call, the gatekeeper records the call statistics for charging.

10.2.3   MS Call Termination

This section describes the message flow for the MS call termination. We assume that the call originated from an H.323 terminal, and that the MS is the called party. The message flow (see Figure 10.16) is described in the following steps:

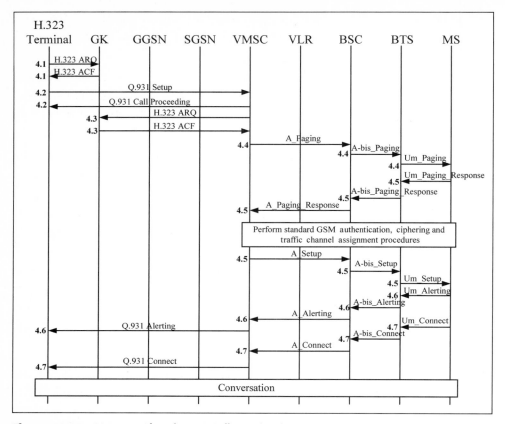

Figure 10.16   Message Flow for MS Call Termination

Step 4.1. The calling party sends the **RAS ARQ** message to the gatekeeper with the called party's address (i.e., the MSISDN of the MS). From the address translation table, the gatekeeper finds the IP address of the GGSN and returns it to the calling party through the **RAS ACF** message.

Step 4.2. The calling party sends the **Q.931 Setup** message to the VMSC through the GGSN (the VMSC is identified by the GGSN through the MS's IMSI). When the GGSN receives the **Setup** packet, it retrieves the PDP context of the MS based on the destination IP address of the packet. The GPRS TEID and the SGSN number of the MS are identified from the PDP context. Then the GGSN routes the packet to the VMSC through the SGSN. The VMSC sends the **Q.931 Call Proceeding** message back to the H.323 terminal as described in Step 2.4.

Step 4.3. The VMSC exchanges the **RAS ARQ** and **ACF** message pair with the gatekeeper as described in Step 2.5.

Step 4.4. The VMSC sends the **A Paging** message to the BSC. The BSC sends the **A-bis Paging** message to the BTS. The BTS then pages the MS.

Step 4.5. Upon receipt of the paging response from the MS, the network (VMSC and VLR) performs the standard GSM traffic channel assignment and ciphering procedures. The VMSC sends the **A Setup** message to the BSC. The BSC sends the **A-bis Setup** message to the BTS. The BTS forwards this setup instruction to the MS.

Step 4.6. The MS rings and sends the **Um Alerting** message to the BTS. The BTS forwards this signal to the VMSC through the BSC. The VMSC sends the **Q.931 Alerting** message to the H.323 terminal. The H.323 terminal generates the ringback tone.

Step 4.7. The called party answers the call and the MS sends the **Um Connect** message to the BTS. The BTS forwards this message to VMSC, which results in the **Q.931 Connect** message delivered to the calling party.

At this point, the call is established and the conversation begins.

10.2.4 Inter-System Handoff

This section describes inter-system handoff for vGPRS. We assume that the serving VMSC (old VMSC) and the target VMSC (new VMSC) are connected to different SGSNs (see Figure 10.17). In this figure, Path (1)↔(2) is the

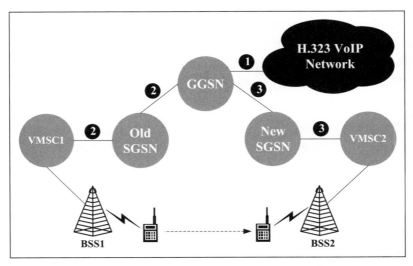

Figure 10.17 The Routing Path before and after Inter-System Handoff

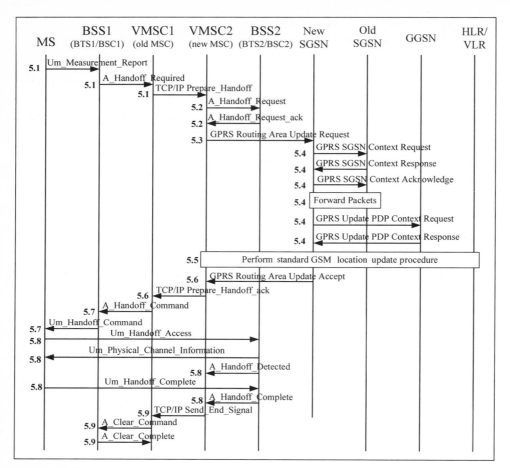

Figure 10.18 The Message Flow for Inter-System Handoff

routing path before the handoff, and Path (1)↔(3) is the routing path after the handoff. The inter-system handoff message flow in Figure 10.18 is described in the following steps:

Step 5.1. The MS periodically measures the radio link qualities of the surrounding BTSs, and sends the measurement report to the Base Station System (BSS; BTS plus BSC). Suppose that the MS connects to BSS1. BSS1 forwards the measurement report to VMSC1. VMSC1 determines that handoff is required. Since the target (new) BSS (BSS2 in our example) is connected to a different VMSC (VMSC2 in our example), VMSC1 sends the **Prepare Handoff** message to VMSC2. Note that if the new MSC is a standard GSM MSC, then the message is **MAP Prepare Handoff**. If the new MSC is a VoIP MSC, then the message is a TCP/IP message that provides the same information as its MAP counterpart. Further-more, the TCP/IP **Prepare Handoff** message delivers the MM and PDP contexts of the MS to the new VMSC.

Step 5.2. VMSC2 checks whether BSS2 is available to accommodate the handoff. If so, the following steps are executed.

Step 5.3. VMSC2 sends the **Routing Area Update Request** message to the new SGSN, which asks the new SGSN to switch the routing path (from (2) to (3) in Figure 10.17).

Step 5.4. The new SGSN communicates with the old SGSN to obtain the MM and PDP contexts for the MS. It also requests the old SGSN to forward the packets destinated at the MS. The new SGSN then exchanges the GPRS **Update PDP Context** message pair with the GGSN to switch the routing path.

Step 5.5. The standard GSM location update is performed to modify the location information in the HLR and the VLR.

Step 5.6. The new SGSN informs VMSC2 that the routing path is success-fully switched and the MS can hand off to BSS2.

Step 5.7. VMSC1 asks the MS to move to BSS2.

Step 5.8. The MS communicates with BSS2 to switch the radio link.

Step 5.9. When the handoff is complete, VMSC2 sends the TCP/IP **Send End Signal** message to VMSC1. VMSC1 deletes the MM and PDP contexts of the MS, and asks BSS1 to reclaim the radio resources used by the MS.

In the preceding inter-system handoff procedure, if both VMSC1 and VMSC2 are connected to the same SGSN, then Steps 5.3, 5.4, and 5.6 are not executed. Furthermore, in the vGPRS handoff, no H.323 signaling is required. That is, the gatekeeper and the H.323 terminal (the other call party) are not involved in the handoff.

10.2.5 Comparing vGPRS and 3GPP TR 21.978

This subsection compares vGPRS with the 3GPP TR 21.978 approach [3GP00a]. The following issues are discussed:

PDP Context Activation. In vGPRS, after an MS is attached to the GPRS network, a PDP context is activated. In GPRS, the PDP context must be activated before a routing path can be established between the GGSN and the MS. In vGPRS, when a call (either incoming or outgoing) to the MS arrives, the call path can be quickly established because the PDP context is already activated (see Step 2.2, Section 10.2.2, and Step 4.2, Section 10.2.3). 3GPP TR 21.978 takes a different approach. In this approach, after the MS has registered to the H.323 gatekeeper, the PDP context is deactivated. The PDP context must be activated again when there are call activities to the MS. Thus, the SGSN and the GGSN do not need to maintain the PDP contexts of MSs when they are idle. In this scenario, however, the PDP context must be established and deactivated for every phone call. 3GPP TR 21.978 does not provide details about how the PDP context can be activated for phone calls. We briefly discuss the 3GPP TR 21.978 call path setup as follows.

For MS call origination, the PDP context is activated as described in Step 1.3, Section 10.2.1. For MS call termination, a network-initiated PDP context activation is required. As pointed out in [ETS 00b], to perform this task, a static PDP address is required (which may not be practical for a large-scaled network). Furthermore, IMSI is used by the GGSN (through the GPRS PDP context activation) and the gatekeeper (through **MAP Send Routing Information** and other MAP messages) to obtain MS information from the HLR. This implies that the H.323 gatekeeper should memorize the IMSIs of the vGPRS users. Since IMSI is considered confidential to the GPRS network operator, this approach may not work if the GPRS network and the H.323 network are owned by different service providers.

When an H.323 terminal makes a call to the MS, the gatekeeper must find the IP address of the GGSN, and the IMSI of the MS based on the dialed MSISDN. The gatekeeper then requests the GGSN to perform PDP context activation. After PDP context activation, the routing path between the GGSN and the MS is established, and the network can set up the call to the MS. Clearly, the call setup time is longer in this approach. vGPRS registration and call procedures can be easily modified to deactivate the PDP contexts when the MSs are idle. However, this approach may significantly increase the call setup time and is not considered in the vGPRS implementation.

Modifications to the Existing Networks. To receive VoIP service in the 3GPP TR 21.978 approach, the MS must be an H.323 terminal. Also, the gatekeeper must equip with GSM MAP because it needs to communicate with the HLR. In vGPRS, however, standard GSM and GPRS MSs can enjoy VoIP service. Furthermore, the gatekeeper is a standard H.323 gatekeeper, which only communicates with the GGSN using the

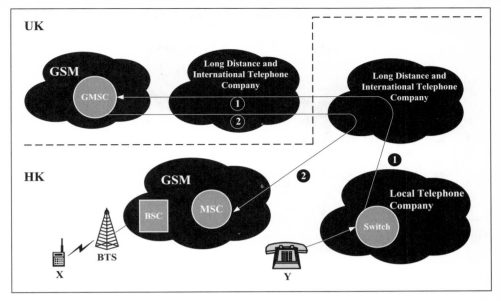

Figure 10.19 Tromboning Phenomenon in GSM

standard H.323 protocol. However, vGPRS introduces a new component, VMSC, that will replace the existing MSC. Since VMSC is a router-based softswitch, this component is anticipated to be much cheaper than the existing MSC.

Tromboning Elimination. In GSM, when a GSM user X in one country (say, the U.K.) roams to another country (say, Hong Kong) and someone Y in Hong Kong makes a phone call to X, it will result in two international calls, as described in [Cho97]. As illustrated in Figure 10.19, when Y in Hong Kong dials X's MSISDN, the call is first routed to X's gateway MSC (GMSC), which is located in the U.K. (Figure 10.19 (1)). After the GMSC has queried the HLR and the VLR (not shown in this figure), the location of X is identified, and a trunk is connected back to Hong Kong (Figure 10.19 (2)). Thus, the call setup results in two international calls.

If the local telephone company in Hong Kong connects to the VoIP network (many local telephone companies are evolving into this configuration now), vGPRS can eliminate this tromboning situation; in other words, the call from Y to X will be a local phone call. When X roams from the U.K. to Hong Kong, it registers at the gatekeeper of Hong Kong. As shown in Figure 10.20, when Y at Hong Kong makes a phone call, the local telephone company first routes the call to the H.323 gateway

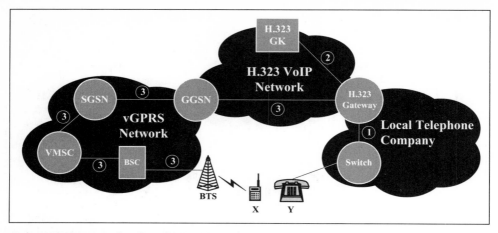

Figure 10.20   Tromboning Elimination in vGPRS

through VoIP service (Figure 10.20 (1)). The gateway checks with the gatekeeper to see whether the entry for X can be found in the address translation table (Figure 10.20 (2)). If so, the call setup follows the procedure described in Section 10.2.3 (Figure 10.20 (3)), which results in a local phone call, and the tromboning situation is avoided. However, whether X is not found in the gatekeeper, the gatekeeper will instruct Y to connect to the international telephone network as a normal PSTN call. Note that 3GPP TR 21.978 cannot eliminate tromboning. To make a local call setup as illustrated in Figure 10.20, the Hong Kong's gatekeeper needs to communicate with the U.K.'s HLR to set up the call. This implies that the Hong Kong gatekeeper needs to keep X's IMSI that is confidential to U.K.'s HLR. Such confidential information sharing is not allowed in a real business model.

10.3   Concluding Remarks

This chapter described how signaling protocols such as MGCP and H323 are utilized in non-all-IP mobile networks to offer VoIP services. We first introduced GSM-IP, a VoIP service for GSM. A new MGCP package named GSM-IP was introduced to support media gateways connected to standard GSM BTSs. Based on the signaling protocol translation mechanism in the MGCP signaling gateway, we described how to interwork the MGCP elements with HLR, VLR, and MSC in the GSM network. Then we presented the message flows for registration, call origination, call delivery, call release, and inter system handoff procedures for the GSM-IP service. From the message flows we designed, we showed the feasibility of integrating GSM and the MGCP-based VoIP system without introducing any new MGCP protocol primitives.

Section 10.2 described vGPRS, a VoIP service for GPRS, which allows standard GSM and GPRS MSs to receive VoIP service. In this approach, a new network element called VoIP Mobile Switching Center (VMSC) is introduced to replace standard GSM MSC. The vGPRS approach is implemented using standard H.323, GPRS, and GSM protocols. Thus, existing GPRS and H.323 network elements are not modified. We described the message flows for vGPRS registration, call origination, call release, call termination, and intersystem handoff procedures. We also showed that for international roaming, vGPRS can effectively eliminate the tromboning situation (two international trunks in call setup) for an incoming call to a GSM roamer.

10.4   Questions

1. Describe the media gateway (MG), the signaling gateway (SG), and the media gateway controller (MGC). Which protocols are executed between these nodes?

2. Show how short message service (SMS) can be implemented in GSM-IP. Do you need to add new packages in the MGCP Connection Model? (See Chapter 1 for details of SMS.)

3. How can a QoS mechanism be implemented in GSM-IP? Which MGCP command is responsible for QoS setup?

4. Please implement a GSM-IP system using H.323. Can GSM-IP and/or vGPRS be implemented in SIP (see Chapter 12)?

5. Consider a vGPRS call session between an MS and an H.323 terminal. Suppose that the H.323 user hangs up the phone first. Draw the call-release message flow.

6. In vGPRS, how do SGSN and GGSN support VoIP QoS? (Hint: See the QoS model described in Chapter 4.)

7. Draw the vGPRS location update message flow when the MS movement involves the crossing of two SGSNs.

8. Redraw the vGPRS inter-system handoff message flow in Figure 10.18 such that both VMSC1 and VMSC2 are connected to the same SGSN.

9. Draw the registration message flow due to MS movement in vGPRS. Point out the steps that are different from the standard GSM registration procedure.

10. Compare vGPRS with the 3GPP TR 21.978 approach in terms of MS call termination.

11. Show the message flow for call setup from a GSM-IP user to a vGPRS user through the Internet.

## CHAPTER 11

Multicast for Mobile Multimedia Messaging Service

Existing 2G systems support *Short Message Service* (*SMS*), which allows mobile subscribers to send and receive simple text messages (e.g., up to 140 bytes in GSM). In the 2.5G systems (e.g., GPRS) and the 3G systems (e.g., UMTS), *Multimedia Messaging Service* (*MMS*) is introduced to deliver messages of sizes ranging from 30K bytes to 100K bytes [3GP04c, 3GP05n]. The content of an MMS can be text (just like SMS), graphics (e.g., graphs, tables, charts, diagrams, maps, sketches, plans, and layouts), audio samples (e.g., MP3 files), images (e.g., photos; see Figure 11.1), video (e.g., 30-second video clips), and so on [Nov01]. Figure 11.2 illustrates an abstract view of the MMS architecture, described here:

- In this architecture, the MMS user agent (Figure 11.2 (a)) resides in a Mobile Station (MS) or an external device connected to the MS, which has an application-layer function to receive the MMS.

- The MMS can be provided by the MMS value-added service applications (Figure 11.2 (b)) connected to the mobile networks or by the external servers (e.g., email server, fax server; see Figure 11.2 (d)) in the IP network.

- The MMS server (Figure 11.2 (c)) stores and processes incoming and outgoing multimedia messages.

Figure 11.1 An Example of Photo MMS

- The MMS relay (Figure 11.2 (e)) transfers messages between different messaging systems, and adapts messages to the capabilities of the receiving devices. It also generates charging data for billing purposes. The MMS server and the relay can be separated or combined.

- The MMS user database (Figure 11.2 (f)) contains user subscriber data and configuration information.

- The mobile network can be a *WAP* (*Wireless Application Protocol*) [OMA04] based 2G, 2.5G, or 3G system (Figure 11.2 (g)).

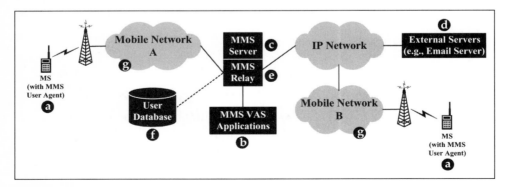

Figure 11.2 Multimedia Messaging Service Architecture

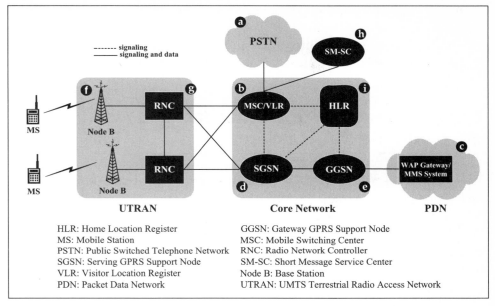

Figure 11.3 UMTS Network Architecture (with SM-SC and WAP Gateway)

Connectivity between different mobile networks is provided by the Internet protocol.

We consider UMTS as the mobile network for MMS. For illustration purposes, Figure 11.3 redraws Figure 2.1. In Figure 11.3, the MMS server/relay connects to the GGSN through a WAP gateway. In UMTS, short messages are delivered through the control plane of the CS domain. The short message is issued from a message sender (e.g., an MS or an input device) to a Short Message Service Center (SM-SC; Figure 11.3 (h)). As described in Chapter 1, the SM-SC is connected to a specific Mobile Switching Center (MSC; Figure 11.3 (b)) called the *Short Message Service Gateway MSC* (*SMS GMSC*). The SM-SC may connect to several mobile networks, and to several SMS GMSCs in a mobile network. Following the UMTS roaming protocol, the SMS GMSC locates the current MSC of the message receiver by querying the Home Location Register (HLR; Figure 11.3 (i)), and forwards the message to that MSC. Then the MSC broadcasts the message to the UMTS Terrestrial Radio Access Network (UTRAN; Figure 11.3 (g)), and the corresponding Node Bs (Figure 11.3 (f)) page the destination MS. Messages can be stored either in the SIM card or in the memory of the *mobile equipment* for display on the standard screen of the MS.

Conversely, multimedia messages can be delivered through either the user plane of the PS or the CS domain. Without loss of generality, we assume that multimedia messages are transmitted over the user plane of the PS domain. In the existing MMS architectures, the mechanisms for MMS unicast and broadcast are well defined. However, no efficient multicast mechanism has been proposed in the literature. In this chapter, we describe efficient multicast mechanisms for messaging.

11.1 Existing Multicast Mechanisms for Mobile Networks

This section describes several approaches for mobile broadcasting and multicasting. In order to track the MSs, the cells (i.e., the coverage area of Node Bs) in the UMTS service area are partitioned into several LAs (in the CS domain) and RAs (in the PS domain). As described in Chapter 2, an RA is typically a subset of an LA. To simply our discussion, this chapter assumes that an RA is equivalent to an LA. To deliver services to an MS, the cells in the group covering the MS are paged to establish the radio link between the MS and the corresponding Node B. The location change of an MS is detected as follows: The Node Bs periodically broadcast their cell identities. The MS listens to the broadcast cell identity, and compares it with the cell identity stored in the MS's buffer. If the comparison indicates that the location has been changed, then the MS sends the location update message to the network. The major task of mobility management is to update the location of an MS when it moves from one LA (RA) to another. The location information is stored in the UMTS mobility databases such as the HLR, the Visitor Location Register (VLR; Figure 11.3 (b)), and the SGSN (Figure 11.3 (d)). In the CS domain, the LA of an MS is tracked by the VLR, and every VLR maintains the information of a group of LAs. In the PS domain, the RA of an MS is tracked by the SGSN, and every SGSN maintains the information of a group of RAs.

Several approaches have been proposed to provide GSM/UMTS broadcast and multicast services. They are described as follows:

Approach I. GSM voice group call service [3GP05i]: This approach can be used to support MMS when the voice calls are replaced by multimedia messages. The GSM voice group call service is provided through a broadcast mechanism. Specifically, the call is delivered to all LAs when a voice call is destined to the multicast members. Every LA is paged, even when no multicast member is in that area.

Approach II. iSMS (see Chapter 1): In this approach, multicast is achieved by sending a message to every individual member in the multicast list. If n members are in an LA, then the same message is sent n times to this LA.

Approach III. GSM/UMTS short message multicasting based on multicast tables [Lin 02a]: In this approach, the short messages are only delivered to the LAs where the multicast members currently reside, and the LAs broadcast the messages to these MSs. The LAs without multicast members do not need to establish the communication link for short message transmission.

Approach III utilizes the existing GSM/UMTS short message architecture as shown in Figure 11.4. In this figure, there are three VLRs in the GSM/UMTS network: VLR1, VLR2, and VLR3. VLR1 covers location areas LA1 and LA2. VLR2 covers location areas LA3 and LA4. VLR3 covers location areas LA5 and LA6.

- To perform multicast, the message sender first issues a short message to the SM-SC, and the SM-SC sends the message to the SMS GMSC associated with the multicast group (Figure 11.4 (1)).

- Then the SMS GMSC queries the HLR to identify the MSCs where the multicast members currently reside (Figure 11.4 (2)) and forwards the message to these MSCs (Figure 11.4 (3)).

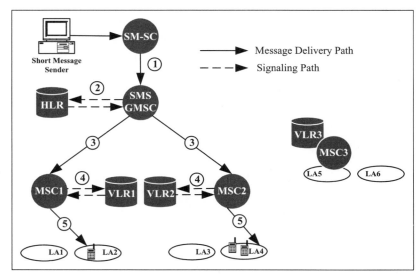

Figure 11.4 Short Message Multicast Architecture (for Approach III)

- Upon receipt of the short message, the MSCs query the corresponding VLRs to identify the LAs where the multicast members currently reside (Figure 11.4 (4)) and page these LAs to establish the radio links (Figure 11.4 (5)).

In Figure 11.4 the message delivery path for SMS multicast is $(1) \rightarrow (3) \rightarrow (5)$.

Two types of tables are utilized in this multicast mechanism. A table MC_H is implemented in the HLR to maintain the addresses of the VLRs and the numbers of multicast members residing in the VLRs. A table MC_V is implemented in every VLR to store the identities of the LAs and the numbers of multicast members in these LAs. In Figure 11.4, there is one multicast member in LA2 and two multicast members in LA4. Thus, we have

$$MC_H[VLR1] = 1, \quad MC_H[VLR2] = 2, \quad \text{and} \quad MC_H[VLR3] = 0$$

For VLR1,

$$MC_V[LA1] = 0 \quad \text{and} \quad MC_V[LA2] = 1$$

For VLR2,

$$MC_V[LA3] = 0 \quad \text{and} \quad MC_V[LA4] = 2$$

For VLR3,

$$MC_V[LA5] = 0 \quad \text{and} \quad MC_V[LA6] = 0$$

By using these tables, the locations of the multicast members are accurately recorded, and the short messages are only delivered to the LAs in which the multicast members currently reside. Details of the location update and message delivery procedures are given in Section 11.2. Note that the PS-domain MMS multicasting cannot be realized by using this mechanism because the GSM/UMTS SMS architecture is only implemented in the CS domain. In Section 11.3, we will describe an efficient MMS multicast mechanism for the UMTS PS domain, which minimizes the number of multimedia messages sent to the RAs.

11.2   The SMS Multicast Approach III

This section describes the CS-domain GSM/UMTS location update and multicast procedures in Approach III, and shows how the multicast tables MC_H and MC_V are maintained through these procedures.

11.2.1 Location Tracking of the Multicast Members

We assume that exactly one MSC is associated with a VLR. This one-MSC-per-VLR configuration is a typical implementation in the existing GSM/UMTS systems. Two types of movement are considered: inter-LA (intra-VLR) movement and inter-VLR movement. In *inter-VLR movement*, the old and new LAs are connected to different MSCs and thus different VLRs. Assume that the MS of LA3 moves into LA4, where LA3 and LA4 belong to VLR1 and VLR2, respectively. The location update message flow is given in Figure 11.5. The steps are described as follows:

Step 1. A location update request message is sent from the MS to MSC2. MSC2 sends the message **MAP Update Location Area** to VLR2.

Step 2. Since the MS is a new visitor to VLR2, VLR2 does not have a VLR record of the MS. According to the message received from MSC2 at Step 1, VLR2 identifies the address of the previous VLR (i.e., VLR1).

Step 3. VLR2 sends the message **MAP Send Identification** to VLR1. The message provides information for VLR1 to retrieve the International

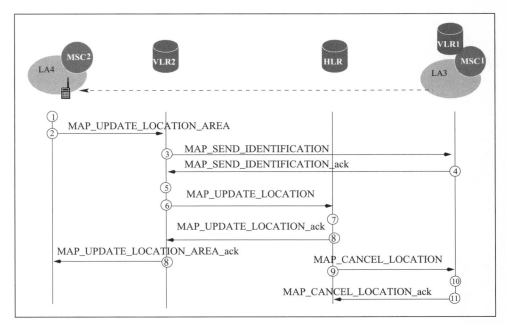

Figure 11.5   Registration for a Multicast Member in Approach III (Inter-VLR Movement)

Mobile Subscriber Identity (IMSI) of the MS in the database. IMSI uniquely identifies the HLR of the MS.

Step 4. The IMSI is sent back from VLR1 to VLR2. VLR2 creates a VLR record for the MS, updates the Location Area Identifier (LAI) and the MSC fields of the VLR record, and derives the HLR address of the MS from the MS's IMSI.

Step 5. $MC_V[LA4]$ (in VLR2) is incremented by 1.

Step 6. VLR2 sends the **MAP Update Location** message to the HLR. By using the received IMSI, the HLR identifies the MS's record. The MSC number field and the VLR address field of the record are updated.

Step 7. $MC_H[VLR1]$ is decremented by 1, and $MC_H[VLR2]$ is incremented by 1.

Step 8. An acknowledgment is sent back to VLR2, and then to the MS.

Step 9. The HLR sends the **MAP Cancel Location** message to VLR1. The obsolete record of the MS in VLR1 is deleted.

Step 10. $MC_V[LA3]$ (in VLR1) is decremented by 1.

Step 11. VLR1 acknowledges the cancel location operation.

In this procedure, Steps 1–4, 6, 8, 9, and 11 are defined in the standard UMTS/GSM specifications. Steps 5, 7, and 10 are executed if the MS is a multicast member. For inter-LA (intra-VLR) movement, only Steps 1, 2, 6, and 10 in Figure 11.5 are executed.

From the descriptions of the above procedure, it is apparent that the tables MC_Hs and MC_Vs accurately record the multicast members distributed in the LAs of a UMTS network.

11.2.2 Mobile Multicast Message Delivery

This subsection describes how messages are multicast by using the multicast tables in Approach III. The procedure is described in the following steps (Figure 11.6):

Step 1. The SM-SC sends a multicast message to the SMS GMSC.

Steps 2 and 3. Through the message **MAP Send Routing Info For SM**, the SMS GMSC requests the routing information from the HLR. The HLR

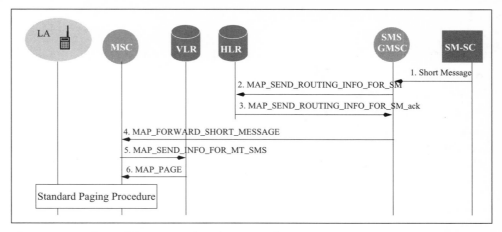

Figure 11.6 The Multicast Procedure in Approach III

searches the multicast table MC_H. If $MC_H[VLR_i] > 0$, then the Mobile Station Roaming Number (MSRN) for the VLR_i is returned from the HLR to the SMS GMSC through **MAP Send Routing Info For SM Ack**. The MSRN is used to identify the destination MSC of the message.

Step 4. The SMS GMSC delivers the multicast message to the destination MSCs (based on the MSRNs received from the HLR) by sending **MAP Forward Short Message**. In Figure 11.4, the multicast message is sent to MSC1 and MSC2.

Steps 5 and 6. Every destination MSC sends **MAP Send Info For MT SMS** to its VLR to obtain the subscriber-related information. When the VLR receives this message, it searches the multicast table MC_V to identify the LAs in which the multicast members reside. These location areas LA_j satisfy the condition $MC_V[LA_j] > 0$. In Figure 11.4, such an LA is LA2 in VLR1, and is LA4 in VLR2. A micro procedure Check_Indication in the VLR is invoked to verify the data value of the message. If the tests are passed, the VLR requests the MSC to page LA_j.

Step 7. The MSC broadcasts the message to the multicast members in the LAs following the standard GSM/UMTS paging procedures. The multicast members listen and receive the message broadcast in the LAs.

From the above message delivery procedure, it is clear that only the LAs with multicast members will be paged for multicasting. The LAs without multicast members will not be paged.

11.3 The MMS Multicast Approach IV

Approach IV is implemented in the UMTS PS domain, which utilizes the existing *Cell Broadcast Service* (*CBS*) architecture [3GP03b]. Figure 11.7 illustrates an example of the CBS architecture. This example consists of two SGSNs: SGSN1 and SGSN2. SGSN1 covers routing areas RA1 and RA2. SGSN2 covers routing areas RA3, RA4, RA5, and RA6. We assume that radio network controller RNC1 (covering RA1 and RA2) connects to SGSN1. Both RNC2 (covering RA3 and RA4) and RNC3 (covering RA5 and RA6) connect to SGSN2.

> **Step 1.** The multimedia message is first delivered from the message sender to the *Cell Broadcast Entity* (*CBE*; see (1) in Figure 11.7).
>
> **Step 2.** The CBE sends the message to the *Cell Broadcast Center* (*CBC*; see (2) in Figure 11.7).
>
> **Steps 3 and 4.** The CBC determines the RAs that should receive the multimedia message (Figure 11.7 (3)), and forwards the message to the corresponding RNCs (Figure 11.7 (4)).
>
> **Step 5.** Then the RNCs multicast the multimedia message to the multicast members (Figure 11.7 (5)).

In Figure 11.7, the message delivery path for MMS multicast is $(1) \rightarrow (2) \rightarrow (4) \rightarrow (5)$. Like Approach III, a multicast table MC_C is implemented in the CBC to maintain the identities of the RAs and the number of multicast

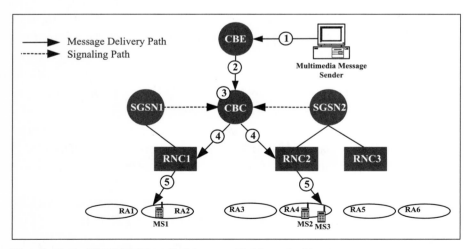

Figure 11.7 UMTS Multimedia Message Multicast Architecture

Table 11.1 Signaling message format defined in the interface between the CBC and the SGSN

MESSAGE TYPE	ADDRESS FIELD 1	ADDRESS FIELD 2
Attach Indication	Current RA	—
Attach Response	—	—
Detach Indication	Current RA	—
Detach Response	—	—
RA Update Indication	Current RA	Previous RA
RA Update Response	—	—

members in these RAs. In Figure 11.7, there is one multicast member in RA2 and two multicast members in RA4. Thus, we have

$$MC_C[RA1] = 0, \quad MC_C[RA2] = 1, \quad MC_C[RA3] = 0,$$

$$MC_C[RA4] = 2, \quad MC_C[RA5] = 0, \quad \text{and} \quad MC_C[RA6] = 0 \quad (11.1)$$

In order to accurately record the current locations of multicast members, we define a new signaling interface between the CBC and the SGSN. This interface can be based on the Internet protocol or Mobile Application Part (MAP) described in Chapter 8. Table 11.1 shows the signaling message format in this interface. Six message types are defined:

- Attach Indication/Response
- Detach Indication/Response
- RA Update Indication/Response

Address Field 1 and Address Field 2 specify the addresses of the MS's current and previous RAs, respectively. In the attach, detach, and location update procedures, the SGSN informs the CBC of the multicast member's current location through these signaling messages. Detailed message flows are described in the following subsections.

11.3.1 Location Tracking of the Multicast Members

This section provides a simplified description of the UMTS/GPRS attach, detach, and location update procedures, and shows how the multicast table MC_C is modified through these procedures. The complete description of the standard UMTS mobility management procedures is given in Chapter 2.

Attach for a Multicast Member

When an MS powers on and attaches to the UMTS PS domain, the standard UMTS/GPRS attach procedure is performed to inform the network of the MS's presence. We use the multicast member MS1 in Figure 11.7 as an example. In this example, MS1 attaches to SGSN1 with the message flow shown in Figure 11.8, which consists of the following steps:

Step 1. MS1 initiates the attach procedure by sending the **Attach Request** message to SGSN1.

Step 2. The authentication function is performed between MS1, SGSN1, and the HLR as described in Section 9.1.

Step 3. Through the standard UMTS RA update procedure, SGSN1 informs the HLR of the MS1's current location and obtains the MS1's user profile from the HLR.

Step 4. SGSN1 informs the CBC of the MS1's RA identity (i.e., RA2 in Figure 11.7) by sending the **Attach Indication** message. The CBC increments $MC_C[RA2]$ by 1. Then it acknowledges SGSN1 by sending the **Attach Response** message.

Step 5. SGSN1 and MS1 exchange the **Attach Accept** and the **Attach Complete** message pair to indicate that the attach procedure is complete.

In the above procedure, Steps 1–3 and 5 are defined in the standard UMTS specifications (see Section 2.5). Step 4 is executed if the mobile user is a multicast member. Note that the update of the multicast table is done in

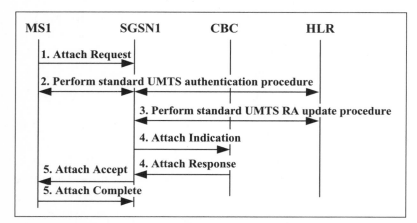

Figure 11.8 Attach Procedure for a Multicast Member

microseconds, which can be ignored, in comparison to the delays of message exchanges in the normal attach procedure.

Detach for a Multicast Member

After PS detach is executed, the MS will not receive the GPRS-based services. We use the multicast member MS1 in Figure 11.7 as an example to illustrate the detach procedure. In this example, MS1 resides at RA2, which is covered by SGSN1. Simplified steps of the detach procedure are given as follows (see Figure 11.9):

Step 1. MS1 detaches from the UMTS PS domain by sending the **Detach Request** message to SGSN1.

Step 2. Upon receipt of the MS1's detach request, SGSN1 and the GGSN exchange the **Delete PDP Context Request** and **Delete PDP Context Response** message pair to deactivate the MS1's Packet Data Protocol (PDP) context. SGSN1 sends the **Purge MS** message to the HLR. This message indicates that SGSN1 has deleted the MS1's Mobility Management (MM) and PDP contexts. The HLR acknowledges with the **Purge MS Ack** message. Then Steps 3 and 4 are executed in parallel.

Step 3. SGSN1 sends the **Detach Indication** message to the CBC to indicate that MS1 in RA2 has been detached from the network. The CBC decrements $MC_C[RA2]$ by 1. Then it replies with the **Detach Response** message to SGSN1.

Step 4. If the MS1's detach is not caused by power-off, SGSN1 sends the **Detach Accept** message to MS1. At the same time, SGSN1 initiates

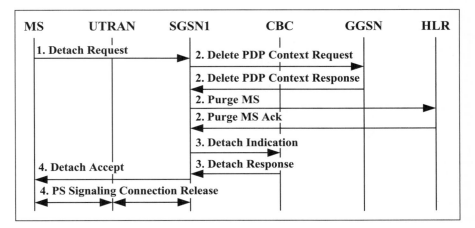

Figure 11.9 Detach Procedure for a Multicast Member

the PS signaling connection release procedure to release the signaling connections between SGSN1 and the UTRAN, and between the UTRAN and MS1.

In the above procedure, Steps 1, 2, and 4 are defined in the standard UMTS specifications (see Section 2.5). Step 3 is executed if the mobile user is a multicast member. The cost of updating the multicast table can be ignored, in comparison to the message delivery costs among the network nodes.

Location Update for a Multicast Member

When an MS moves into a new RA, the RA update procedure is performed. Two types of movements are defined in 3GPP TS 23.060 [3GP05q]: intra-SGSN movement and inter-SGSN movement. In inter-SGSN movement, the previous and current RAs are connected to different SGSNs. We use Figure 11.7 as an example to illustrate the inter-SGSN movement, whereby the multicast member MS1 moves from RA2 to RA3. A simplified description of the location update procedure is given as follows (see Figure 11.10):

Step 1. When detecting the RA location change, MS1 issues the **Routing Area Update Request** message to SGSN2.

Step 2. Through the standard UMTS SGSN context request procedure, SGSN2 obtains the MM and PDP contexts of MS1 from SGSN1.

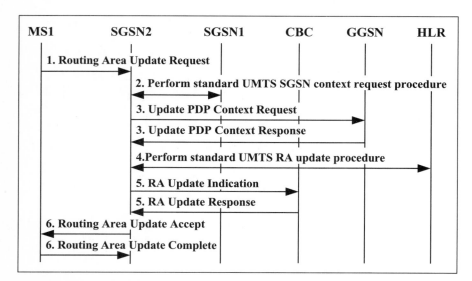

Figure 11.10 Location Update Procedure for a Multicast Member

Step 3. SGSN2 sends the **Update PDP Context Request** message to the corresponding GGSN. With this message, the GGSN PDP context is modified to indicate that SGSN1 has been replaced by SGSN2. The GGSN returns the **Update PDP Context Response** message.

Step 4. The standard UMTS RA update procedure is performed to inform the HLR that the SGSN for MS1 has been changed.

Step 5. Upon receipt of the location update acknowledgment from the HLR, SGSN2 sends the previous RA identity (RA2) and current RA identity (RA3) to the CBC through the **RA Update Indication** message. Then $MC_C[RA2]$ is decremented by 1 and $MC_C[RA3]$ is incremented by 1. The CBC replies with the **RA Update Response** message to SGSN2.

Step 6. Through the **Routing Area Update Accept** and **Routing Area Update Complete** message exchange, SGSN2 informs MS1 that the location update procedure is successfully performed.

In the above procedure, Steps 1–4 and 6 are defined in the standard UMTS specifications (see Section 2.6). Step 5 is executed if the mobile user is a multicast member. In Figure 11.7, before the inter-SGSN movement, the content of the multicast table MC_C is given in Equation (11.1). After the inter-SGSN movement, $MC_C[RA1]$, $MC_C[RA4]$, $MC_C[RA5]$, and $MC_C[RA6]$ remain the same, but $MC_C[RA2]$ and $MC_C[RA3]$ values are updated to be 0 and 1, respectively.

For intra-SGSN movement, only Steps 1, 5, and 6 in Figure 11.10 are executed. In Figure 11.7, if the multicast member MS2 moves from RA4 to RA3, the intra-SGSN location update is performed. Before the intra-SGSN movement, the content of the multicast table MC_C is given in Equation (11.1). After the intra-SGSN movement, the $MC_C[RA1]$, $MC_C[RA2]$, $MC_C[RA5]$, and $MC_C[RA6]$ values remain the same, but the $MC_C[RA3]$ and $MC_C[RA4]$ values are updated to be 1.

From the descriptions of the above procedures, it is apparent that table MC_C accurately records the multicast members distributed in the RAs of an UMTS network.

11.3.2 Mobile Multicast Message Delivery

This section elaborates on how multimedia messages are multicast in Approach IV by using the multicast table MC_C. The procedure is described in the following steps (see Figure 11.7):

Step 1. The multimedia message sender issues the message to the CBE.

Step 2. The CBE forwards the message to the CBC.

Step 3. The CBC searches the multicast table MC_C to identify the routing areas RA_i where the multicast members currently reside (i.e., $MC_C[RA_i] > 0$ in the CBC). In Figure 11.7, $i = 2$ and 4.

Step 4. The CBC sends the multicast message to the destination RNCs (i.e., RNC1 and RNC2 in Figure 11.7) through the **Write Replace** message defined in 3GPP TS 23.041 [3GP03b].

Step 5. The RNCs deliver the multimedia messages to the multicast members in the RAs following the standard UMTS cell broadcast procedure.

In the above procedure, it is clear that the multimedia messages are only delivered to the RAs that contain multicast members.

11.4   Concluding Remarks

This chapter described short message multicast mechanisms (Approaches I, II, and III) in the CS domain, and a multicast mechanism (Approach IV) in the PS-domain MMS. Multicast Approach IV is based on the CBS architecture. The CBC is a standard UMTS network node defined in 3G 23.041 [3GP03b]. We proposed a new interface between the CBC and the SGSN to track the current locations of the multicast members. Then we described location tracking procedures (including attach, detach and location update) for multicast members and the multicast message delivery procedure. The implementation and execution of the multicast table are so efficient that the cost for updating this table can be ignored compared with the standard mobility management procedures. In terms of MMS multicast message delivery cost, Approaches III and IV outperform Approaches I and II. A simple analysis is given below (see also question 1 in Section 11.5 for modeling Approach III):

Signaling Cost for Location Update. The cost of updating the multicast tables can be ignored in Approaches III and IV. The location update costs for Approaches I, II, and III are equal. Approach IV requires two extra messages (e.g., **RA Update Indication** and **RA Update Response**) exchanged between the SGSN and the CBC.

Signaling Cost for Multicast Message Delivery. In Approaches II and III, the SMS GMSC must query the HLR (see Step 2 in Figure 11.4). Furthermore, the destination MSCs must query the VLRs (see Step 4 in

Figure 11.4) in Approach III. Conversely, these signaling costs are not required in Approaches I and IV.

Delivery Cost for Multicast Message. It is clear that the delivery costs for Approaches III and IV are lower than that for Approaches I and II. An analytic model for deriving the multicast cost is given in Section 11.5.

As a final remark, the multicast table mechanism is an ROC patent (205010) and a U.S. pending patent.

11.5 Questions

1. The multicast costs of Approaches I–IV are measured by the number of paging messages sent to the LAs at multicast message delivery. The multicast costs for both Approaches III and IV are similar and it suffices to consider Approaches I–III (i.e., A_I, A_{II} and A_{III}).

   We first model how the multicast members are distributed in the LAs of a UMTS system. Assume that the LAs are classified into l categories. For $1 \leq i \leq l$, there are N_i LAs of class i. Let $\pi_i(j)$ be the probability that there are j multicast members in a class i LA. Probabilities $\pi_i(j)$ are derived as follows: We assume that the multicast members enter a class i LA with rate λ_i. These arrivals can be from other LAs or areas not covered by the UMTS system. It was shown that the aggregate arrivals can be approximated by a Poisson stream. We assume that a multicast member resides at the LA for a period that has a general distribution with mean $1/\mu_i$. From [Lin97a], $\pi_i(j)$ can be derived from the $M/G/\infty$ model. That is, for $j \geq 0$,

$$\pi_i(j) = \left(\frac{\rho_i^j}{j!} \right) e^{-\rho_i} \quad \text{where} \quad \rho_i = \frac{\lambda_i}{\mu_i} \tag{11.2}$$

   Let M be the number of LAs in the UMTS system. That is

$$M = \sum_{i=1}^{l} N_i \tag{11.3}$$

   Consider two random variables K and N. When a multicast message arrives, there are K multicast members in the UMTS system, and these members are distributed among N LAs (where $N \leq M$). It is clear that the expected multicast costs for A_I, A_{II}, and A_{III} are M, $E[K]$, and

$E[N]$, respectively. Let θ_I and θ_{II} be the expected costs for A_I and A_{II} normalized by the cost of A_{III}. Then

$$\theta_I = \frac{M}{E[N]}, \quad \text{and} \quad \theta_{II} = \frac{E[K]}{E[N]} \tag{11.4}$$

Derive θ_I and θ_{II} based on (11.2)–(11.4). (Hint: See the solution in http://liny.csie.nctu.edu.tw/supplementary.)

2. Describe the MMS architecture. Modify this architecture to accommodate multicast approach IV.

3. Describe the SMS architecture for multicast approach III.

4. Can multicast approach III be implemented in iSMS? Can we support this efficient multicast mechanism at the application layer without modifying the GSM/GPRS components?

5. Describe the location update steps for intra-SGSN movement of a multicast MS.

6. Define a special type of PDP context such that Algorithm IV can be implemented by maintaining the multicasting tables in this PDP context. This approach is described in 3GPP 23.246 as the multimedia broadcast/multicast service.

## CHAPTER 12

Session Initiation Protocol

In the UMTS all-IP network, the *Session Initiation Protocol* (*SIP*) [Ros02] is the default protocol for the IP Multimedia Core Network Subsystem. As a standard for Internet telephony published in 1999 by the Internet Society, SIP is a general way for an application to make one computer user aware that another user is online and available for communications; that is, it is the Internet's virtual dial tone. SIP also enables other Internet applications, such as letting a friend's name pop up in the buddy lists of instant messaging software. SIP telephony can also be integrated with applications such as games. Today, the easiest way to make a free Internet phone call is with a network-connected Xbox or by playing a multiplayer online video game [Che05c]. This chapter gives a brief introduction to SIP. Then we use the push mechanism and the prepaid mechanism to illustrate how SIP-based applications can be implemented in UMTS/GPRS.

12.1 An Overview of SIP

As an application-layer signaling protocol over the IP network, SIP is designed for creating, modifying, and terminating multimedia sessions or calls. The SIP message specifies the *Real-Time Transport Protocol/*

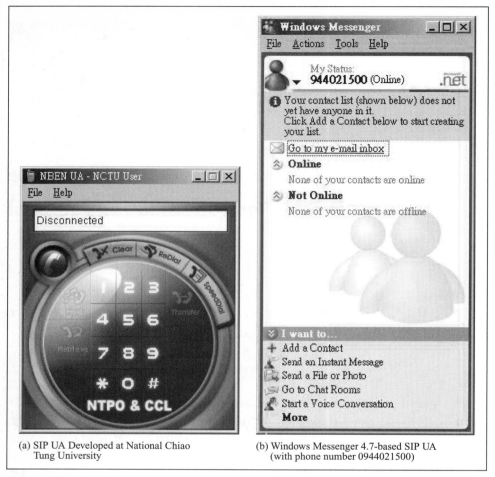

(a) SIP UA Developed at National Chiao
Tung University

(b) Windows Messenger 4.7-based SIP UA
(with phone number 0944021500)

Figure 12.1 User Agents (Softphone)

Real-Time Transport Control Protocol (*RTP/RTCP*), which delivers the data in the multimedia sessions. RTP is a transport protocol on top of UDP, which detects packet loss and ensures ordered delivery. An RTP packet also indicates the packet sampling time from the source media stream. The destination application can use this time stamp to calculate delay and jitter.

Two major network elements are defined in SIP: *user agent* and *network server*. The user agent resides at SIP endpoints (or phones). Figures 12.1 and 12.2 give softphone and hardphone examples of UA. A user agent contains both a *User Agent Client* (*UAC*) and a *User Agent Server* (*UAS*). The UAC (or calling user agent) is responsible for issuing SIP requests, and the

Figure 12.2 User Agent (Leadtek Hardphone)

UAS (or called user agent) receives the SIP request and responds. Six basic types of SIP requests are defined:

REGISTER is sent from a user agent to the registrar (a network server to be defined later) to register the address where the subscriber is located.

INVITE is used to initiate a multimedia session. This request includes the routing information of the calling and the called parties, and the type of media to be exchanged between the two parties.

ACK is sent from a UAC to a UAS to confirm that the final response to an INVITE request has been received.

OPTIONS is used to query the user agent's capabilities, such as the supported media type.

BYE is used to release a multimedia session or call.

CANCEL is used to cancel a pending request (i.e., an incomplete request).

SIP also defines the INFO method [Don00] for transferring information during an ongoing session. For example, The *Dual-Tone Multifrequency* (*DTMF*) digits can be delivered during a VoIP call through the INFO message. We will show another example of INFO usage in Section 12.3.1.

After receiving a request message, the recipient takes appropriate actions and acknowledges with a *SIP response* message. The response message carries a return code indicating the execution result for the request. Examples of the return code are

- 100 Trying: the request is currently being executed
- 180 Ringing: the called party is alerted

- **183 Session Progress** (provisional)

- **200 OK**: the request was executed normally

- **401 Unauthorized**: the client is not authorized to make the request

- **404 Not Found**

- **500 Server Failure**: the request could not be executed because of server internal error

A SIP user is globally and uniquely identified by a *Uniform Resource Identifier (URI)* [Ber98]. The SIP URI is of the format

```
sip:username@hostname:port
```

where

sip is the prefix to indicate that the whole string is a SIP URI,

username is a local identifier of the SIP user on the server hostname,

hostname is the SIP server for the user username, and

port is the transport port to accept the SIP message on the hostname, which typically is "5060".

A SIP URI example is "sip:yjlou@csie.nctu.edu.tw:5060".

A *SIP transaction* consists of a request and one or more responses. The transaction is initiated by a *translation initiator*. The *target* of the transaction may or may not be the *recipient* of the request. For example, in a registration transaction, a UA (the initiator) may register to a SIP registrar (the recipient) for another UA (the target). As another example, in an invite transaction, a UA (the initiator) sets up a call to another UA (the target that is also the recipient). A SIP message consists of the following fields:

Request-URI indicates the URI of the receiver.

To (e.g., To:UA2 <sip:lin@nctu.com> or To:UA2 <sip:lin@140.113.87.40>) contains an optional display name (which could be a person or a system, e.g., UA2) followed by the SIP URI of the transaction target (e.g., sip:lin@nctu.com).

From (e.g., From:UA1 <sip:lin@nctu.com>) contains an optional display name of the transaction initiator (UA1) followed by its SIP URI sip:lin@nctu.com.

Via (e.g., Via:SIP/2.0/UDP 140.113.87.52:5061) contains the version (SIP/2.0) and the transport protocol (UDP) followed by the

IP address (140.113.87.52) and the port number (5061) of the immediate sender.

  Any intermediate server that forwards the SIP message adds a Via field with its address and port number. This field may also be expressed with a domain name (e.g., Via:SIP/2.0/UDP ncku.com:5061).

Contact (e.g., Contact:<sip:wang@ncku.com:5061> or Contact:<sip:lin@140.114.87.52:5061>) indicates the contact where the other party can send subsequent requests.

Content-Length counts the SIP message body in octets. A Content-Length of 0 indicates that there is no message body.

SIP conjuncts with protocols such as *Session Description Protocol* (*SDP*) [Han98] to describe the multimedia information. While RTP transports the voice packets, SDP provides the RTP information such as the network address and the transport port number of the RTP connection. Some of the SDP fields are listed below:

o (e.g., o=lin 625106937 625106937 IN IP4 140.113.87.52) indicates the originator, which contains user name (lin), session Id and version (625106937), network type (IN for Internet), address type (IP4 for IPv4, IP6 for IPv6, and FQDN for domain name), and address (140.113.87.52). The originator of the SDP is the sender of the SIP message.

c (e.g., c=IN IP4 140.113.87.52) indicates the connection information for the media session, which includes the network type (IN), address type (IP4), and connection address (which can be the originator's IP address 140.113.87.52). This field may also be expressed with a domain name (e.g., c=IN FQDN ncku.com).

m (e.g., m=audio 9000 RTP/AVP 0 4 8) indicates the media, which contains the media type (audio), port (9000), protocol (RTP/AVP) and codec number (e.g., 0 for u-law PCM, 4 for GSM, and 8 for a-law PCM [Sch96a]).

SIP supports three types of network servers:

- *registrar*
- *redirect server*
- *proxy server*

To support user mobility, the user agent informs the network of its current location by explicitly registering with a registrar. The registrar is typically co-located with a proxy or redirect server. A SIP UA can periodically register its SIP URI and contact information (which includes the IP address and the transport port accepting the SIP messages) to the registrar. When the SIP UA moves to different networks, the registrar always holds the current contact information of the SIP UA. A registrar may store the contact information in a *location service* database.

A proxy server processes the SIP requests from a UAC to the destination UAS. The proxy server either handles the request or forwards it to other servers, perhaps after performing some translation. For example, to resolve the SIP URI in the INVITE request, the proxy server consults the location service database to retrieve the current IP address and transport port of the SIP UAS. The proxy server then forwards the INVITE message to the SIP UAS.

A redirect server accepts the INVITE requests from a UAC, and returns a new address to that UAC. Similar to a proxy server, a redirect server may query the location service database to obtain the callee's contact information. Unlike the proxy server, the redirect server does not forward the INVITE message. It only returns the contact information to the SIP UAC.

The interaction between the SIP UAC, the SIP UAS, the registrar, the location service database, and the proxy server is illustrated in Figure 12.3, and is described as follows:

Step 1. When a SIP UAS is activated, the SIP UAS registers its SIP URI to the registrar by sending the REGISTER message.

Step 2. The registrar stores the contact information in the location service database.

Step 3. The registrar generates the 200 OK message and returns it to the SIP UAS.

When a SIP UAC attempts to call a SIP UAS, Steps 4–10 below are executed (see Figure 12.3).

Step 4. The SIP UAC first sends the INVITE request to a proxy server, which is pre-configured by the SIP UAC. The INVITE message contains the SIP URI of the SIP UAS and the SDP that describes the RTP information of the SIP UAC. The RTP information includes the IP address and port number for receiving voice data at the SIP UAC.

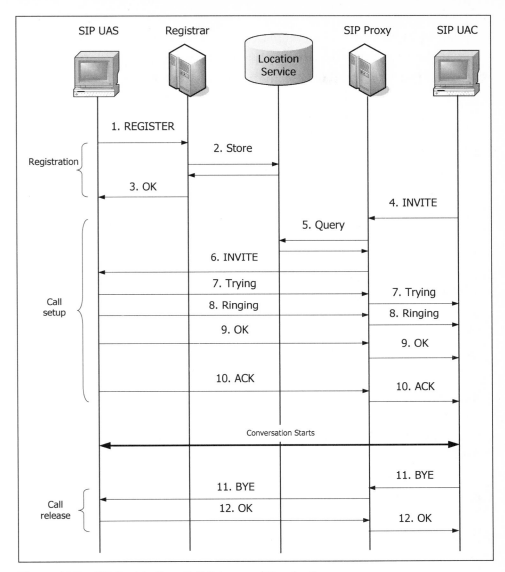

Figure 12.3 SIP Registration, Call Setup, and Termination

Step 5. To resolve the SIP URI in the **INVITE** request, the proxy server may query the location service database to obtain the contact information of the SIP UAS.

Step 6. Then the proxy server forwards the **INVITE** message to the SIP UAS.

Step 7. Upon receipt of the INVITE request, the SIP UAS replies with a 100 Trying response to indicate that the call is in progress. This message is received by the SIP UAC through the proxy server.

Step 8. The SIP UAS plays an audio ringing tone to alert the called user that an incoming call arrived, and sends the 180 Ringing response to the SIP UAC through the proxy server. The SIP UAC plays an audio ringback tone to the calling user.

Step 9. When the called user picks up the handset, the SIP UAS sends the final 200 OK response to the SIP UAC. The OK message includes the SDP that describes the RTP information of the SIP UAS, such as the IP address and the transport port used by the SIP UAS.

Step 10. Upon receipt of the OK response, the SIP UAC sends the ACK request to acknowledge the SIP UAS.

At this point, the conversation starts. The SIP UAS sends the RTP packets to the SIP UAC according to the parameters described in the SDP of the INVITE message. The SIP UAC sends the RTP packets to the SIP UAS according to the parameters described in the SDP of the final OK message.

Assume that the SIP UAC terminates the session after the conversation. The following steps in Figure 12.3 are executed:

Step 11. The SIP UAC sends the BYE request to the SIP UAS through the proxy server.

Step 12. The SIP UAS responds with the OK message to confirm the request, and the session is terminated.

12.2   SIP-based GPRS Push Mechanism

Based on the SIP protocol described in Section 12.1, we elaborate on a push mechanism for GPRS. Figure 12.4 illustrates the SIP-based push architecture. The GPRS network is shown in Figure 12.4 (A). In several commercial GPRS implementations, an MS is dynamically assigned a private IP address [IET94]. Therefore, a *Network Address Translator (NAT)* (Figure 12.4 (14)) is required to translate the IP addresses of the packets delivered between the public address realm (the external data network) and the private address realm (the GPRS network).

Another issue in GPRS is that if an MS does not activate the PDP context for a specific service, the network cannot "push" this service to the MS. To resolve the above issues, this section describes a SIP-based push

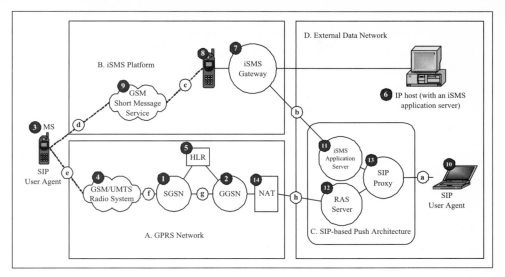

Figure 12.4 GPRS, iSMS and SIP-based Push Architecture

mechanism for GPRS supporting private IP addresses. This solution does not modify existing GPRS components. We utilize the iSMS service platform (see Chapter 1) to implement the push mechanism. iSMS is an operator-independent platform that integrates the IP network with the Short Message Service (SMS) in mobile telephone systems. Through iSMS (Figure 12.4 (B)), an IP host in the external data network (Figure 12.4 (6)) can offer Internet services to an MS (Figure 12.4 (3)). Specifically, messages are created by an iSMS application server run on the IP host, and then sent to the iSMS gateway (see Figure 12.4 (7)). The iSMS gateway is connected to a GSM modem (Figure 12.4 (8)) that delivers the messages to the MS through the short message service (Figure 12.4 (9)). Note that (9) and (4) in Figure 12.4 typically utilize the same radio system. In the iSMS platform, no components in the GSM are modified. The iSMS gateway can be implemented by an off-the-shelf high-reliability PC or workstation connected to an MS.

The idea behind the push mechanism is simple. When the VoIP calling party in the IP network (Figure 12.4 (10)) initiates a call to an MS, the iSMS application server (Figure 12.4 (11)) issues a short message to the MS through the iSMS gateway. This short message instructs the MS to activate the PDP context for the VoIP service. Therefore, the network-requested PDP context activation is performed without requiring the GGSN to interact with the HLR.

At National Chiao Tung University (NCTU), we have implemented a *SIP-based Push Architecture (SPA*; see Figure 12.4 (C)) that uses the SIP described in Section 12.1 for VoIP signaling. In this architecture, a SIP proxy

(Figure 12.4 (13)) connects to a SIP user agent (Figure 12.4 (10)) in the external data network and an MS (with another SIP user agent) through the GPRS network. An iSMS application server (Figure 12.4 (11)) is implemented in the SPA to support the push operation. Note that the port number for SIP applications is pre-defined. Since the NAT server distinguishes the hosts by the port numbers, the fixed port number nature of SIP will result in wrong translation at the NAT server. Therefore, a *Remote Access Service* (*RAS*) server (Figure 12.4 (12)) is implemented to support the tunnel between the SIP proxy and an MS. This tunnel implementation is based on the *Layer Two Tunneling Protocol* (*L2TP*) described in Section 4.4 [IET99b].

Consider the SIP call setup procedure from an IP host in the external data network to an MS. The phone number of the MS is +886936105401 and the fully qualified domain name of the SIP proxy is `fetnet.com`. Before the call termination is initiated, the MS has not activated the VoIP service yet. Steps 1–17 below are executed.

Step 1. The calling party initiates the call to the MS by issuing the SIP **INVITE** message. This message contains the SDP information that provides the RTP network address and transport port number of the calling party. In VoIP, a call party is identified by its IP address. Since the IP address of the MS is dynamically assigned, this address is not available when the call termination is initiated. To resolve this issue, the MS is identified by the telephone number +886936105401 carried by the **INVITE** message using the SIP Request-URI with the following format:

INVITE `sip:+886936105401@fetnet.com`

The above message is routed to the SIP proxy (`fetnet.com`) through path (a) in Figure 12.4. Upon receipt of the **INVITE** message, the SIP proxy instructs the iSMS application server to send a short message to the number +886936105401 (path (b)→(c)→(d) in Figure 12.4). This short message carries the public IP address of the RAS server and a tunnel IP address of the SIP proxy, which will be used by the MS to establish the tunnel to the SIP proxy.

Step 2. The short message triggers the MS to activate the PDP context for VoIP. After the activation, the MS is assigned an IP address from the GGSN (see Chapter 4).

Step 3. By using the RAS IP address and the MS IP address, the MS and the RAS server exchange L2TP messages to establish a tunnel between the MS and the SIP proxy (path (e)·→(f)↔(g)↔(h) in Figure 12.4). After the tunnel is established, the MS is assigned a tunnel IP address from the RAS server. Note that this tunnel IP address is different from the MS IP address assigned by the GGSN.

Step 4. Using the established tunnel, the MS sends its tunnel IP address and telephone number +886936105401 to the SIP proxy. This tunnel IP address and the phone number are saved in the SIP proxy. When the SIP proxy receives a SIP message with the destination phone number +886936105401, it will forward the message by using the tunnel IP address of the MS.

Step 5. The SIP proxy modifies the **INVITE** message received at Step 1. Specifically, the connection information field and the transport port field of the SDP are modified to the IP address and the port number of the SIP proxy. Therefore, the RTP packets will be routed from the MS to the calling party through the SIP proxy. The SIP proxy then forwards this message to the MS.

Step 6. Upon receipt of the modified **INVITE** message, the MS answers the call by sending the SIP **OK** response back to the SIP proxy through the tunnel established in Step 3. The SDP of this message contains the RTP information of the MS, which is also modified by the SIP proxy, similar to that in Step 5.

Step 7. The SIP proxy forwards the **OK** message to the calling party. The calling party confirms this session by sending the SIP **ACK** message to the MS through the SIP proxy. At this point, the conversation begins. The RTP packets are routed through the path (e)↔ (f)↔(g)↔(h)↔(a).

12.3   SIP-based VoIP Prepaid Mechanism

In this section, we describe the NCTU prepaid system, a SIP-based mechanism to handle prepaid calls. Unlike the previous VoIP prepaid approaches that require modifications to endpoints [Cis04, Woo04], we show that this prepaid mechanism can be easily integrated with the VoIP platform without modifying the existing network nodes. Figure 12.5 illustrates the SIP-based prepaid mechanism. In this figure, the *Remote Authentication Dial-In User Service (RADIUS)* protocol [IET00] enables centralized *Authentication, Authorization, and Accounting* (AAA) functions for network access. Two network nodes are related to the prepaid mechanism:

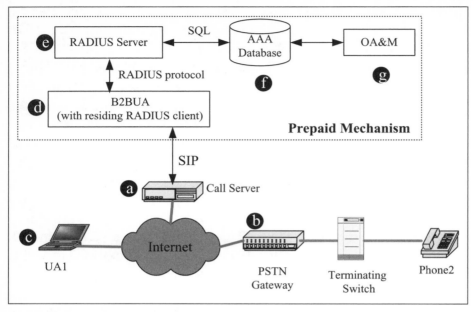

Figure 12.5   NCTU Prepaid System Architecture

- The *PSTN Gateway* (Figure 12.5 (b)) supports interworking between the VoIP network and the PSTN. Figure 12.6 illustrates a 32-line PSTN gateway developed by the Industrial Technology Research Institute (ITRI), which allows IP phones to reach PSTN phones.

- The *Call Server* (Figure 12.5 (a)) is a SIP proxy with the registrar functions. Figure 12.7 shows the administrative console of a Call Server. The console monitors all SIP UAs registered in this Call Server.

Figure 12.6   PSTN Gateway Developed in the ITRI

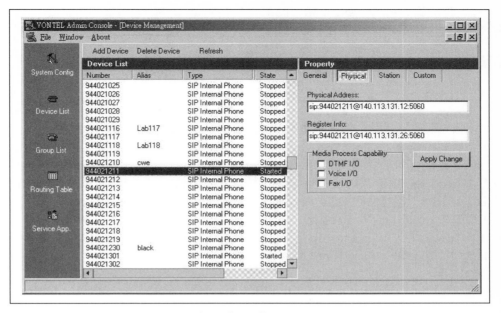

Figure 12.7   Administrative Console of the Call Server

The prepaid system includes four components:

- **RADIUS client**, residing in a SIP-based *Back-to-Back User Agent* (*B2BUA*; see Figure 12.5 (d)) can process and exchange SIP messages between the call parties, and terminates a prepaid call when the *authorized session time* for the call is expired. Before a prepaid call starts, the RADIUS client asks the RADIUS server for authorization. When the prepaid call terminates, it instructs the RADIUS server to log the session accounting information.

- **RADIUS server** (Figure 12.5 (e)) authorizes prepaid requests and responds to the RADIUS client with authorized information (such as maximum call time). It also processes the RADIUS accounting messages and stores the accounting record in the nonvolatile AAA database.

- **AAA database** (Figure 12.5 (f)) is a *Structured Query Language* (*SQL*)-based database, which is queried by the RADIUS server to retrieve user information. The AAA database also acts as the accounting database, storing the user *Call Detail Records* (*CDRs*) for the prepaid users.

- **Operations, Administration, and Maintenance (OA&M)** system (Figure 12.5 (g)) can be accessed from web browsers over the *Secure*

Socket Layer (*SSL*) protocol. Through OA&M, an administrator can view and/or modify user information and browse the prepaid CDRs generated by the prepaid system.

By utilizing the B2BUA technique, the prepaid mechanism is inserted in a VoIP call by breaking the signaling session into two subsessions. Consider the scenario in which a SIP user agent (UA1) attempts to establish an IP-PSTN call to a PSTN phone (Phone2). The prepaid B2BUA breaks the SIP session between UA1 and the PSTN gateway into one subsession between UA1 and the B2BUA (Subsession 1) and another subsession between the B2BUA and the PSTN gateway (Subsession 2). The B2BUA triggers the RADIUS **Accounting Request** message with the Status "start" or "stop" when it receives SIP messages such as **INVITE/200 OK/ACK** or **BYE/200 OK**, and terminates the prepaid call session when the prepaid credit of a user is depleted.

To support prepaid service in a VoIP network, some telephone numbers are reserved for prepaid call services. When the Call Server receives a SIP request from a prepaid user, it forwards the request to the B2BUA for authorization. After authorization, the B2BUA sends the authorized request to the Call Server. The Call Server then sets up the call to the called party. Conversely, if the Call Server receives SIP requests from non-prepaid users, it sets up the calls without involving the prepaid mechanism. Based on the prepaid user phone numbers, the prepaid functions can be easily achieved by reconfiguring the call routing rules in the Call Server. Suppose that a prepaid phone number has the prefix "0944021", and the B2BUA resides at the host `prepaid.com`. An example of the reconfiguring routing rules in the Call Server (based on the *SIP Express Router* [Ipt05]) is shown below:

```
# Alternate Routing for A Prepaid Call
if (search("From:.*sip:0944021[0-9][0-9][0-9]@") &&
    # If this is a prepaid call
    !search("Call-ID:.*@prepaid.com") &&
    # and this request is to be authorized
    # by the prepaid mechanism
    (method=="INVITE" || method=="CANCEL" ||
    method=="BYE" || method=="ACK"))
{
    log(1,"Prepaid-User");
    rewritehost("prepaid.com");
    forward(prepaid.com,5060);
    break;
};
```

The preceding configuration ensures that the request messages generated from a prepaid user are forwarded to the prepaid B2BUA for authorization.

12.3.1 Prepaid Call Setup

Simplified prepaid call setup and force-termination message flows are illustrated in Figure 12.8. In a prepaid call setup, the SIP messages are exchanged among the Call Server, the B2BUA, and the PSTN gateway. The SS7 messages are exchanged between the Terminating Switch and the PSTN gateway. The call setup message flow in Figure 12.8 (Steps 1–14) is described as follows:

Step 1. UA1 sends the Call Server the **INVITE** message with `Request-URI` addressing to Phone2. According to the phone number configuration in the Call Server, the calling party phone number is investigated and UA1 is considered a prepaid user. The Call Server forwards this message to the B2BUA for authorization.

Step 2. When the B2BUA receives the **INVITE** message, its residing RADIUS client sends the RADIUS **Access Request** message (including the prepaid phone number) to the RADIUS server.

Step 3. The RADIUS server retrieves the prepaid user's record from the AAA database. If no record is found, the RADIUS server replies with the RADIUS **Access Reject** message, indicating that the user is not an authorized prepaid user. Otherwise, the RADIUS server replies with the RADIUS **Access Accept** message that contains the available prepaid credit in the `Session-Timeout` attribute representing the maximum quota of money or time.

Step 4. When the B2BUA receives the **Access Accept** message, it generates another **INVITE** message (for Subsession 2) to Phone2. The address of the B2BUA is filled in the `Contact` header field. This message is sent to the PSTN gateway through the Call Server.

Steps 5 and 6. The PSTN gateway generates the ISUP **Initial Address Message** (IAM) to the Terminating Switch at the PSTN. Phone2 then starts ringing. When the called party picks up the phone, the ISUP **Answer Message** (ANM) is sent from the Terminating Switch to the PSTN gateway. The message indicates that the called party has answered the call.

Step 7. The PSTN gateway generates a **200 OK** message. The address of the PSTN gateway is filled in the `Contact` header field. This message is routed to the Call Server and then forwarded to the B2BUA.

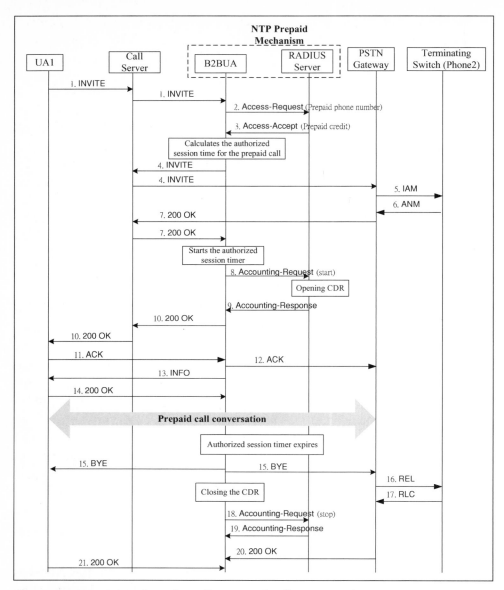

Figure 12.8 Message Flows for Call Setup and Call Force-Termination

Steps 8 and 9. Upon receipt of the final response, the B2BUA starts an authorized session timer with the value based on the Session-Timeout attribute (obtained in Step 3). The RADIUS client sends a RADIUS **Accounting Request** message with the Status "start," indicating the beginning of a prepaid call session. The RADIUS server creates an accounting CDR in the AAA database (via the SQL command

INSERT) and acknowledges the RADIUS client with a RADIUS **Accounting Response** message.

Step 10. The B2BUA generates a **200 OK** message to UA1. The address of the B2BUA is set in the `Contact` header field. This response is routed to the Call Server and then forwarded to UA1.

Steps 11 and 12. Upon receipt of the **200 OK** message, UA1 sends the **ACK** message to the B2BUA. The B2BUA forwards the **ACK** message to the PSTN gateway. At this point, the PSTN gateway opens a UDP port for RTP so that the voice packets sent from UA1 can pass through the PSTN gateway. The prepaid call session is established.

Step 13. The B2BUA sends the SIP **INFO** message to UA1. This message indicates the available prepaid credit.

Step 14. When UA1 receives the **INFO** message, it replies with the final response **200 OK** message.

After the call is connected, the prepaid credit is decremented by the prepaid system.

12.3.2 Forced Termination of a Prepaid Call

When one call party terminates the prepaid call, the **BYE** message is sent to the B2BUA directly. Then the B2BUA triggers the **Accounting Request** message with the Status "stop" to close the CDR. If the prepaid credit is exhausted before the call is complete, the call is forced to terminate by the prepaid system (Steps 15–21 illustrated in Figure 12.8):

Step 15. When the authorized session timer (started in Step 8) is timed out, the B2BUA sends the **BYE** messages to both UA1 and the PSTN gateway.

Steps 16 and 17. The PSTN gateway sends the ISUP **Release** (REL) message to the Terminating Switch. The Terminating Switch replies with the ISUP **Release Complete** (RLC) message. At this point, the PSTN gateway closes the UDP port for the RTP connection. Therefore, subsequent voice packets sent from UA1 cannot pass through the PSTN gateway.

Steps 18 and 19. The B2BUA sends the **Accounting Request** message with the Status "stop" and Terminate-Cause "Session-Timeout" to the RADIUS server, indicating forced termination of the prepaid call. The RADIUS server responds with the **Accounting Response** message after the accounting information (including remaining prepaid credit) is stored in the AAA database (via the SQL command **UPDATE**).

Steps 20 and 21. When the B2BUA receives the **200 OK** messages from UA1 and the PSTN gateway, the call is terminated.

Note that the prepaid model in this section assumes that a prepaid call is established between an IP phone and a PSTN phone. Calls among IP phones are typically free (e.g., Skype), and are not considered in this prepaid model.

12.4 Concluding Remarks

This chapter introduced SIP. We showed how SIP supports user mobility, call setup, and call release. Based on the SIP protocol, we illustrated how the push mechanism and the prepaid mechanism can be implemented in GPRS/UMTS.

In GPRS, the push feature is not supported. That is, an MS must activate the PDP context for a specific service before the external data network can push this service to the MS. An example is VoIP call termination (incoming call) to an MS. However, maintaining a PDP context without actually using it will significantly consume network resources. Therefore, it is desirable to devise a GPRS push mechanism that only activates a PDP context when it is needed. By using the iSMS platform, we described a SIP-based push mechanism for SIP call termination of GPRS supporting private IP addresses. A major advantage of this approach is that no GPRS/GSM nodes need to be modified.

As another SIP application example, we described a SIP-based prepaid mechanism to handle the prepaid calls in a VoIP system. Integration of this prepaid mechanism with the existing VoIP platform can be easily achieved by reconfiguring the Call Server.

12.5 Questions

1. Describe the six basic commands of SIP. Is this command set sufficient to support mobile telecommunications applications (e.g., mobility management)?

2. Describe the registrar, redirect server, and proxy server.

3. Describe the SIP registration transaction. Is the target of the transaction also the recipient of the request?

4. Compare SIP with H.323 and MGCP, described in Chapter 10.

5. Describe the SDP fields o, c, and m.

6. After a phone conversation is finished, the call can be disconnected using procedures that depend on who hangs up first. In the *caller model*, if the calling party hangs up the phone, the call is immediately released. If the called party hangs up the phone first, the call is suspended. After the call is suspended, if the calling party hangs up, the call is terminated. Conversely, if the called party picks up the phone again, the call is resumed. Show how the caller model (for call release) is implemented in SIP.

7. Why is L2TP tunneling required to implement the SIP-based GPRS push mechanism?

8. Show how a prepaid GPRS user can be supported through the architecture shown in Figure 12.5.

9. To support prepaid calls among IP phones, should the architecture shown in Figure 12.5 be modified? What potential problems may occur?

10. In Section 12.3, the prepaid users are distinguished from the postpaid users through the phone numbers. When a prepaid user decides to change from prepaid to postpaid, how can he/she keep the original phone number?

11. The INFO message can be used to indicate that the user's prepaid account balance is near zero. Show how this feature can be implemented using the SIP-based prepaid mechanism architecture in Figure 12.5.

12. Show how to integrate the push mechanism in Section 12.2 with the prepaid mechanism in Section 12.3. Does this integration make sense?

13. Show how a SIP phone can be used for eavesdropping, as described in Section 9.3.

## CHAPTER 13

Mobile Number Portability

Number portability is a network function that allows a subscriber to keep a "unique" telephone number. Imposed by the National Regulatory Authority and agreed upon by different network operators, number portability is one of three important mechanisms to enhance fair competition among telecommunications operators and to improve customer service. The other two mechanisms for fair competition are equal access and network unbundling.

Three types of number portability have been discussed: location portability, service portability, and operator portability:

- With *location portability*, a subscriber may move from one location to another without changing his/her telephone number. This type of portability is already implied in mobile phone service.

- With *service portability*, a subscriber may keep the same telephone number when changing telecommunications services. In the U.S., service portability between fixed telephone service and mobile phone service is implementable because both services follow the "NPA-NXX-XXXX" telephone number format. In Taiwan, the service code "09" for mobile service is distinguished from area codes of fixed telephone service. As a result, service portability cannot be made available in Taiwan unless the numbering plan is modified.

■ With *operator portability*, a subscriber may switch telecommunications operators without changing his/her telephone number.

In most countries, location portability and service portability are not enforced, and only operator portability is implemented. The reason is twofold. First, operator portability is considered essential for fair competition among operators, while location portability and service portability are typically treated as value-added services. Second, implementation and operation costs can be significantly reduced if service portability and location portability are not considered.

Many countries, including Australia, China, Hong Kong, Japan, Taiwan, the United Kingdom, the U.S., and numerous European countries, have implemented or are in the process of implementing fixed-network number portability. In these countries, the implementation schedule for mobile number portability typically follows that of fixed-network number portability [ACA99, IDA00b]. Survey studies by OFTEL (Office of Telecommunications, United Kingdom) and DGT/Taiwan indicate that most mobile operators are not enthusiastic about implementing number portability. They questioned whether there is a real demand for mobile number portability and whether it would provide significant benefits. However, number portability must be enforced because it is considered a mechanism that will help a new operator or CLEC (Competitive Local Exchange Carrier) compete with the existing operator, or ILEC (Incumbent Local Exchange Carrier). Some mobile operators also claim that the absence of number portability may not deter customers from switching operators. In the U.S., for example, *called-party-pays policy* is exercised, whereby mobile subscribers typically pay for the air-time usage and mobility for both incoming and outgoing calls. In order not to receive undesirable calls, customers are unlikely to distribute their numbers widely. From this perspective, number portability may not be an important factor in a customer's decision to change mobile operators. Thus, in the U.S., "most mobile operators fought number portability kicking and screaming, and expressed amazement that the FCC would do such an evil thing to them" [Ame97].

Conversely, in Taiwan or the United Kingdom, *calling-party-pays* policy is exercised, whereby the mobile subscribers only pay for outgoing calls, and the incoming calls are paid by the callers. In this scenario, mobile customers, especially business people who have high mobility (such as sales-people, plumbers, electricians, and builders), are likely to widely distribute their numbers. Furthermore, compared with fixed-network telephone numbers, few mobile numbers are published in telephone directories. Therefore, the benefits of number portability for mobile customers are greater than they are for fixed-network customers. According to a U.K. survey, without number portability, only 42% of corporate subscribers are willing to change mobile operators. This percentage would increase to 96% if number portability were introduced. In-Stat MDR polled 1,050 mobile business users and found that only 6% said they were likely to churn in the next 12 months, while 36.6% said they might. On the other hand, 52% said they were more likely to churn if number portability were introduced [Lyn02]. Changing telephone numbers becomes a barrier to switching mobile operators, and has turned out to be a major reason why mobile operators are against mobile number portability. As pointed out, mobile service providers know that with the hold of a unique number, customer loyalty will be even harder to keep [Che99].

A recent OFTEL analysis shows that there will be a net gain to the United Kingdom economy of 98 million £ with the introduction of mobile number portability. NERA's analysis indicated that by introducing mobile number portability, the net benefit for Hong Kong's economy will range from HK$1,249 million to HK$1,467 million. Thus, it was concluded that to improve a country's economy, the government should enforce mobile number portability.

This chapter introduces mobile number portability. We discuss number portability mechanisms, the costs incurred by number portability, and cost recovery issues. We first define basic number portability terms. Originally, a telephone number is assigned to a mobile network. This network is called the *number range holder (NRH)* network. The *subscription network* is the network with which a mobile operator has a contract to implement the services for a specific mobile phone number. Originally, the NRH network is the subscription network of the customer. For example, if a mobile phone number is ported from mobile operator A to mobile operator B, then network A is called the *donor network* or *release network*, and network B is called the *recipient network*. Before the porting process, network A is the subscription network. After porting, network B is the subscription network. The "moved" number is referred to as a *ported number*. Note that the ported number indicates the routing information to the NRH network.

13.1 Number Portability for Mobile Telecommunications Networks

Although most mobile operators are not enthusiastic about implementing mobile number portability, they cannot avoid the impact of fixed-network number portability. When a Mobile Station (MS) originates a call to a ported number in the fixed network, the originating Mobile Switching Center (MSC) needs to route the call to the correct destination by using fixed-network number portability solutions. Alternatively, the MSC may direct the call to a switch in the fixed network, which then routes the call to the recipient switch. In this case, the mobile operator should reimburse the fixed-network operator for extra routing costs.

Before describing mobile number portability, we point out that an MS is associated with two numbers: the *directory* number and the *identification* number. In a UMTS circuit-switched domain, the Mobile Station ISDN Number (MSISDN) is the directory number, which is dialed to reach the MS. In other words, the MSISDN is the telephone number of the MS. The International Mobile Subscriber Identity (IMSI) is the identification number that uniquely identifies an MS in the mobile network. The IMSI is used to authenticate/identify the MS during mobile network access such as location update and call origination, which is hidden from the mobile user. When a mobile user switches operators, a new MSISDN and IMSI pair is assigned to the user. When mobile number portability is introduced, a porting mobile user would keep the MSISDN (the ported number) while being issued a new IMSI. In other words, IMSI shall not be ported. When the ported number is no longer used, the number is returned to the NRH network. Note that lawful interception will be possible on a ported MSISDN.

For mobile systems based on ANSI IS-41 [Lin 01b, EIA97], the identification number and the directory number are referred to as the *Mobile Identification Number (MIN)* and the *Mobile Directory Number (MDN)* respectively. Mobile operators typically assume that both the MIN and the MDN have the same value and are used interchangeably. The MIN/MDN is of the format "NPA-NXX-XXXX", where the first six digits "NPA-NXX" identify the home system of the MS. Without this home network identification, roaming is not possible. The MDN is used as the calling party number parameters in signaling and billing records. If mobile number portability is introduced, the MIN will be different from the MDN. In such a case, using the MIN as the calling party number will result in misrouting in services such as automatic callback and calling number/calling name presentation. Similarly, using MDN for location update will result in errors when performing the registration procedure. Thus, to support portability, separation of the MIN and MDN is required for the IS-41-based systems. This means that extra costs will be incurred to modify mobile software in the MSC, the HLR, the VLR, the billing system, and so on.

Following the above discussion, the impact of number portability on mobile network is considered in three aspects:

- Location Update: The identification number (IMSI or MIN) is used in the location update procedure. Since the assignment of this number is not affected by the introduction of number portability, location update is not affected by portability except that MIN/MDN separation is required for the IS-41-based systems.

- Mobile Call Origination: As mentioned in the beginning of this section, to originate a call to a ported number, the MSC needs to be equipped with a number portability routing mechanism.

- Mobile Call Termination: To deliver or terminate a call to a ported mobile number, the standard mobile call termination procedure must be modified to accommodate the portability mechanism.

The U.S. introduces number portability to mobile operators in two phases. In phase 1, mechanisms for mobile to (ported) fixed-network calls are implemented. In phase 2, the MIN/MDN separation, as well as mobile call termination mechanism, is implemented.

13.2 Call Routing Mechanisms with Number Portability

In mobile service, the network tracks the location of every MS. The location information is stored in two mobile databases, the HLR and the VLR. To deliver a call to an MS, the databases are queried for routing information for the MSC where the MS resides. Figure 13.1 illustrates a simplified UMTS circuit-switched call termination procedure in which the interaction between the HLR and the VLR is omitted (a more complete flow is illustrated in Figure 9.6). The message flow is described in the following steps:

Step GSMCT.1. When the calling party dials the MSISDN of a mobile station MS2, the call is routed to the Gateway MSC (GMSC) of MS2 using the ISUP Initial Address Message (IAM).

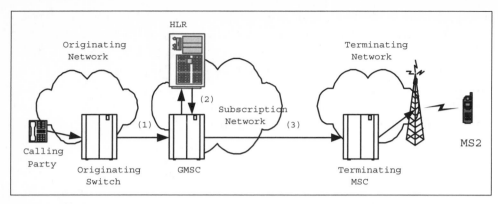

Figure 13.1   A Simplified UMTS Circuit-Switched Call Termination Procedure without Number Portability

Step GSMCT.2. The GMSC queries the HLR to obtain the Mobile Station Roaming Number (MSRN; the address of the terminating MSC where MS2 resides).

Step GSMCT.3. Based on the MSRN, the IAM message is routed to the destination MSC, and the call is eventually set up.

In Figure 13.1, the terminating network (where the MS resides) may be different from its subscription network. Call termination to the MS must be routed to the GMSC at the subscription network due to the following restrictions:

Restriction 1. The GMSC must be in the call path for the provision of special features and services as well as for billing.

Restriction 2. The originating switch does not have the capability to query the HLR database, which must be done by the GMSC through the Mobile Application Part (MAP) *C interface* (the protocol between the GMSC and the HLR).

To support mobile number portability, call termination in Figure 13.1 should be modified. In 3GPP TS 23.066 [3GP04d], two approaches are proposed to support number portability call routing:

- *Signaling Relay Function* (SRF)-based solution
- *Intelligent Network* (IN)-based solution

Both approaches utilize the *Number Portability Database* (*NPDB*), which stores the records of the ported numbers. The record information includes the following:

- The ported MSISDN
- The status (active or pending)
- The time stamps (when the ported number record is created, activated, disconnected, and modified)
- The NRH mobile operator
- The subscription operator
- The routing information

The routing information includes several addresses to support applications such as switch-based services (i.e., *Custom Local Area Signaling Service*, or *CLASS*), calling cards, and short message service. For non-ported numbers, no records will be maintained in the NPDB. The call routing mechanisms are described in the following subsections.

13.2.1 The SRF-based Solution for Call-Related Signaling

The SRF-based solution utilizes the MAP protocol. The SRF node is typically implemented on the Signal Transfer Point (STP) platform (see Section 8.1). Three call setup scenarios have been proposed for the SRF-based approach: *direct routing*, *indirect routing*, and *indirect routing with reference to the subscription network*. These scenarios are elaborated as follows:

Direct Routing Scenario (DR): The mobile number portability query is performed in the originating network, which is basically the same as the *all-call-query* approach in fixed-network number portability (see Chapter 15 in [Lin 01b]). All call-related messages for ported and non-ported subscribers are acknowledged with appropriate routing information in order to route the calls to their subscription networks. Figure 13.2 (a) illustrates a DR call from a mobile station MS1 to a ported mobile station MS2. The following steps are executed:

Step DR.1. MS1 dials the MSISDN of MS2. An ISUP **IAM** message is routed from the originating MSC to the originating GMSC (the GMSC in the originating network). As mentioned in **Restriction 2**, only the GMSC is equipped with the MAP C interface to communicate with the HLR/SRF.

Step DR.2. The originating GMSC issues the MAP **Send Routing Information** message to the SRF.

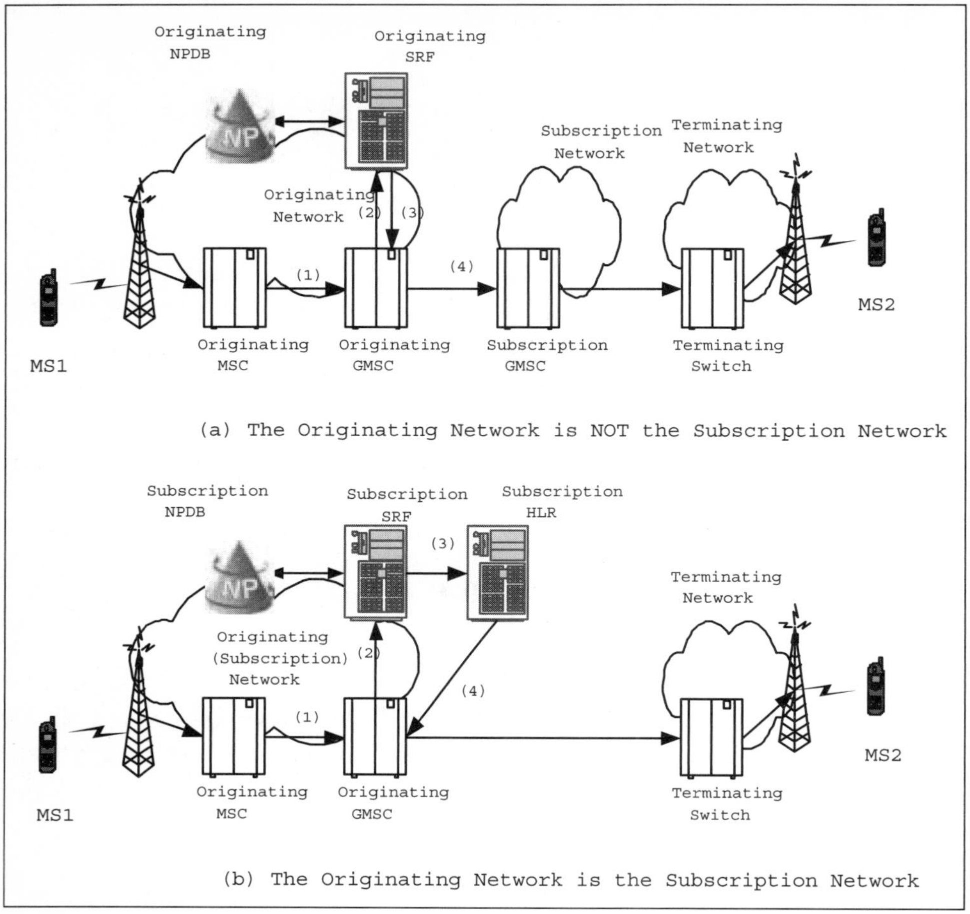

(a) The Originating Network is NOT the Subscription Network

(b) The Originating Network is the Subscription Network

Figure 13.2   SRF-based Directed Routing (DR)

Step DR.3. By consulting the NPDB (possibly through the MAP or *Intelligent Network Application Part (INAP)*), the SRF obtains the subscription network information of MS2 and forwards the information to the originating GMSC.

Step DR.4. The originating GMSC then routes the IAM message to the GMSC of MS2 in the subscription network. After this point, the call is set up following the standard call termination procedure described in Steps GSMCT.2 and 3.

In Step DR.3, the SRF provides the *Routing Number (RN)* to the originating GMSC. The RN consists of a RN prefix plus the MSISDN of the called party. The RN prefix points to the subscription GMSC, which

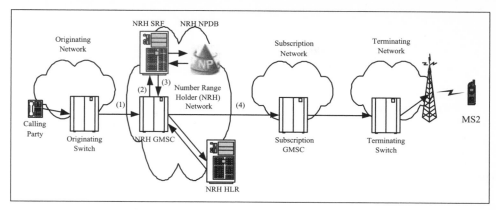

Figure 13.3 SRF-based Indirect Routing (IR-I)

may also provide the HLR address of the called party. (Note that the subscription networks may have several HLRs, and the HLR address cannot be simply identified by the MSISDN.) If so, the subscription GMSC can access the subscription HLR directly in Step GSMCT.2. If the prefix does not provide the HLR information, then the subscription GMSC must utilize the SRF to route the **Send Routing Information** to the HLR. Details provided by the RN prefix may be constrained by issues such as security (of the subscription network) and length limit [ETS02]. In Germany, the routing prefix format is Dxxx where D is a hex digit and x is a decimal digit.

If the originating network is the subscription network of MS2, then as illustrated in Figure 13.2 (b), in Steps 3 and 4, the SRF sends the **Send Routing Information** message to the subscription HLR, and the HLR returns the MSRN of MS2 to the originating GMSC (which is also the GMSC of MS2), and the call setup proceeds to Step GSMCT.3.

Indirect Routing Scenario (IR-I): The mobile number portability query is performed in the NRH network, which is similar to *onward routing* (remote call forwarding) in fixed-network number portability [Lin 01b]. All call-related messages for ported subscribers are acknowledged with appropriate routing information in order to route the call to the subscription network. Figure 13.3 illustrates the IR-I call setup to a ported mobile station MS2 with the following steps:

Step IR-I.1. The calling party dials the MSISDN of MS2, and the IAM message is routed to the NRH GMSC of MS2.

Step IR-I.2. The NRH GMSC queries the SRF using the MAP **Send Routing Information** message.

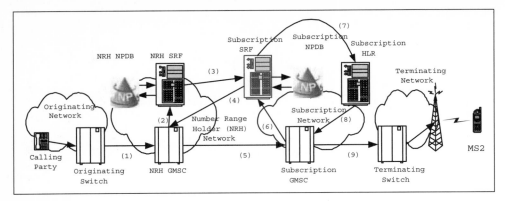

Figure 13.4 SRF-based Indirect Routing with Reference to the Subscription Network (IR-II)

Step IR-I.3. By consulting the NPDB, the SRF obtains the subscription network information of MS2 (the RN prefix points to the subscription GMSC) and forwards the information to the NRH GMSC.

Step IR-I.4. The NRH GMSC then routes the IAM message to the GMSC of MS2 in the subscription network. After this point, the call is set up by following Steps GSMCT.2 and 3. As mentioned in the DR scenario, if the RN information provided in Step IR-I.3 does not point out the HLR location, then the subscription SRF is queried in Step GSMCT.2.

Indirect Routing with Reference to the Subscription Network Scenario (IR-II): The mobile number portability query is performed in the NRH network. All call-related signaling messages for ported subscribers are relayed to the subscription network. Figure 13.4 illustrates the IR-II call setup to a ported mobile station MS2 with the following steps:

Step IR-II.1. The calling party dials the MSISDN of MS2, and the IAM message is routed to the NRH GMSC of MS2.

Step IR-II.2. The NRH GMSC queries the SRF using the MAP **Send Routing Information** message.

Step IR-II.3. By consulting the NPDB, the NRH SRF identifies that the called party's MSISDN is ported out. The NPDB provides the RN prefix pointing to the SRF in the subscription network. The NRH SRF relays the **Send Routing Information** request to the subscription SRF.

Step IR-II.4. By consulting the NPDB, the subscription SRF identifies the GMSC of MS2, and returns the routing information to the NRH GMSC.

Step IR-II.5. The NRH GMSC then routes the IAM message to the GMSC of MS2 in the subscription network.

Steps IR-II.6–8. These steps are the same as Step GSMCT.2 except that the HLR is queried indirectly through the SRF.

Step IR-II.9. After the NRH GMSC has obtained the MSRN of MS2, Step GSMCT.3 is executed.

The SRF-based DR is utilized when the originating network has the GMSC that can query the SRF, and a routing mechanism exists for the originating switch (that connects to the calling party) to access the GMSC:

- For a mobile-to-mobile call, this scenario incurs the lowest cost. IR-I is basically the same as the onward routing proposed in fixed-network number portability (see Chapter 15 in [Lin 01b]).

- For a fixed-to-mobile call, this scenario is recommended so that the fixed networks do not need to make any modifications due to the introduction of mobile number portability.

IR-II is typically used for international call setup where the NRH network and the subscription network are in different countries. IR-II is preferred for international calls because the originating network/country does not have the ported number information for the subscription network's country. To exercise DR, the originating NPDB should contain records for all ported numbers in the portability domain. Conversely, the NRH NPDB in IR-I and IR-II only needs to contain the numbers ported out of the NRH network, and the subscription NPDB only needs to contain the numbers ported in the subscription network.

13.2.2 The SRF-based Solution for Noncall-Related Signaling

For non-call-related signaling such as the Short Message Service (SMS) delivery and *Call Completion on Busy Subscriber* (*CCBS*), no voice trunk setup is involved, and the message flows are different from that in Section 13.2.1. We describe the SMS for the SRF-based direct and indirect routing scenarios:

Short Message Service Direct Routing Scenario (SMS-DR). Figure 13.5 illustrates a DR short message to a ported mobile station MS2. The following steps are executed:

Step SMS-DR.1. The SM-SC issues the short message to the SMS GMSC in the same interrogating network using a proprietary

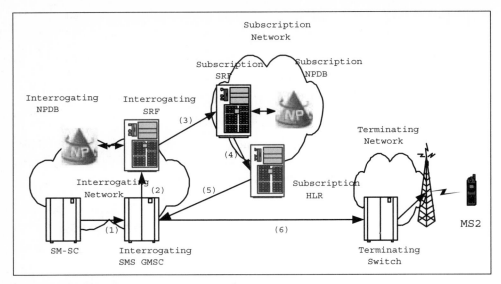

Figure 13.5 SRF-based SMS-Direct Routing (SMS-DR)

interface. In most GSM implementations, the SM-SC is co-located with the SMS GMSC. The term "interrogating network" means that the network will interrogate the HLR for a non-call-related signaling message.

Step SMS-DR.2. The SMS GMSC queries the SRF in the interrogating network.

Step SMS-DR.3. By consulting the NPDB, the interrogating SRF identifies that the called party's MSISDN is ported. The interrogating SRF relays the routing query message to the SRF in the subscription network.

Step SMS-DR.4. By consulting the NPDB, the subscription SRF identifies the HLR of MS2, and forwards the routing query message to the HLR.

Step SMS-DR.5. The HLR returns the routing information similar to the MSRN of MS2 to the interrogating SMS GMSC.

Steps SMS-DR.6. The interrogating SMS GMSC forwards the short message following the standard SMS delivery procedure described in Chapter 1.

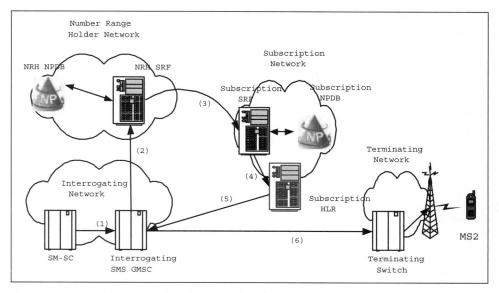

Figure 13.6 SRF-based SMS-Indirect Routing (SMS-IR)

Short Message Service Indirect Routing Scenario (SMS-IR). Figure 13.6 illustrates an indirect-routed short message to a ported mobile station MS2 with the following steps:

Step SMS-IR.1. Like Step SMS-DR.1, the SM-SC issues the short message to the SMS GMSC in the same interrogating network.

Step SMS-IR.2. The SMS GMSC queries the NRH SRF.

Step SMS-IR.3. By consulting the NPDB, the NRH SRF identifies that the called party's MSISDN is ported out. The NRH SRF relays the routing query message to the SRF in the subscription network.

Steps SMS-IR.4–6. These steps are similar to Steps SMS-DR.4.–6.

Note that the major difference between SMS-DR and SMS-IR is the execution of Step 2. In direct routing, the interrogating network has its own SRF to forward the routing query message to the subscription HLR.

13.2.3 The IN-based Solution for Call-Related Signaling

There are two IN-based solutions for querying the NPDB: ETSI Core INAP and ANSI IN Query [3GP04d]. The IN solution is implemented in the Service

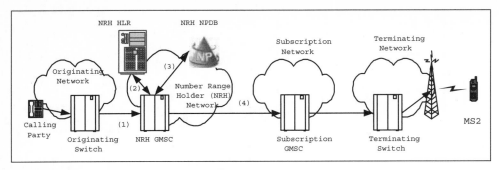

Figure 13.7   IN-based Query on HLR Release (QoHR)

Control Point (SCP; see Chapter 8). The major differences between the IN and the SRF solutions are as follows:

- Any switch equipped with the IN protocol can query the NPDB. In the SRF solution, only GMSC equipped with the MAP C interface can query the SRF (see **Restriction 2**).

- The IN approach does not support the non-call-related signaling messages. Note that the IN approach can support non-call-related signaling messages if the message sender first performs an IN query and uses the routing number for routing the signaling message. But this is usually not exercised in current number portability implementations.

To route the calls, three scenarios have been proposed for the IN-based number portability solutions. *Originating call Query on Digit analysis (OQoD)* is similar to direct routing in the SRF-based approach, except that the originating switch can directly query the NPDB using the IN protocol. *Terminating call Query on Digit analysis (TQoD)* is similar to indirect routing (IR-I) in the SRF-based approach. The third scenario of the IN-based approach is called *Query on HLR Release (QoHR)*. The message flow of QoHR is illustrated in Figure 13.7 and the steps are described as follows:

Step QoHR.1. The calling party dials the MSISDN of MS2. The originating network routes the call to the GMSC of the NRH network.

Step QoHR.2. The NRH GMSC first queries its HLR for the routing information. If MS2 is ported, the NRH HLR replies with the "Unknown Subscriber" error, which triggers the NRH GMSC to query the NPDB.

Step QoHR.3. Using the INAP **InitialDP** message, the NRH GMSC queries the NPDB with the MSISDN of MS2. The NPDB returns the routing number pointing to the subscription network through the INAP **Connect**

message. If no entry is found in the NPDB, the INAP **Continue** message is returned.

Step QoHR.4. The NRH GMSC sends the **IAM** message to the subscription GMSC. Then Steps GSMCT.2 and 3 are executed for call setup. If the RN prefix obtained in Step QoHR.3 does not provide the HLR information, then the subscription GMSC may not be able to identify the HLR from the MSISDN [3GP04d]. To resolve this problem, a function similar to SRF may need to be implemented in the subscription GMSC to find out the HLR for the ported-in MSISDN.

If the MSISDN of MS2 is not ported, then the NRH network is the same as the subscription network, and Steps QoHR.3 and 4 are not executed. It is clear that the routing cost for OQoD (direct routing) is lower than that for TQoD (indirect routing). If MS2 is not ported, the routing cost for QoHR is lower than that for OQoD. If MS2 is ported, the result reverses. To exercise OQoD, the originating NPDB should contain records for all ported numbers in the portability domain. Conversely, the NRH NPDB in QoHR and TQoD only needs to contain the numbers ported out of the NRH network, and the subscription NPDB only needs to contain the numbers ported in the subscription network. Therefore, if the population of ported subscribers is small (say, less than 30%), QoHR is a preferred mobile number portability solution as compared with OQoD and TQoD.

13.3 Number Porting and Cost Recovery

When a number is ported from the donor operator to the recipient operator, the NPDBs of all operators in the portability domain may need to be updated. Number porting is an off-line administrative process that can be performed centrally by a neutral third party or distributed among the participating mobile operators. This section describes the number porting mechanisms, and then discusses the cost recovery issues.

13.3.1 Number Porting Administration

We use the North America *Number Portability Administration Center* (*NPAC*) and Hong Kong *Central Ticketing System* (*CTS*) models to illustrate number porting administration [Fos03]. Mandated by regulators and service providers, North America NPAC is administered by a neutral third

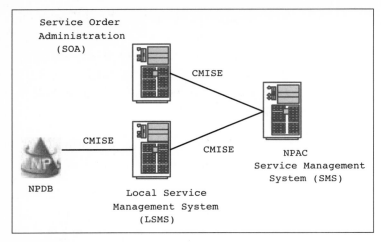

Figure 13.8 Connectivity between the NPAC and a Mobile Operator

party that has a fixed-term contract. When the contract expires, the new NPAC is selected through open bidding, and system ownership is transferred to the new NPAC. The NPAC functions include service provider data administration, subscription data administration, audit administration, resource accounting, billing and cost apportionment, and so on. The NPAC is designed to support various types of number portability, and is developed according to standardized functional requirements and interface specifications that are maintained in the public domain. Note that the NPAC supports the master database and is not involved in individual call setups. Figure 13.8 illustrates the connectivity between the NPAC and a mobile operator. In this figure, the facilities of a mobile operator are *Service Order Administration (SOA)*, *Local Service Management System (LSMS)*, and the NPDB. The interface among these components is *Common Management Information Service Element (CMISE)*. The SOA connects to the NPAC service management system for service order processing. An SOA transaction with the NPAC must complete within 2 seconds or less. The LSMS connects to the NPAC service management system for database updates. The services offered by the NPAC service management system are available 99.9% of the time, and the unavailability of scheduled services should not exceed 2 hours per month. After a record is activated in the NPAC service management system, this change should be broadcast to the LSMSs within 60 seconds. The NPAC should respond to both the SOA and the LSMS within 3 seconds. 95% of the NPAC-to-LSMS transactions must occur at a rate in accordance with the performance improvement plan. Every NPDB connects

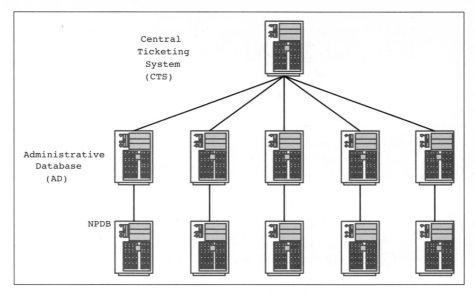

Figure 13.9   Connectivity between the CTS and Mobile Operators

to the corresponding LSMS for accessing the record of each active ported number. Therefore, switch routing information and network element identification are kept in the NPDB. When a number is ported, the NPAC service management system updates the LSMSs (and therefore the NPDBs and the HLRs) of the participating operators.

In Hong Kong's number porting administration, a *Central Ticketing System (CTS)* is shared by all operators, as illustrated in Figure 13.9. The CTS connects to several *Administrative Databases (ADs)* owned by or leased to the mobile operators. Every AD is connected to the NPDBs of a mobile operator. To port a number, the recipient operator issues a request to the CTS. The CTS approves a limited number of porting requests (5,000–10,000 requests per day) to ensure that the subscribers will not change operators too frequently. The ADs transfer a service order between donor and recipient operators, and notify other network operators when the service order is confirmed. Therefore, the porting process is performed in the distributed manner among the ADs. The ADs also update the NPDBs during the daily cutover window. Table 13.1 lists all possible scenarios for administrative actions in number porting. For example, in number porting scenario III (where neither the donor network nor the recipient network is the NRH network), the databases of the donor network are modified as follows: The HLR record of the ported number is deleted, and the subscription operator field of the

Table 13.1 Administrative actions in number porting

SCENARIO	DONOR HLR	SUBSCRIPTION HLR	ALL NPDB
I. Donor=NRH*	delete	add	add
II. Subscription=NRH	delete	add	delete
III. Neither	delete	add	update

*NRH: the number range holder network.

NPDB record is set to the recipient network. Note that the order of the sequence for the administrative actions to be performed both within a network and by different network operators is significant with respect to prevention of disruption in service to the mobile subscriber and prevention of looping calls between networks during the porting process.

13.3.2 Costs of Number Portability

To support number portability, the following costs are incurred: initial system setup costs, customer transfer costs, and call routing costs.

Initial system setup includes the costs for number portability system development, network management, line testing (for fixed networks), operator services, billing information, exchanges overlay, maintenance, and support. Initial system setup costs were estimated to be £8.01 million for British Telecom's mobile number portability [Oft97a]. In-Stat/MDR predicted that mobile number portability implementation would cost the U.S. wireless industry up to US$1 billion in setup costs [Lyn02].

Customer transfer costs or per-line setup costs are incurred when moving a ported number from the donor operator to the recipient operator. The costs include the following:

- Closing down an old account
- Opening a new account
- Coordinating physical line switching (for fixed networks)

The transfer cost for a mobile ported number is estimated to be £19. Some number portability studies considered the customer transfer costs as a part of number portability overhead. It is clear that a major part of the costs exists even if number portability is not implemented. However, extra coordination between the number range holder, the donor, and the recipient operators is required to transfer a ported number as described in

Section 13.3.1. Furthermore, issues such as "what to do when a customer who leaves bad debt to the donor operator moves to the recipient operator with the ported number" should be carefully resolved. In-Stat/MDR predicted that mobile number portability implementation will cost the U.S. wireless industry up to US$500 million in annual running costs [Lyn02].

There has been a long argument regarding who should cover the costs for number portability. Some argue that the ported customers should cover the costs for number portability. Others believe that the number portability costs should be borne by all telecommunications users, for the following reasons:

- Number portability should be treated as a "default function" instead of an added-value service. Since all users have the opportunity to reach ported numbers, they all benefit from number portability.

- If the cost for a ported number is significantly higher than the cost for a non-ported number, then users will be discouraged from utilizing the number portability feature. If so, the original goal to provide fair competition and service quality improvement will not be achieved. Depending on telecommunications policies, different countries may make different decisions for cost recovery [IDA00a].

Most studies suggest that operators should bear their own costs for initial system setup. For customer transfer costs, the donor operator should bear the cost for closing down an account, and the recipient operator should bear the cost for opening a new account. In indirect routing, the subscription operator needs the NRH operator's assistance to transfer the ported customers. Thus, requiring reimbursement from the recipient or originating operators to the donor operator is reasonable. The recipient operator could charge the customers for importing their numbers. For fixed-network number portability in the U.S., the FCC 3rd Report & Order (CC 95-116) pointed out that this cost could be recovered via a monthly end-user charge. Monthly charges range from US$.23 (Verizon) to US$.43 (US West) or US$.48 (Sprint Local) per line. Donor operators are not anticipated to charge customers who move out with ported numbers. In the North America NPAC model, the participating operators should cover the costs of NPAC. For fixed-network number portability in the U.S., the FCC 3rd Report & Order (CC 95-116) identified the cost for NPAC recurring and nonrecurring administration, and it is shared by all fixed-network operators based on regional end-user revenue. In Canada, Telecom Order 97-1243 set the ported number transaction charge at C$5.00.

Telecom Order 98-761 Stentor proposed a query charge of C$1.05 per 1,000 queries [Fos03].

13.4 Concluding Remarks

This chapter introduced mobile number portability. Based on [Lin03b], we described and analyzed number portability routing mechanisms and their implementation costs. We first described the SRF-based solution for call-related and noncall-related routing. Then we described the IN-based solution for call-related routing. In these routing mechanisms, if the population of ported subscribers is small, the NPDB can be integrated with the MSC. Typically, a NPDB contains millions of entries. The throughput of most SRF/NPDB products is up to several thousands of transactions per second. To implement the routing mechanisms, the government needs to specify the routing number plan. In both the United Kingdom and Spain, the SRF approach is used. In both Hong Kong and Australia, the operators can choose either SRF or IN approaches. In the above four countries, the network interface is standard ISUP. In the U.S., an advanced IN approach is used, and the network interface is ANSI ISUP with LRN enhancement. In Portugal, the IN approach is used, and the network interface is standard ISUP with query on release. In most of these countries and Germany as well, the originating networks are responsible for querying the NPDB.

Cost recovery issues for number portability were discussed in this chapter from a technical perspective. We should point out that rules for cost recovery also depend on business and regulatory factors, which vary from country to country. Several surveys for number portability have been conducted by OFTEL [Oft97a, Oft97b], NERA [Att98], DGT/Taiwan [Yu 99], and OVUM [Hor98]. An excellent overview of number portability can be found in [Fos03]. Service portability between fixed telephone service and mobile phone service is discussed in [NAN98]. MDN/MIN issues for mobile number portability are elaborated in [NAN98].

As a final remark, mobile number portability may affect the existing services. For example, it is difficult to provision the prepaid services and friendly tariff (e.g., special tariff D1 to D1) when mobile number portability is exercised. For further reading, refer to the following web sites for the most up-to-date information:

- Office of the Telecommunications Authority (Hong Kong): www.ofta.gov.hk/mnp/main.html

- Office of Telecommunications (U.K.): www.oftel.gov.uk (search with "mobile number portability")

- Number Portability Administration Center: www.npac.com

- Ported Communications: www.ported.com

- FCC: www.fcc.gov/wcb/

- NeuStar: www.neustar.com

- North American Number Plan Administration: www.nanpa.com

13.5   Questions

1. Describe three types of number portability. Which type of number portability is required for fair competition?

2. Explain the following terms: number range holder network, subscription network, donor network, recipient network, and ported number.

3. What are the differences between fixed network number portability and mobile number portability?

4. Describe the three routing mechanisms for the SRF-based approach. What are the advantages and disadvantages of these mechanisms?

5. In mobile number portability what is "routing number"? Which network component provides this number?

6. Compare the direct routing and the indirect routing scenarios for SMS delivery to a ported number.

7. Describe how prepaid SMS can be supported by the SRF-based solution.

8. What are the major differences between the SRF and the IN-based number portability approaches?

9. Describe OQoD, TQoD, and QoHR. What are the advantages and disadvantages of these approaches?

10. Illustrate the message flows for OQoD and TQoD.

11. What number portability routing approach should be selected to support international incoming calls to a ported number?

12. Describe the number porting administration mechanisms based on NPAC and CTS, respectively. Compare these two approaches.

13. What regulation is needed when a customer who left bad debt to the donor operator moves to the recipient operator with the ported number?

14. Can we support service number portability between prepaid and post-paid mobile services? What potential issues need to be resolved?

## CHAPTER 14

Integration of WLAN and Cellular Networks

In Bibliography Section of this book, please refer to, 3GPP TR 22.934 [3GP03a] which conducts a feasibility study on cellular system and *Wireless LAN* (*WLAN*) interworking that extends mobile services to the WLAN environment. In this interworking, WLAN serves as an access technology to the cellular system, which scales up the coverage of mobile services. Six scenarios were proposed for incremental development of cellular and WLAN interworking. Each scenario enhances interworking functionalities over the previous scenarios, as illustrated in Table 14.1. The service and operational capabilities of each scenario are described as follows:

- **Scenario 1** provides common billing and customer care for both WLAN and mobile operators. That is, a customer receives a single monthly billing statement combining both mobile and WLAN services. The customer also consults the same customer care center about problems with both services.

- **Scenario 2** re-uses cellular-access control and charging mechanisms for WLAN services. The WLAN customers are authenticated by the mobile core network without introducing a separate authentication procedure. In addition, the roaming mechanism between the cellular system and the WLAN is supported. In this scenario, users can access

Table 14.1   Interworking scenarios and service capabilities

Service Capabilities Scenario	1	2	3	4	5	6
Common Billing	Yes	Yes	Yes	Yes	Yes	Yes
Common Customer Care	Yes	Yes	Yes	Yes	Yes	Yes
Cellular-based Access Control	No	Yes	Yes	Yes	Yes	Yes
Cellular-based Access Charging	No	Yes	Yes	Yes	Yes	Yes
Access to Mobile PS Services	No	No	Yes	Yes	Yes	Yes
Service Continuity	No	No	No	Yes	Yes	Yes
Seamless Service Continuity	No	No	No	No	Yes	Yes
Access to Mobile CS Services with Seamless Mobility	No	No	No	No	No	Yes

traditional Internet services but cannot access mobile services (such as *Circuit Switched* (CS) voice and GPRS data services) through the WLAN.

- **Scenario 3** allows a customer to access mobile *Packet Switched* (*PS*) services over the WLAN. The PS services include Short Message Service (SMS; see Chapter 1), Multimedia Messaging Service (MMS; see Chapter 11), and IP Multimedia Core Network Subsystem (IMS) Service (see Chapter 15). Customers equipped with both a WLAN card and a cellular (e.g., UMTS or GPRS) module can simultaneously but independently access WLAN and cellular networks.

- **Scenario 4** allows a customer to change access between cellular and WLAN networks during a service session. The system is responsible for reestablishing the session without user involvement. Service interruption during system switching is allowed in this scenario. Quality of Service (QoS) is a critical issue for service continuity. Since cellular and WLAN networks have different capabilities and characteristics, the user would gain different QoS grades in different networks. Therefore, QoS adaptation is required during system switching.

- **Scenario 5** provides seamless service switching (that is, handoff) between the cellular system and the WLAN. Techniques must be developed to minimize the data lost rate and delay time during switching so that the customer does not experience significant interruption during handoff.

- **Scenario 6** supports mobile CS services in the WLAN environment. The seamless continuity feature described in Scenario 5 is also required to support CS services when customers roam between different networks.

Our survey with several mobile service providers indicates that the Scenario 3 features are essential for commercial operation of cellular/WLAN interworking in the first stage deployment. Depending on the business strategies, the Scenario 4 features may or may not be deployed in the long-term commercial operation. Scenarios 5 and 6 are typically ignored because the benefits of the extra features might not justify the deployment costs. This chapter describes a UMTS and WLAN interworking solution called WLAN-based GPRS Support Node (WGSN). The features of WGSN are described in Section 14.1. The design and implementation of WGSN components are elaborated in Section 14.2. The WGSN authentication and network access procedures are given in Section 14.3. The WGSN push mechanism is analyzed in Section 14.4.

14.1 The WGSN Approach

WLAN-based GPRS Support Node (WGSN) interworks UMTS with WLAN to support Scenario 3 features. This section describes the architecture and the features of WGSN.

14.1.1 WGSN Network Architecture

Figure 14.1 illustrates the inter-connection between the UMTS and a WLAN network through WGSN. The UMTS network (Figure 14.1 (1)) provides mobile PS services. The WLAN network (Figure 14.1 (2)) provides access to the Internet. The customers are allowed to roam between the two networks as long as the Mobile Station (MS) is equipped with both a cellular module (for example, GPRS module) and a WLAN card. The WLAN radio network includes 802.11-based *Access Points (APs)* that provide radio access for the MSs. The WGSN acts as a gateway between the PDN and the WLAN node, which obtains the IP address for an MS from a *Dynamic Host Configuration Protocol (DHCP)* [IET99a] server and routes the packets between the MS and the external *Packet Data Network (PDN)*. The WGSN node communicates with the HLR to support GPRS/UMTS mobility management following 3GPP TS 23.060 [3GP05q] as described in Chapter 2. Therefore, the WLAN authentication and network access procedures are exactly the same as that for GPRS/UMTS.

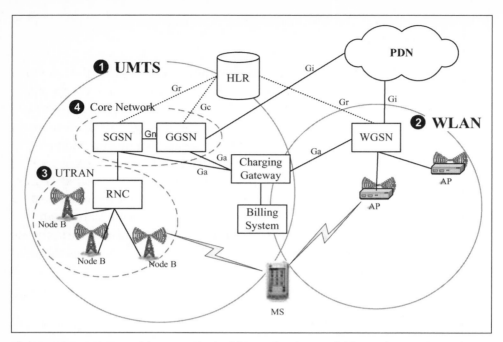

Figure 14.1 WGSN Architecture (dashed lines: signaling; solid lines: data and signaling)

The WGSN node integrates both SGSN and GGSN functionalities. Like an SGSN, the WGSN communicates with the HLR through the *Gr* interface. Conversely, like a GGSN, the WGSN communicates with the external PDN via the Gi interface. Therefore, for other GPRS/UMTS networks, the WGSN node and the corresponding WLAN network are considered as a separate GPRS network. The WGSN node can be plugged in any UMTS core network without modifying the existing UMTS nodes (such as SGSNs and GGSNs). To integrate the billing system for both UMTS and WLAN, WGSN communicates with the charging gateway using the same UMTS protocols; that is, the GPRS Tunneling Protocol (GTP') implemented in the Ga interface (see Chapter 7) or by *File Transfer Protocol* (*FTP*). To access the WGSN services, the MS must be either a UMTS-WLAN dual-mode handset or a laptop/Personal Data Assistant (PDA) equipped with both a WLAN *Network Interface Card* (*NIC*) and a GPRS/UMTS module.

14.1.2 WGSN Features

Based on the seven interworking aspects listed in 3GPP TS 22.934 [3GP03a], we describe the features implemented in WGSN [Fen04]:

Service aspects: WGSN provides general Internet access and *Voice over IP* (*VoIP*) services based on *Session Initiation Protocol* (*SIP*; see Chapter 12). Since a *Network Address Translator* (*NAT*) is built into the WGSN node, the VoIP voice packets delivered by the *Real-Time Transport Protocol* (*RTP*) connection [Sch96b] cannot pass through the WGSN node. This issue is resolved by implementing a *SIP Application Level Gateway* (*ALG*) [3CO00] in the WGSN node, which interprets SIP messages and modifies the source IP address contained in these SIP messages.

In UMTS, an MS must activate the *Packet Data Protocol* (*PDP*) context for VoIP service before a caller from the external PDN can initiate a phone call to this MS. Also, for both UMTS and WLAN, a SIP *User Agent* (*UA*) must be activated in an MS before it can receive any incoming VoIP call. Therefore, a *SIP-based Push Center* (*SPC*) is implemented in the WGSN node to provide MS terminated SIP services. The SPC is implemented on iSMS, the SMS-based IP service platform described in Chapter 1. The SPC also provides a push mechanism through WLAN for a WGSN user who does not bring up the SIP UA. Therefore, the SIP terminated services (for example, incoming VoIP calls) can be supported in WGSN.

Access control aspects: WGSN follows standard UMTS access control for users to access WLAN services, where the existing UMTS Subscriber Identity Module (SIM) card and the subscriber data (user profile) records in the HLR are utilized. Therefore, the WGSN customers do not need a separate WLAN access procedure, and maintenance of customer information is simplified. User profiles for both UMTS and WLAN are combined in the same database (that is, the HLR).

Security aspects: WGSN utilizes the existing UMTS authentication mechanism (see Section 9.1). That is, the WLAN authentication is performed through the interaction between an MS (using a UMTS SIM card) and the Authentication Center (AuC). Therefore, WGSN is as secure as existing cellular networks. We do not attempt to address the WLAN encryption issue. It is well known that WLAN based on IEEE 802.11b is not secure. For a determined attack, *Wired Equivalent Privacy* (*WEP*) is not safe, which only makes a WLAN network more difficult for an attacker to intrude. The IEEE 802.11 Task Group I has been investigating the current 802.11 *Media Access Control Address* (*MAC*) security. WGSN follows the resulting solution.

Roaming aspects: WGSN provides roaming between UMTS and WLAN. We utilize the standard UMTS mobility management mechanism described in Chapter 2 without introducing any new roaming procedures.

Terminal aspects: A terminal for accessing WGSN is installed with a *Universal IC Card (UICC)* reader (a smart card reader implemented as a standard device on the Microsoft Windows platform). The UICC reader interacts with the UMTS SIM card (i.e., the UICC containing the SIM application) to obtain authentication information for the WGSN attach procedure.

Naming and addressing aspects: WGSN user identification is based on the *Network Access Identifier (NAI)* format [Abo99], following the 3GPP recommendation. Specifically, the International Mobile Subscriber Identity (IMSI) is used as WGSN user identification.

Charging and billing aspects: The WGSN acts as a router, which can monitor and control all traffics for the MSs. The WGSN node provides both offline charging and online charging (for pre-paid services) based on the *Call Detail Records (CDRs)* delivered to the charging gateway.

Besides the six aspects listed above, WGSN also provides automatic WLAN network configuration recovery. A WGSN MS can be a notebook, which is used at home or at the office with different network configurations. The network configuration information includes IP address, subnet mask, default gateway, WLAN *Service Set Identifier (SSID)*, et cetera. When the MS enters the WGSN service area, its network configuration is automatically reset to the WGSN WLAN configuration if the MS is successfully authenticated. The original network configuration is automatically recovered when the MS detaches from the WGSN. This WGSN functionality is especially useful for those users who are unfamiliar with network configuration setup.

14.2   Implementation of WGSN

This section describes the implementation of WGSN. We first introduce the protocol stack among MS, AP, WGSN, and HLR. Then we elaborate on the WGSN components for the WGSN network node and the MS. Figure 14.2 illustrates the WGSN protocol stack. In this figure, the lower-layer protocol between the MS and the WGSN node is IP over 802.11 radio (through WLAN AP). In the control plane, standard GPRS Mobility Management (GMM) defined in 3GPP TS 23.060 [3GP05q] (see Chapter 2) is implemented on top

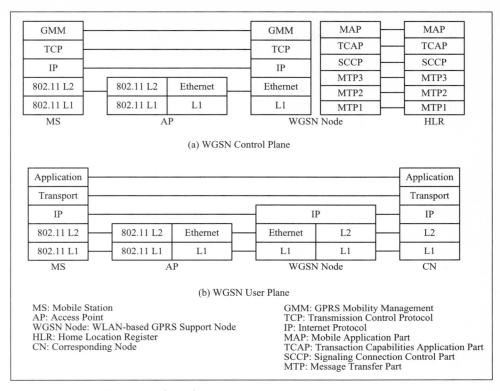

Figure 14.2 WGSN Protocol Stack

of TCP/IP between the MS and the WGSN node. The standard UMTS Gr interface is implemented between the WGSN node and the HLR through the Signaling System Number 7 (SS7)-based Mobile Application Part (MAP) protocol. The layers of the SS7 protocol include Message Transfer Part (MTP), Signaling Connection Control Part (SCCP), and Transaction Capabilities Application Part (TCAP), described in Chapter 8.

The WGSN node communicates with the charging gateway through the IP-based GTP protocol described in Chapter 7. In Section 14.5, the TCP/IP layers in the control plan will be replaced by *Extensible Authentication Protocol / EAP over LAN* (*EAP/EAPOL*) [IET03, IEE01]. EAP/EAPOL operates over the 802.11 MAC layer, which allows authentication of an MS before it is assigned an IP address. Therefore, the IP resource of the WGSN system can be managed with better security. Also, between the WGSN node and the HLR, the lower-layer SS7 protocols (i.e., MTP and SCCP) can be replaced by the IP-based *Stream Control Transmission Protocol* (*SCTP*) to support all-IP architecture. See Chapter 8 for the SCTP description.

The WGSN user plane follows the standard IP approach. That is, the MS and the WGSN node interact through the Internet protocol. The MS communicates with a host in the external PDN using the transport layer over IP. In the user plane, the WGSN node serves as a gateway between the WLAN network and the external PDN. The WGSN MS must be either a UMTS-WLAN dual-mode handset or a laptop/PDA equipped with both a WLAN NIC and a UMTS/GPRS module. The UICC reader (which can be contained in the UMTS/GPRS module or a separate smart card reader) communicates with the standard SIM card to obtain the authentication information required in both the cellular network and the WLAN. The WGSN UICC reader is implemented as a standard device on the Microsoft Windows platform. The WGSN software modules are implemented on the Windows 2000 and XP OS platforms for notebooks, and on WinCE for PDAs. A WGSN client is implemented to carry out tasks in the control plane. Several SIP user agents are implemented for SIP-based applications in the user plane. The modules for the WGSN client are described as follows:

SIM Module: (Figure 14.3 (1)) As in UMTS, a WGSN user is authenticated using the UMTS/GPRS SIM card before the user can access the WLAN network. Through the UICC smart card reader, the SIM module retrieves the SIM information (including IMSI, SRES and Kc; see Section 9.1) and forwards the information to the GMM module.

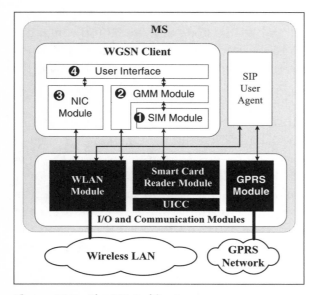

Figure 14.3   The MS Architecture

GMM Module: (Figure 14.3 (2)) Based on the SIM information obtained from the SIM module, the GMM module communicates with the WGSN node to perform MS attach and detach operations. The authentication action is included in the attach procedure.

NIC Module: (Figure 14.3 (3)) The network configurations of different WLANs may vary. The NIC module dynamically sets up appropriate network configurations when a WGSN user moves across different WLAN networks. WGSN utilizes DHCP for IP address management. The WGSN MS must obtain a legal IP address and the corresponding network configuration through the DHCP lease request. Conversely, when the MS terminates a WGSN connection, it should send the IP release message to the WGSN node, and the IP address is reclaimed for the next WGSN user. The NIC module then recovers the original network configuration for the MS. If the MS is abnormally terminated, the NIC module cannot immediately recover the network configuration. Instead, the NIC module offers a Window OS program called the *WGSN Service*. When the MS is restarted, this service will check if the network configuration has been recovered. If not, the configuration previously recorded by the NIC module is used.

User Interface: (Figure 14.3 (4)) A user interacts with the WGSN system through the MS user interface. As illustrated in Figure 14.4, the user

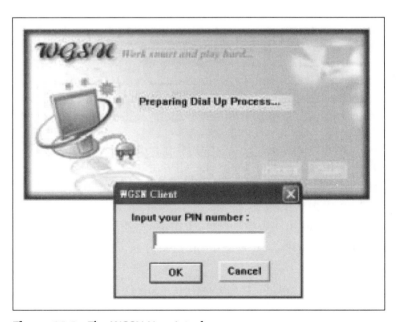

Figure 14.4   The WGSN User Interface

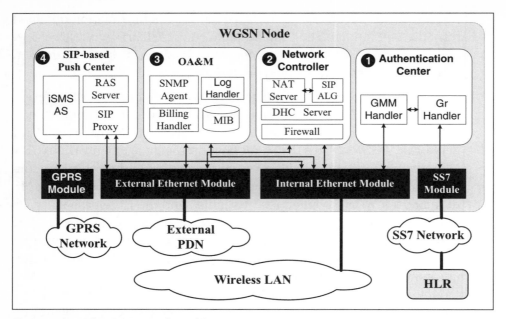

Figure 14.5 The WGSN Node Architecture

enters the GSM /GPRS pin number to initiate the WGSN connection. Like the usage of the user interface of a GPRS handset, the pin number can be disabled. Based on the received command, the corresponding modules are instructed to carry out the desired tasks. During a WGSN session, the user interface indicates the status of the execution and displays the elapsed time of the WGSN connection.

On the network side, the WGSN node is implemented on the Advantech Industrial Computer platform S-ISXTV-141-W3. The black boxes in Figure 14.5 illustrate the WGSN communication modules, which include the following:

- A SS7 module for communications with the HLR (through the SS7 network). In this module, the MTP, SCCP, and TCAP layers (see Figure 14.2 (a)) are based on Connect7 2.4.0-Beta version software developed by SS8 Networks Cooperation

- An internal Ethernet module for communications with the WLAN APs

- An external Ethernet module for communications with the external PDN

- A GPRS module for communications with the MS (through the GPRS network)

The software architecture of the WGSN node includes four major components:

Authentication Center (Figure 14.5 (1)) consists of the GMM and the Gr Handlers. Through the internal Ethernet module, the GMM Handler receives the GMM messages from the WGSN MS, and dispatches the corresponding tasks to the other WGSN modules (the details are described in Steps 4 and 8 of the attach procedure of Figure 14.6 in Section 14.3). The Gr Handler implements the standard GMM primitives for the Gr interface [3GP05q]. Through the SS7 module, the Gr Handler interacts with the HLR for MS network access and authentication. Specifically, it obtains an array of authentication vectors, including a random number (RAND), a signed result (SRES), and an encryption key (Kc) (see Section 9.1) from the GPRS authentication center (which may or may not be co-located with the HLR). Each time the WGSN MS requests authentication, the Gr Handler uses an authentication vector to carry out the task as specified in 3GPP TS 33.102 [3GP04g]. Furthermore, when an MS detaches, the Gr Handler should inform the HLR to update the MS status. The WGSN MAP service primitives (such as **MAP Send Authentication Info** and **MAP Purge MS**) are implemented on MAP version 1.4 software developed by Trillium Digital Systems, Inc.

Network Controller (Figure 14.5 (2)) provides the following functions for Internet access:

- IP address management: A DHCP server is implemented in the WGSN node to distribute private IP addresses to the MSs. A NAT server performs address translation when the IP packets are delivered between the private (WLAN) and public (external PDN) IP address spaces.

- Internet access control: The WGSN node only allows authenticated users to access Internet services. Unauthorized packets will be filtered out by the firewall.

- SIP application support: To support SIP-based applications under the NAT environment, the WGSN node implements a SIP ALG that modifies the formats of SIP packets so that these packets can be delivered to the WGSN MSs through the WGSN node.

Operation, Administration, and Maintenance (OA&M; see Figure 14.5 (3)) controls and monitors individual WGSN user traffic. WGSN utilizes *Simple Network Monitoring Protocol* (*SNMP*) as the network management protocol. With *Management Information Base* (*MIB*), every managed network element is represented by an object with an identity and several attributes. An SNMP agent is implemented in the WGSN node, which interacts with the managed network element through SNMP. For example, the traffic statistics of an AP can be accessed by the OA&M (through the corresponding MIB object) and displayed in a web page using *Multi Router Traffic Grapher (MRTG)* [MRT]. The SNMP agent can also detach an MS through the MIB object of the MS. The Log Handler is implemented in the OA&M to record all events occurring in the WGSN node. The Billing Handler generates CDRs. This handler communicates with the billing gateway through the GTP' protocol or FTP.

SIP-based Push Center (SPC; see Figure 14.5 (4)) provides a push mechanism for GPRS networks that support private IP addresses. Since GPRS significantly consumes the MS power, a mobile user typically turns on GSM but turns off GPRS unless he/she wants to originate a GPRS session. In this case, services cannot be pushed to the users from the network side. For an MS that is CS attached but PS detached, the SPC can push a SIP request to the MS through a SMS application server called the iSMS Application Server (see Chapter 1). The SPC also provides a push mechanism through the WLAN for a WGSN user who does not bring up the SIP UA. Therefore, the SIP terminated services (e.g., incoming VoIP calls) can be supported in WGSN. The SPC is basically the same as the SPA described in Section 12.2. We will elaborate on the SPC in Section 14.4.

14.3 Attach and Detach

In the WGSN attach and detach procedures, the message flows between the WGSN node and the HLR are the same as that between the SGSN and the HLR in UMTS. The message flows between the MS and the WGSN node are specific to the WGSN network, which are not found in UMTS. The attach procedure is illustrated in Figure 14.6, and consists of the following steps:

Step 1. When the WGSN user brings up the MS user interface, the SIM module is invoked to configure the smart card reader and (optionally) request the user to input the *Personal Identification Number (PIN)* number. The card reader authenticates the user through the pin number, just like a GPRS mobile phone.

Step 2. The MS NIC module is invoked to store the current WLAN network configuration. To obtain the network configuration of WGSN, the

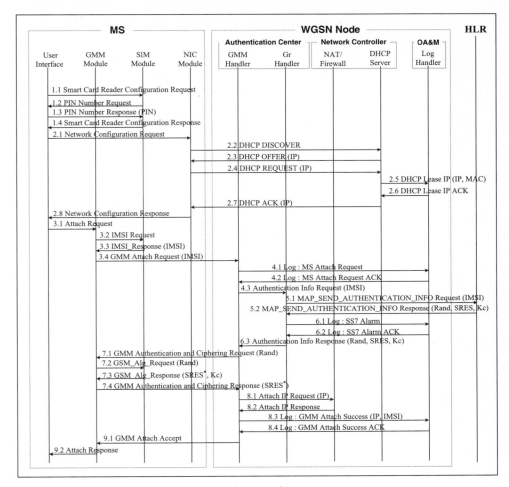

Figure 14.6 Message Flow for the Attach Procedure

MS broadcasts the **DHCP Discover** message on its subnet to look for a DHCP server. The DHCP server in the WGSN node replies to the MS with the **DHCP Offer** message, which includes an available IP address. Then the MS sends the **DHCP Request** message to the DHCP server and asks for the usage of an available IP address contained in the **DHCP Offer** message. If the DHCP server accepts the request, it reports the IP lease event to the Log Handler and sends the MS the **DHCP Ack** message with network configuration parameters. Finally, the MS NIC module sets up the new network configuration.

Step 3. The MS GMM module is invoked to perform the attach operation. The GMM module first obtains the IMSI from the SIM module. Then it sends the **GMM Attach Request** message (with the parameter IMSI) to the WGSN node.

Step 4. When the GMM Handler of the WGSN node receives the attach request, it reports this event to the Log Handler, and sends the authentication information request to the Gr Handler.

Step 5. The Gr Handler sends the **Send Authentication Info Request** message (with the argument IMSI) to the HLR. The HLR returns the authentication vector (RAND, SRES, Kc) through the **Send Authentication Info Response** message.

Step 6. The WGSN Gr Handler issues the **SS7 Alarm** message to the Log Handler, and the event is logged. The Gr Handler returns the authentication vector to the GMM Handler.

Step 7. The GMM Handler sends the **GMM Authentication and Ciphering Request** message (with the parameters IMSI and RAND) to the GMM module of the MS. The GMM module passes the random number (RAND) to the SIM module, and the SIM module computes the signed result (SRES*) and the encryption key (Kc) based on the received RAND and the authentication key (Ki) stored in the SIM card. These results are returned to the GMM module. The GMM module returns the computed SRES* to the GMM Handler of the WGSN node using the **GMM Authentication and Ciphering Response** message (with the parameters IMSI and SRES*). The GMM Handler compares SRES with SRES*. If they match, the authentication is successful.

Step 8. The GMM Handler sends the **Attach IP** message to the firewall, which will allow the packets of this IP address to pass the WGSN node. Then the GMM Handler reports to the Log Handler that the attach is successful (with the corresponding IMSI and IP address).

Step 9. The GMM Handler sends the **GMM Attach Accept** message to the GMM module of the MS, and the GMM module passes the **Attach Response** message to the user interface. At this point, the attach procedure is completed.

The WGSN connection can be detached by the MS or by the network (the WGSN OA&M). The message flow for MS-initiated detach is illustrated in Figure 14.7, and the steps are described as follows:

Step 1. When the user presses the detach button in the user interface, the GMM module is invoked to send the **GMM Mobile Originated Detach**

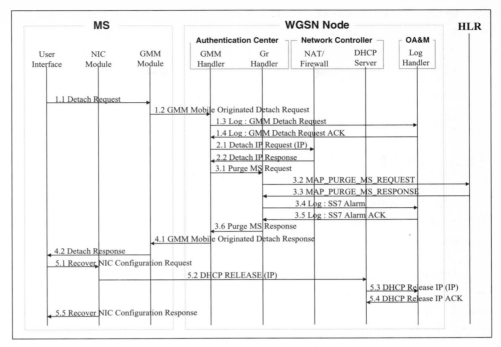

Figure 14.7 Message Flow for the MS-Initiated Detach Procedure

Request message (with parameters IMSI and IP address) to the GMM Handler of the WGSN node. The GMM Handler reports this detach event to the Log Handler.

Step 2. The GMM Handler sends the detach IP request to the firewall. From now on, the packets of this IP address will be filtered out by the firewall of the WGSN node.

Step 3. The GMM Handler invokes the Gr Handler to send the **MAP Purge MS Request** message (with the parameters IMSI and the SSN address of the WGSN node) to the HLR. The HLR updates the MS status in the database and sends the **MAP Purge MS Response** message to the Gr Handler. The Gr Handler reports this event to the Log Handler.

Step 4. Through the **Mobile Originated Detach Response** message, the GMM Handler informs the MS GMM module that the detach operation is complete.

Step 5. The MS NIC module is instructed to recover the original network configuration. It sends the **DHCP Release** message to the DHCP server in the WGSN node. The DHCP server reclaims the IP address and reports this event to the Log Handler. The NIC module then recovers the original network configuration (which was saved in Step 2 of the attach procedure in Figure 14.6).

The message flow for the network-initiated detach procedure is similar to that illustrated in Figure 14.7, and the details are omitted.

14.4 WGSN Push Mechanism

To reduce the power consumption and computation complexity of a WGSN MS, most WGSN applications are not activated at the MS until the user actually accesses them. This approach does not support "always-on" or MS terminated services such as incoming VoIP calls. To address this issue, WGSN implements a push mechanism called *SIP-based Push Center* (*SPC*) that is similar to the SPA in Section 12.2. Suppose that a SIP VoIP caller in the external PDN issues a call request to a WGSN user. This request is first sent to the WGSN node. The WGSN node checks if the SIP UA of the called MS is activated. If so, the request is directly forwarded to the called MS. If not, the WGSN node sends a message to the MS to activate the corresponding SIP UA. After the SIP UA is brought up, the call request from the caller is then delivered to the SIP UA, following the standard SIP protocol. Note that the VoIP call model is typically handled by the SIP UAs or a call server (or softswitch) that controls the call setup process and indicates whether the called party is busy or idle. The WGSN SPC is only responsible for pushing the SIP requests to the MSs where the SIP UAs are not activated. For every SIP UA (an MS may have several SIP UAs for different applications), a status record is maintained in the SPC. A four-state *Finite State Machine* (*FSM*) is associated with the record. These states are as follows:

State 0: The SPC has not initiated the activation process.

State 1: The SPC has initiated the activation process, and one incoming call is waiting for setup.

State 2: The SPC has initiated the activation process. No incoming call is waiting for setup.

State 3: The SIP UA is active.

The incoming call waiting for setup at **State 1** is referred to as the *outstanding call*. There is at most one outstanding call during the SIP UA activation

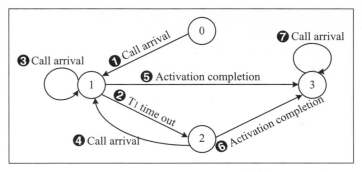

Figure 14.8 The FSM State Transition Diagram for the SPC Status Record

process. The state transition diagram for the FSM of a status record is illustrated in Figure 14.8 and the details are given below:

Transition 1: An incoming call request arrives at **State 0**. The SPC sends a message to activate the SIP UA. The SPC sets a timer $T1$. The FSM moves to **State 1**.

Transition 2: The timer T1 expires at **State 1**. The SPC drops the current incoming call request by sending a timeout message to the caller. The FSM moves to **State 2**.

Transition 3: An incoming call arrives at **State 1**. The SPC drops this call request (because the called MS is already engaged in an outgoing call setup). The FSM moves to **State 1**.

Transition 4: An incoming call request arrives at **State 2**. This call becomes the outstanding call. The SPC sets the timer T1. The FSM moves to **State 1**.

Transition 5: When the SPC receives the activation complete message from the called MS at **State 1**, the SPC forwards the outstanding call request to the SIP UA of the called MS, following the standard SIP protocol. The FSM moves to **State 3**.

Transition 6: When the SPC receives the activation complete message from the called MS at **State 2**, the FSM moves to **State 3**.

Transition 7: When the SPC receives a call request at **State 3**, it directly forwards the call request to the SIP UA of the called MS, following the standard SIP protocol.

It is clear that for every SIP UA, the FSM eventually moves to **State 3**. There is exactly one outstanding call at **State 1**, and there is no outstanding call at

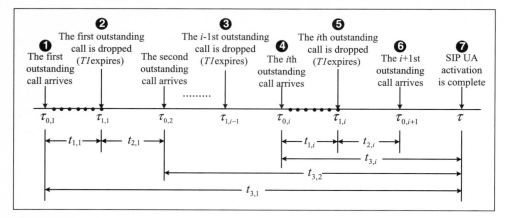

Figure 14.9 The Timing Diagram I (A dot (".") represents dropping of an incoming call immediately after it arrives at the SPC)

State 2. During the state transition, an incoming call is "lost" if either Transition 2 or Transition 3 occurs. Consider the timing diagram in Figure 14.9. In this figure, the bullets in the periods $t_{1,i}$ represents the incoming call losses due to Transition 3. The bullets at $\tau_{1,i}$ represent the outstanding call drops due to Transitions 2. Suppose that the first outstanding call arrives at the SPC at time $\tau_{0,1}$ when the FSM is at **State 0** (Figure 14.9 (1)). At time τ, the SPC received the activation complete message from the called MS (which indicates that the SIP UA has been activated; see Figure 14.9 (7)). The ith outstanding call arrives at the SPC at time $\tau_{0,i}$ (see Figure 14.9 (4)), which means that the previous $i - 1$ outstanding calls were timed out and lost). Based on Figure 14.9, [Fen04] analyzed the restrictions on WGSN transmission delay so that the system can accommodate most SIP-based terminated calls (i.e., the incoming calls) to a WGSN user during the SIP UA activation process (see also question 1 in Section 14.7).

14.5 IEEE 802.1X-based Authentication

The IEEE 802.1X standard specifies authentication and authorization for an IEEE 802 LAN [IEE01]. IEEE 802.1X has been widely adopted for mobile devices to access WLAN. Furthermore, if WLAN is integrated with a cellular network (such as GSM or UMTS), the SIM (in the mobile device) and AuC are utilized together with IEEE 802.1X for authentication. An example of WLAN and cellular integration (in terms of authentication) is illustrated in Figure 14.10. In this figure, the AuC provides security data management for mobile subscribers. The AP provides radio access to mobile devices.

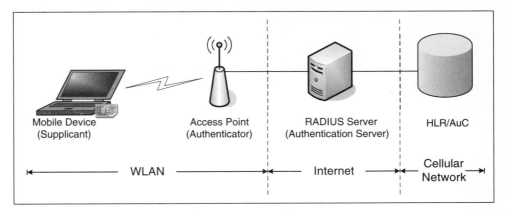

Figure 14.10 A WLAN and Cellular Integration Environment

Before a mobile device is authenticated, the AP only allows this mobile device to send the IEEE 802.1X authentication message. When the *Remote Authentication Dial-In User Service* (*RADIUS*) server receives an authentication request from the mobile device, it retrieves the authentication information of the mobile device from the HLR/AuC. After the HLR/AuC returns the authentication information, the RADIUS server authenticates the mobile device following the standard GSM/UMTS authentication procedure described in Section 9.1.

Figure 14.11 illustrates the IEEE 802.1X protocol stack. In this figure, the mobile device to be authenticated is called a *supplicant*. The server (typically a RADIUS server) performing authentication is called the *authentication server*. The *authenticator* (e.g., a wireless access point) facilitates authentication between the IEEE 802.1X supplicant and the authentication server. IEEE 802.1X utilizes the *Extensible Authentication Protocol* (*EAP*) to support multiple authentication mechanisms based on the challenge-response paradigm [IET04a]. The IEEE 802.1X supplicant encapsulates the EAP packets in *EAP over LAN* (*EAPOL*) frames before they are transmitted to the

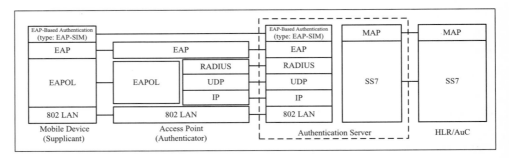

Figure 14.11 IEEE 802.1X Protocol Stack.

authenticator. Upon receipt of an EAPOL frame, the authenticator decapsulates the EAP packet from the EAPOL frame. Then the EAP packet is sent to the authentication server using the RADIUS protocol [IET00]. Implemented on top of UDP, RADIUS provides mechanisms for per-packet authenticity and integrity verification between the authenticator and the authentication server.

IEEE 802.1X authentication for the WLAN and cellular integration network has been investigated in [Ahm03, Sal04]. These studies focused on the design of the network integration architectures, and proposed IEEE 802.1X authentication procedures for the integration network. In Section 14.2, the mobile device of the WGSN must obtain an IP address before it is authenticated by the HLR/AuC. This section describes the IEEE 802.1X authentication that enhances WGSN security by allowing a mobile device to be authenticated before it is assigned an IP address.

In our solution, the WLAN and cellular integration network in Figure 14.10 employs EAP-SIM authentication, which is an EAP-based authentication protocol utilizing the GSM SIM [IET04b]. The authentication server (which is implemented in the WGSN node) communicates with the HLR/AuC to obtain the GSM authentication information through the MAP implemented on top of the SS7 protocol described in Chapter 8. In the EAP-SIM authentication, the MAP is responsible for retrieving the GSM authentication information in the HLR/AuC.

14.5.1 Related Protocols for IEEE 802.1X Authentication

This subsection describes the protocols used in IEEE 802.1X authentication, including Extensible Authentication Protocol (EAP), EAP over LAN (EAPOL), and Remote Authentication Dial-In User Service (RADIUS).

> **Extensible Authentication Protocol (EAP):** The authentication server authenticates the supplicant using the *Extensible Authentication Protocol (EAP)*. Figure 14.12 illustrates the EAP header format.

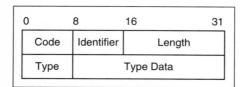

Figure 14.12   EAP Header Format

The `code` field identifies the type of an EAP packet. There are four EAP codes:

- The EAP-Request code is used by the authenticator to challenge the supplicant.

- The supplicant replies to the authenticator's challenge with the EAP-Response code.

- The EAP-Success and the EAP-Failure codes are used by the authenticator to notify the supplicant of the authentication outcome.

The `identifier` field is used for matching a response with the corresponding request.

The `length` field indicates the size of the EAP packet, including the `code`, the `identifier`, the `length`, the `type`, and the `type data` fields.

The `type` field specifies the authentication method. The `type data` field conveys the information used in the authentication process. These two fields appear only in the EAP-Request and EAP-Response packets. Two types, Identity and SIM, are used in our implementation.

An EAP-Request packet with Identity type is sent from the authenticator to the supplicant to request the identity of the supplicant. An EAP-Response packet with the Identity type is sent from the supplicant to the authenticator to provide the identity of the supplicant.

For a SIM-type EAP packet, the `type data` field is filled with the EAP-SIM authentication messages, to be elaborated in Section 14.5.2.

EAP over LAN (EAPOL): In the IEEE 802.1X supplicant, EAP packets are encapsulated in the EAPOL frames. Figure 14.13 illustrates the fields in an EAPOL frame. The `Ethernet type` field is assigned the value `0x888E` to indicate that the Ether frame is an EAPOL frame. The `protocol version` field identifies the supported EAPOL version. For IEEE 802.1X authentication, the `protocol version` field of an

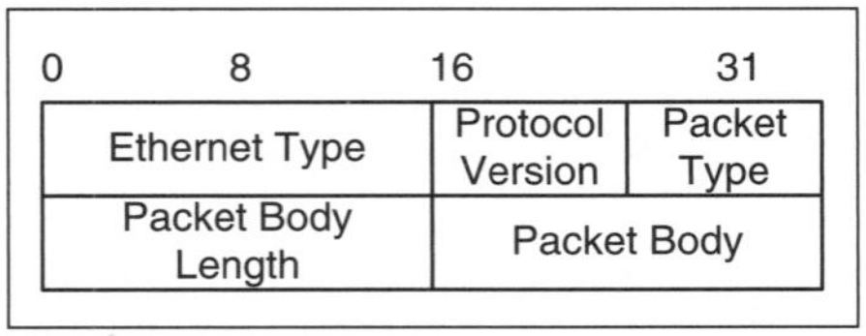

0	8	16	31
Ethernet Type		Protocol Version	Packet Type
Packet Body Length		Packet Body	

Figure 14.13 EAPOL Frame Format

EAPOL frame is set to the value `0x01`. The `packet body length` field indicates the packet body field length in octets.

The `packet type` field indicates the EAPOL type of a packet. Examples of EAPOL types are

- EAPOL-Start
- EAPOL-Packet and
- EAPOL-Logoff

EAPOL-Start indicates the initiation of IEEE 802.1X authentication. The `packet body` field of this frame is empty. EAPOL-Packet encapsulates an EAP packet for IEEE 802.1X authentication. EAPOL-Logoff indicates that the supplicant has logged off and the authenticated association between the supplicant and the authenticator is terminated. Like EAPOL-Start, the `packet body` field of EAPOL-Logoff is empty.

Remote Authentication Dial-In User Service (RADIUS): Based on the EAP packets received from the supplicant, the authenticator communicates with the authentication server via the RADIUS protocol. The authenticator acts as a RADIUS client and the authentication server acts as a RADIUS server. Figure 14.14 illustrates the RADIUS packet format.

The `code` field identifies the type of a RADIUS packet. The RADIUS codes include:

- Access-Request
- Access-Accept

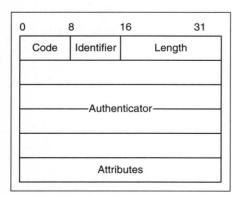

Figure 14.14 RADIUS Packet Format

- Access-Reject and
- Access-Challenge

The Access-Request packet is sent from a RADIUS client to the RADIUS server. This code determines whether a user is allowed to access the requested services.

If the Access-Request packet is accepted, then the RADIUS server sends a packet with the code Access-Accept. Otherwise, the Access-Reject packet is returned. If the RADIUS server attempts to send the supplicant a challenge requiring a response, then it sends Access-Challenge.

The `identifier` field is used to match a reply with the corresponding request.

The `length` field indicates the length of the RADIUS packet.

The `authenticator` field is 16 octets in length. The value of the authenticator field is "request authenticator" in an Access-Request packet and "response authenticator" in Access-Accept, Access-Reject, or Access-Challenge packets. The value "request authenticator" is a random number encapsulated in the Access-Request packet. The value "response authenticator" is a one-way MD5 hash which takes the Access-Request packet as input.

The `variable-length attribute` field contains the list of the attributes for the requested service. In IEEE 802.1X authentication, the RADIUS Access-Request packet encapsulates an EAP-Response message in the EAP-Message attribute, and the RADIUS Access-Challenge packet encapsulates an EAP-Request message in the EAP-Message attribute. An Access-Accept packet encapsulates an EAP-Success message, and an Access-Reject packet encapsulates an EAP-Failure message.

14.5.2 SIM-based IEEE 802.1X Authentication

This section describes the SIM-based IEEE 802.1X authentication procedure. The authentication message flow is illustrated in Figure 14.15, which consists of the following steps:

Step 1. The mobile device (the supplicant) sends the EAPOL-Start packet to the AP to initiate the IEEE 802.1X authentication.

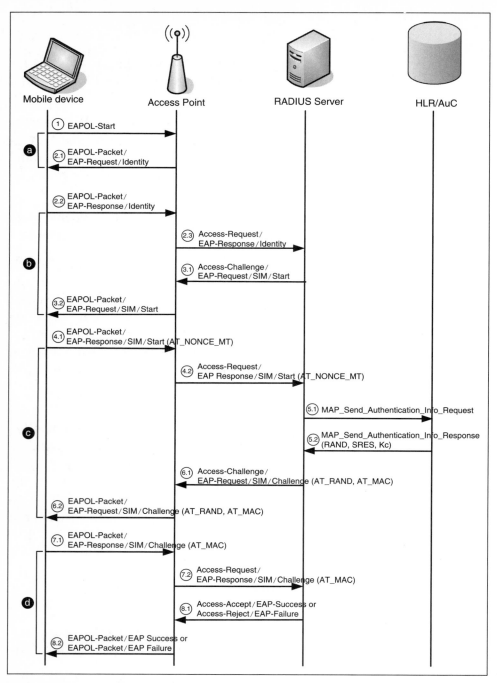

Figure 14.15 IEEE 802.1X Authentication Message Flow

Step 2. The AP requests the identity of the mobile device through the EAP-Request message with type Identity. The identity data is an NAI user@realm, where the user portion identifies the user and the realm portion is used for message routing. In EAP-SIM, the IMSI with the "1" prefix is utilized in the user portion (e.g., user = 1466012400260692 where the last 15 digits represent the IMSI of the mobile user). The domain name of the RADIUS server is utilized in the realm portion (e.g., wgsn.com).

When the AP receives the EAP-Response/Identity message from the mobile device, it encapsulates this message in the Access-Request packet. Then the packet is sent to the RADIUS server.

Step 3. Upon receipt of the Access-Request packet, the RADIUS server conducts the EAP-SIM authentication with the mobile device (the supplicant). Specifically, the RADIUS server generates an Access-Challenge packet that encapsulates EAP-Request/SIM/Start in the EAP-Message attribute and sends it to the AP. EAP-Request/SIM/Start is an EAP-Request message whose type is SIM and whose type data is an EAP-SIM message Start. This message requests the mobile device to initiate the EAP-SIM authentication. The AP decapsulates the EAP-Request from the Access-Challenge packet, and delivers it to the mobile device by an EAPOL packet.

Step 4. The mobile device responds to the RADIUS server with the EAP-Response/SIM/Start message containing a random nonce AT_NONCE_MT. The random nonce will be used to generate the encryption key for data transmission between the mobile device and the RADIUS server after the IEEE 802.1X authentication.

Step 5. To obtain the authentication information of the mobile device, the RADIUS server sends the **MAP Send Authentication Info Request** message (with the argument IMSI) to the HLR/AuC. The HLR returns the authentication vector (RAND, SRES, Kc) through the **MAP Send Authentication Info Response** message, where RAND is a random number generated by the HLR/AuC. By exercising the GSM authentication algorithm $A3$, both the SIM module and the HLR/AuC use the RAND and the secret key, Ki, to produce a signed result, SRES (see Step 7). The RADIUS server will authenticate the mobile device by comparing the signed result, SRES*, generated by the SIM module with the SRES generated by the HLR/AuC (see Step 8). If they are equal, the mobile device is successfully authenticated and an encryption key, Kc*, is produced by the GSM encryption/decryption key generation algorithm *A8* (using Ki and RAND as inputs).

Step 6. The RADIUS server sends the EAP-Request/SIM/Challenge (with the parameter RAND, which is encapsulated as AT_RAND) to the mobile device. To ensure the integrity of the challenge message, the message contains a *Message Authentication Code (MAC)*, AT_MAC. Detailed usage of MAC can be found in [IET04b], and will not be elaborated here.

Step 7. After verifying AT_MAC has been received from the RADIUS server, the mobile device passes the random number, RAND, to the SIM module to perform GSM authentication. The SIM module computes its signed result, SRES*, and the encryption key, Kc*, based on the received RAND and the authentication key, Ki, stored in the SIM card. Then the mobile device sends SRES* (encapsulated in AT_MAC) to the RADIUS server through the EAP-Response/SIM/Challenge message.

Step 8. The RADIUS verifies AT_MAC and compares the SRES* calculated by the mobile device with the SRES received from the HLR/AuC. If the values are identical, the RADIUS server notifies the AP that authentication is successful through the EAP-Success message (encapsulated in the RADIUS Access-Accept packet). The AP passes the EAP-Success message to the mobile device. At this point, the mobile device is allowed to access the network through the AP. If the signed results are not the same, the RADIUS server notifies the AP that the authentication failed through the EAP-Failure message (encapsulated in the RADIUS Access-Reject packet).

14.5.3 EAPOL Timers

In the IEEE 802.1X supplicant (mobile device), three EAPOL timers are defined: *startWhen* (associated with message pair (a) in Figure 14.15), *authWhile* (associated with message pairs (b), (c), and (d) in Figure 14.15), and *heldWhile* (associated with Step 8.2 in Figure 14.15 if the client fails the authentication):

- **startWhen:** When the IEEE 802.1X supplicant initiates the authentication, it sends EAPOL-Start to the authenticator and starts the startWhen timer. If the supplicant has not received any response from the authenticator after this timer expires, it resends EAPOL-Start. The supplicant gives up when it sends EAPOL-Start for n_1 times. In the IEEE 802.1X specification [IEE01], the default n_1 value is 3. The default value of the startWhen timer is 30 seconds.

- **authWhile:** Every time the supplicant sends an authentication message (Steps 2.2, 4.1, and 7.1 in Figure 14.15), it starts the authWhile timer. If the supplicant does not receive any response from the authenticator after this timer has expired, the supplicant sends an EAPOL-Start message to restart the authentication procedure. The supplicant gives up after it has consecutively sent EAPOL-Start for n_2 times. The default n_2 value is 3. The default value of the authWhile timer is 30 seconds.

- **heldWhile:** If the IEEE 802.1X authentication fails, the supplicant has to wait for a period heldWhile before it restarts the authentication procedure. The default value of the heldWhile timer is 60 seconds.

Selection of the EAPOL timer values is not trivial. If the timer value is too large, it will take a long time before the mobile device detects the failure of the network (e.g., RADIUS server failure). If the timer value is too small, the timer may expire before the mobile device receives the response message. In this case, the mobile device needs to restart the authentication process due to false failure detection.

Table 14.2 shows the expected *Round-Trip Times* (*RTTs*) of message exchanges measured from the WGSN implemented at National Chiao Tung University. These measurements do not experience waiting delays due to queuing at the network nodes (i.e., AP, RADIUS server, and HLR/AuC). In our measurement, the mobile device and the AP are located in one subnet. The RADIUS server and the HLR are located in another subnet. The data rate of the fixed network is 100Mbps. It is observed that the RTT of a message exchange between the mobile device and the RADIUS server is much shorter than that of a message exchange between the mobile device and the HLR/AuC. This significant RTT discrepancy is due to the fact that accessing the HLR/AuC is much more time-consuming than accessing the RADIUS server. This phenomenon is especially true when the HLR/AuC is fully loaded by cellular user accesses and when the RADIUS server and the HLR/AuC are located in different cities or countries. To reduce the false failure detection probability without non-necessary timer timeout delay, the values of the startWhen timer and authWhile timers should not be identical for all message exchanges in the IEEE 802.1X authentication. For example, the authWhile timer for (c) in Table 14.2 should be much longer than that for (b) and (d).

Table 14.2 Expected round-trip times for EAP-SIM authentication messages

EVENTS OCCURRING AT THE MOBILE DEVICE	MESSAGE PATH	ASSOCIATED TIMER	RTT (SEC.)
Send the EAPOL-Start message. Start the startWhen timer. Receive the EAPOL-Start message. Stop the startWhen timer. ((a) in Figure 14.15)	mobile device and AP	startWhen	0.005
Send the EAP request identity message. Start the authWhile timer. Receive the EAP-SIM Start message. Stop the authWhile timer. ((b) in Figure 14.15)	mobile device and RADIUS server	authWhile	0.013
Send the EAP-SIM Start message Start the authWhile timer. Receive the EAP-SIM Challenge message. Stop the authWhile timer. ((c) in Figure 14.15)	mobile device and HLR/AuC	authWhile	1.087
Send the EAP-SIM Challenge message. Start the authWhile timer. Receive the EAP Success/ Failure message. Stop the authWhile timer. ((d) in Figure 14.15)	mobile device and RADIUS server	authWhile	0.013

14.6 Concluding Remarks

This chapter described WGSN, a solution for integrating cellular and WLAN networks. We described how the cellular/mobile mechanisms are re-used for WLAN user authentication and network access without introducing new procedures and without modifying the existing cellular network components. We described the WGSN features and showed how they are designed and implemented. Then we focused on the WGSN push mechanism. In [Fen04], we proposed an analytic model to investigate the restrictions on WGSN transmission delay so that the system can accommodate most SIP-based terminated calls (i.e., the incoming calls) to a WGSN user during the SIP UA activation process. The study indicates that, for example, if the expected SIP UA activation delay is 1 second, then it is appropriate to set the $T1$

timeout period as 8 or 16 seconds. We also investigated the expected number of lost calls for SPC. Our study indicates that, for example, if a WGSN user receives an incoming call every 20 minutes, then a small number of lost calls is expected if the elapsed time for SIP UA activation is less than 19 seconds. Finally, we showed how IEEE 802.1X authentication can be integrated in WGSN.

14.7 Questions

1. The expected number $E[N]$ of lost calls during a SIP UA activation process can be derived by defining the following periods:

    (a) The $T1$ timer for the ith outstanding call expires at $\tau_{1,i}$ (Figure 14.9 (2), (3), and (5)). The timeout period is $T_{1,i} = \tau_{1,i} - \tau_{0,1}$, which has the density function $f_{t_{1,i}}(t) = \mu e^{-\mu t}$.

    (b) The incoming call arrivals are a Poisson process with rate λ. Therefore, the inter call arrival times are exponentially distributed with mean $\frac{1}{\lambda}$. The ith outstanding call is timed out at time $\tau_{1,i}$, and the $i + 1$st outstanding call arrives at time $\tau_{0,i+1}$ (Figure 14.9 (6)). From the residual life theorem (see question 2 in Section 9.6) and the memoryless property of the Exponential distribution, the period $t_{2,i} = \tau_{0,i+1} - \tau_{1,i}$ has the Exponential distribution with the density function $f_{t_{2,i}}(t) = \lambda e^{-\lambda t}$.

    (c) The SIP UA activation time is $t_{3,1} = \tau - \tau_{0,1}$, which has the density function $f_{t_{3,1}}(t) = \gamma e^{-\gamma t}$.

    Based on the above assumption, derive $E[N]$. (Hint: See the solution in http://liny.csie.nctu.edu.tw/supplementary.)

2. Describe the six WLAN and cellular integration scenarios. Which scenario is implemented in WGSN?

3. Show how QoS adaption (Scenario 4) can be achieved between UMTS-WLAN system switching. (Hint: See Section 4.5).

4. Describe the service aspect of WGSN. To provide Scenario 4 features, which part of the WGSN should be enhanced?

5. The NIC module of a WGSN MS can recover the network configuration. Describe this feature and explain why it is useful.

6. Describe the major components of the WGSN node. Which component is responsible for authentication if we accommodate the IEEE 802.1X protocols?

7. Draw the MS push message flow for SPC based on the SIP message flow for the SPA described in Section 12.2. Why is the push mechanism required in WGSN?

8. Compare the WGSN attach and detach procedures with the 3GPP versions described in Section 2.5.

9. Draw the message flow for the network-initiated detach procedure for WGSN.

10. Redraw the WGSN protocol stack in Figure 14.2 when IEEE 802.1X protocols are accommodated in WGSN.

UMTS All-IP Network

By providing ubiquitous connectivity for data communications, the Internet has become the most important vehicle for global information delivery. The flat-rate tariff structures and low entry cost characteristics of the Internet environment encourage global usage. Furthermore, introduction of the 3G and the Beyond the 3G (B3G) mobile systems has driven the Internet into new markets to support mobile users. As consumers become increasingly mobile, they will demand wireless access to services available from the Internet. Specifically, the mobility, privacy, and immediacy offered by wireless access introduce new opportunity for Internet business. Therefore, mobile networks are becoming a platform that provides leading-edge Internet services.

To integrate IP and wireless technologies, UMTS all-IP architecture has been proposed by the 3GPP [3GP04i, 3GP05r, Bos01]. This architecture evolved from GSM (Second Generation or 2G mobile network), GPRS, UMTS Release 1999 (UMTS R99), and *UMTS Release 2000 (UMTS R00)*. UMTS Release 2000 has been split up into Release 4 and Release 5. Release 4 introduces a next-generation network architecture for the Circuit Switched (CS) domain. Release 5 introduces the IP Multimedia (IM) Core Network Subsystem (IMS) on top of the Packet Switched (PS) domain. Evolution from UMTS R99 to all-IP network has the following advantages. First, mobile networks will benefit directly not only from all existing Internet applications, but also from the huge momentum behind the Internet in terms of the development and introduction of new services. Second, this evolution allows telecommunications operators to deploy a common backbone (for example, IP) for all types of access, and thus to greatly reduce capital and operating costs. Third, the new generation of applications will be developed in an all-IP environment, which guarantees optimal synergy between the ever-growing mobile world and the Internet.

In a UMTS all-IP network, the SS7 transport will be replaced by IP (see Chapter 8), and the common IP technology supports all services, including multimedia and voice services controlled by the Session Initiation Protocol (SIP; see Chapter 12). In UMTS R99, the PS domain supports data services over an enhanced GPRS network. The CS domain mainly supports voice-based services. Conversely, a UMTS all-IP network supports voice applications through the PS domain using SIP. The CS domain call control mechanism in R99 may be re-used to support CS domain services for a UMTS all-IP network.

15.1 All-IP Architecture

There are two options for a UMTS all-IP network. The *option 1* architecture supports PS-domain multimedia and data services. The *option 2* architecture extends the option 1 network by accommodating CS-domain voice services over a packet-switched core network.

15.1.1 Option 1 for All-IP Architecture

The UMTS all-IP network architecture option 1 consists of the following segments (see Figure 15.1):

Radio Access Network (RAN; see Figure 15.1 (a)) can be UMTS Terrestrial Radio Access Network (*UTRAN*) or *GSM Enhanced Data Rates for Global Evolution* (*EDGE*) *Radio Access Network* (*GERAN*). The UTRAN is basically the same as the R99 version (see Chapter 2).

Home Subscriber Server (HSS; see Figure 15.1 (b)) is the master database containing all 3G user-related subscriber data. The HSS consists of the IM functionality (that is, IM user database), a subset of the Home Location Register (HLR) functionality required by the PS domain

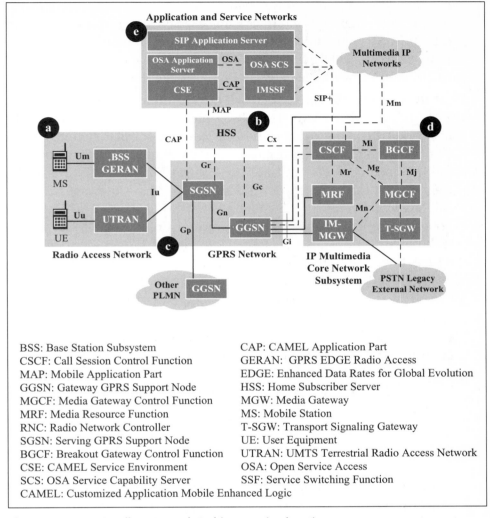

Figure 15.1 UMTS All-IP Network Architecture (option 1)

(that is, 3G GPRS HLR), and a subset of the HLR functionality required by the CS domain (that is, 3G CS HLR) to provide support for call handling entities. We will briefly describe these functionalities in Section 15.2.

GPRS Network (Figure 15.1 (c)) consists of Serving GPRS Support Nodes (SGSNs) and Gateway GPRS Support Nodes (GGSNs) that provide mobility management and session management. The GPRS network should be able to interface with a variety of RANs such as UTRAN and EDGE. The Iu interface between UTRAN and SGSN is IP based.

SGSN and GGSN communicate with HSS through Gr and Gc interfaces, respectively. These two interfaces are based on Mobile Application Part (MAP). SGSN communicates with GGSN through the Gn interface in the same network, and through the *Gp* interface in the different networks. GGSN interacts with the external PDN through the Gi interface.

IP Multimedia (IM) Core Network (CN) Subsystem (IMS; Figure 15.1 (d)) is located between the GGSN and the PDN (specifically, the IP networks). In this subsystem, the Call Session Control Function (CSCF) is a SIP server, which is responsible for call control. Other nodes in the IMS include Breakout Gateway Control Function (BGCF), Media Gateway Control Function (MGCF), IM-Media Gateway Function (IM-MGW), and Transport Signaling Gateway Function (T-SGW). These nodes are typically used in a VoIP network [Col01, Sch99], and will be elaborated in the subsequent sections. Most interfaces among these nodes (that is, Mc, Mg, Mh, Mm, Mr, and Ms) are IP-based gateway control protocols. We will briefly describe these interfaces in Section 15.2.

Application and Service Network (Figure 15.1 (e)) supports flexible services through a service platform. The all-IP network architecture will provide a separation of service control from call/connection control, and the applications are implemented in dedicated application servers that host service-related databases or libraries. As shown in Figure 15.1 (e), the 3GPP defines three possible alternatives to provide flexible and global services:

- Direct SIP+ link between CSCF and SIP Application Server: SIP+ is SIP with extensions for service control, to be defined later. This method will be used by mobile operators to provide new multimedia SIP applications. The SIP application services are either developed by the mobile operators or purchased from trusted third parties.

- SIP+ link between CSCF and *IM-Service Switching Function* (*IM-SSF*) followed by a *Customized Application Mobile Enhanced Logic* (*CAMEL*) *Application Part* (*CAP*) link between IM-SSF and *CAMEL Service Environment* (*CSE*) [3GP05m]: This method will be used by the mobile operator to provide CAMEL services (for example, prepaid service) to the IMS users. Note that similar CS domain services have already been provided via the CAMEL platform.

- SIP+ link between CSCF and Open Service Access (OSA) Service Capability Server (SCS) followed by an OSA link between OSA SCS and OSA Application Server: This method will be used to give third parties controlled access to the operator's network, and enable third parties to run their own applications (in the third-party application servers) using the IM capabilities of the operator's network. See Section 15.4 for the details of OSA.

The designs of Mobile Stations (MSs) or User Equipments (UEs) are strongly influenced by the Internet and content availability. For advanced mobile services and applications, content needs to scale to various display sizes of UEs and requires interoperability to ensure wide end-user acceptance. Thus, a large variety of UEs targeted at different market segments will emerge. Multi-mode and multi-band UEs will be the first step in the transition to 3G and B3G. This migration phase means that initial service quality may not be globally consistent [UMT00a]. In this all-IP network architecture, the GGSN is considered as the border of the network toward the public IP network. The GGSN and MGW together are the network border toward the Public Switched Telephone Network (PSTN) and legacy mobile networks. Note that an MGW is basically a PSTN gateway, described in Section 12.3.

15.1.2 Option 2 for All-IP Architecture

All-IP network option 2 (Figure 15.2) supports R99 CS UEs, which enables the R99 CS and PS domains to evolve independently. Two control elements, the MSC server and the GMSC server, are introduced in option 2. The MSC servers and the HLR functionality in HSS provide an evolution of R99

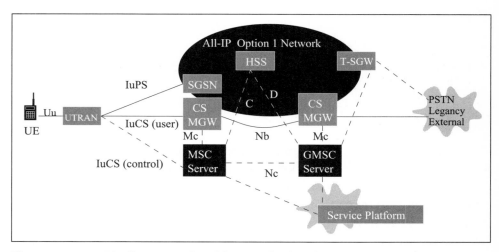

Figure 15.2 UMTS All-IP Network Architecture (option 2)

telephony services. MAP is the signaling interface between HSS and the MSC server (or the Gateway MSC server).

The R99 Iu interface separates transport of user data from control. Evolving from this interface, option 2 UTRAN accesses the core network via a CS-MGW (user plane) separated from the MSC server (control plane). UTRAN communicates with an MSC server using the Radio Access Network Application Part (RANAP) over the Iu interface. The Iu interface between the UTRAN and the CS-MGW is based on the *Iu User Plane* protocol [3GP05l]. Notice that there are one or more CS-MGWs in the option 2 network. If two or more CS-MGWs exist, then they communicate through the *Nb* interface. In our example, there are two CS-MGWs in the all-IP option 2 architecture: One is connected to PSTN and the other is connected to UTRAN via the Iu-CS interface. These two CS-MGWs are responsible for voice format conversion between PS and CS networks.

15.1.3 Partitioning of All-IP Architecture in Horizontal Layers

The all-IP network architecture can also be partitioned horizontally into three layers (see Figure 15.3):

Application and Service Layer consists of service nodes and service platforms such as OSA (same as Figure 15.1 (e)):

Network Control Layer is responsible for control signaling delivery, consists of the MSC server, HSS, CSCF, BGCF, MGCF, SGSN (control plane part), GGSN (control plane part), and T-SGW. In this layer, CSCF, MGCF, the MSC server, and BGCF can serve as call agents.

Connectivity Layer is a pure transport mechanism that is capable of transporting any type of information via voice, data, and multimedia streams. The layer includes MGW, SGSN (user plane part), GGSN (user plane part), and MRF.

In UMTS all-IP option 1, the radio access network (e.g. UTRAN) and the GPRS network together are referred to as the *bearer network*. Through the *Gm* interface (which includes radio, Iu, Gn, and Gi), the bearer network provides transport for signaling (control plane) and data (user plane) exchange between the UE and the CSCF/gateways. Signaling between the bearer network and the PSTN is interworked through CSCF, BGCF, MGCF, and T-SGW. The user plane bearer is connected to the PSTN via an MGW. The bearer

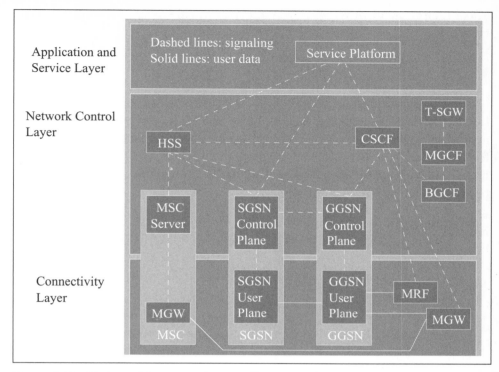

Figure 15.3   Horizontal Structure of UMTS all-IP Network

network nodes (RAN, SGSN, and GGSN) are not aware of the multimedia signaling between the UE and the CSCF. However, the RAN may optimize radio transmission by supporting specific Radio Access Bearers (RABs) for individual flows of the multimedia user plane. These RABs are requested by the UE at Packet Data Protocol (PDP) context activation.

15.2   All-IP Core Network Nodes

This section describes the nodes in the network control layer and the connectivity layer. Among these network nodes, the GGSNs and the SGSNs are basically the same as that in UMTS R99 (see Chapter 2). In all-IP network option 2, an MSC server controls CS services, and an SGSN controls PS bearer services. In the connectivity layer, the MGW uses open interfaces to connect different types of nodes. Like the MSC structure in Figure 15.3, the SGSN control plane handles control layer functions for PS domain communication.

The media gateway provides user plane data transmission in the connectivity layer.

15.2.1 Call Session Control Function

Call Session Control Function (*CSCF*) communicates with HSS for location information exchange, and handles control-layer functions related to application-level registration and SIP-based multimedia sessions. Through the *Mm* interface, the CSCF processes call requests from other VoIP call control servers or terminals in multimedia IP networks. The CSCF consists of the following logical components:

Incoming Call Gateway (ICGW) communicates with HSS to perform routing of incoming calls (see, for example, Step 4 in Section 15.3.4). ICGW may trigger incoming call service (for call screening or call forwarding), and query address handling through the Address Handling component.

Call Control Function (CCF) is responsible for call setup and call-event reports for billing and auditing. It receives and processes application-level registration (see Step 2 in Section 15.3.1), provides a service trigger mechanism (service capabilities features; see Step 4 in Section 15.3.3 or Step 5 in Section 15.3.4) toward application and service networks, and may invoke location-based services related to the serving network. It also checks whether the requested outgoing communication is allowed given the current subscription (see Step 4 in Section 15.3.3). The CCF interacts with the MRF through the *Mr* interface to support multi-party and other services (for example, tones and announcements).

Serving Profile Database (SPD) interacts with HSS in the home network to receive profile information (see Step 7 in Section 15.3.1) for the all-IP network and may store them depending on the *Service Level Agreement* (*SLA*) with the home network. SPD notifies the home network of the initial user's access (includes, for example, the CSCF signaling transport address; see Step 6 in Section 15.3.1).

Address Handling (AH) analyzes and translates (and may modify) addresses (see Step 5 in Section 15.3.3). It supports address portability and alias address mapping (for example, mapping between E.164 number and transport address). It may perform temporary address handling for inter-network routing.

A CSCF can be interrogating, proxy, or serving. The *interrogating CSCF* (I-CSCF) determines how to route mobile terminated calls to the destination

UEs. That is, I-CSCF is the contact point for the home network of the destination UE, which may be used to hide the configuration, capacity, and topology of the home network from the outside world. When a UE attaches to the network and performs PDP context activation, a *proxy CSCF* (P-CSCF) is assigned to the UE. The P-CSCF contains limited CSCF functions (that is, address translation) to forward the request to the I-CSCF at the home network. Authorization for bearer resources in a network is performed by a P-CSCF within that network (see Step 7 in Section 15.3.3). By exercising the application-level registration described in Section 15.3.1, a *serving CSCF* (S-CSCF) is assigned to serve the UE. Through the Gm interface, this S-CSCF supports the signaling interactions with the UE for call setup and supplementary services control (for example, service request and authentication). The S-CSCF provides SPD and AH functionalities to the UE. Details of proxy, interrogating, and serving CSCFs will be elaborated in Section 15.3.

15.2.2 Home Subscriber Server

Home Subscriber Server (*HSS*) keeps a master list of features and services (*user profile* information; for example, user identities, subscribed services, numbering, and addressing information) associated with a user, and maintains the location of the user. The HSS provides the HLR functionality required by the PS and CS domains, and the IM functionality required by the IMS to support the network entities (for example, SGSN, GGSN, MSC server, and CSCF) actually handling calls. In other words, the HSS serves as terminations of the following functionalities:

MAP Termination. The HSS enhances HLR to support the all-IP network. It stores mobility management information, and inter-system location information. Like the R99 HLR, the HSS communicates with the SGSN/GGSN in the PS domain and the GMSC server/MSC server in the CS domain through Gr/Gc and C/D, respectively. Unlike the R99 HLR, both interfaces for HSS are MAP transported over IP.

Addressing Protocol Termination. The HSS provides logical-name to transport-address translation for answering Domain Name Server (DNS) queries (see Step 4 in Section 15.3.4).

Authentication and Authorization Protocol Termination. The HSS may also generate, store, and manage security data and policies used in the IMS. CSCF-UE security parameters are sent from the HSS to the CSCF, which allows the CSCF and the UE to communicate in a trusted and secure way. The HSS stores the all-IP network service profiles and service mobility or S-CSCF-related information for the UEs. The HSS also provides the S-CSCF with the service parameters of UEs (for example, supplementary service parameters, application server address, triggers, and so on; see Step 7 in Section 15.3.1). The HSS communicates with the CSCF using Cx that can be implemented based on the Diameter [3GP05a].

15.2.3   Other Network Nodes

This subsection describes BGCF, MGCF, MSC server, and T-SGW in the network control layer. We also elaborate on MRF and MGW in the connectivity layer.

Breakout Gateway Control Function (BGCF) is responsible for selecting the appropriate PSTN breakout point based on the received SIP request from the S-CSCF. If the BGCF determines that the breakout occurs in the same network, then the BGCF selects an MGCF, which is responsible for interworking with the PSTN. If the breakout is in another network, then depending on the configuration, the BGCF forwards this SIP request to another BGCF or an MGCF in the selected network.

Media Gateway Control Function (MGCF) is the same as the media gateway controller in a Voice over IP (VoIP) network (see Section 10.1), which controls the connection for media channels in an MGW. MGCF communicates with CSCF through SIP over the *Mg* interface. It selects the CSCF depending on the routing number for incoming calls from legacy networks (see Step 3, Section 15.3.4). MGCF should support different call models.

MSC (Gateway MSC) Server supports the Media Gateway Control Protocol (MGCP) or H.248 to handle control layer functions related to CS domain services at the borders between the radio access network and the UMTS all-IP core network, and between the PSTN and the UMTS all-IP core network. It comprises the call control and mobility control parts of UMTS R99 MSC. An MSC server and associated MGW can be implemented as a single node that is equivalent to an R99 MSC. For call setup, the GMSC server communicates with the MSC server through the ISDN User Part (ISUP) protocol (see Chapter 8) in the *Nc* interface.

Transport Signaling Gateway Function (T-SGW) serves as the PSTN signaling termination point and provides the PSTN/legacy mobile network to IP transport level address mapping. T-SGW maps call-related signaling from/to the PSTN on an IP bearer and sends it to/from the MGCF (see Step 11 in Section 15.3.3) or the GMSC server. T-SGW does not interpret the messages in the MAP or the ISUP layer. (Details of the signaling gateway are also given in Section 10.1).

Media Resource Function (MRF) performs multiparty call, multimedia conferencing, tones, and announcements functionalities. Through the *Mr* interface, the MRF communicates with the S-CSCF for service validation of multiparty/multimedia sessions.

Media Gateway Function (MGW) provides user plane data transport in the UMTS core network. The MGW terminates bearer channels from PSTN/legacy mobile networks and media streams from a packet network, for example, Real-Time Transport Protocol (RTP) streams in an IP network (see Chapter 12). The implementation of UMTS MGW should be consistent with existing/ongoing industry protocols/interfaces. Through the *Mc* interface (which is fully compliant with H.248 [ITU00]), the MGW interacts with the MGCF, the MSC server, and the GMSC server (see Figures 15.1 and 15.2) for resource control. Two MGWs can communicate through the *Nb* interface (see Figure 15.2), where the transport for the user plane can be RTP/UDP/IP or AAL2/ATM.

In the option 2 network architecture, the MGW also interfaces UTRAN with the all-IP core network over the Iu interface. The MGW provides flexible connection handling that supports different call models and different media processing through different Iu options for CS services (AAL2/ATM based as well as RTP/UDP/IP based). Media processing includes media conversion, bearer control, and payload processing (for example codec, echo canceller, and conference bridge). The MGW bearer control and payload processing capabilities will also need to support mobile-specific functions such as serving radio network system relocation/handoff and anchoring.

15.3   Registration and Call Control

In an all-IP network, the bearer network mobility management (for both CS and PS domains) follows UMTS R99. Specifically, mobility management procedures in the all-IP CS and PS domains are based on the *Location Area Identifier* (*LAI*) and *Routing Area Identifier* (RAI), respectively. A mechanism to convert the formats of identities among the all-IP network, R99, and 2G is needed. This section describes application-level registration, CS and PS call origination, and PS call termination. For PS call origination and termination, we consider scenarios in which one party is the UE and the other party is in the PSTN. The call agents involved in these scenarios include an S-CSCF (for the UE) and an MGCF (for the call party in the PSTN). SIP is the signaling protocol for multimedia session setup, modification, and tear-down. Although SIP is used with any transport protocol, the media is normally exchanged by using RTP.

15.3.1 Application-Level Registration

In an all-IP network, a UE conducts two types of registration:

- In *bearer-level registration*, the UE registers with the GPRS network following the standard UMTS routing area update or attach procedures (see Chapter 2). After bearer-level registration, the UE can activate PDP contexts in the GPRS network. Bearer-level registration and authentication are required to support GPRS-based services.

- To offer IM services, *application-level registration* must be performed in the IM subsystem after bearer-level registration. In application-level registration, an S-CSCF is assigned to the UE. Specifically, the HSS interacts with the I-CSCF to determine the S-CSCF. This action is referred to as *CSCF selection* [3GP04i].

Before application-level registration is initiated, the UE must have performed bearer-level registration to obtain an IP address and to discover the P-CSCF. The P-CSCF discovery can be achieved, for example, through the Dynamic Host Configuration Protocol (DHCP) server that provides the UE with the domain name of the P-CSCF and the address of a DNS that can resolve the P-CSCF name. The UE stores the P-CSCF address that will be used for mobile-originated signaling (see Step 1 in Section 15.3.3). The application-level registration procedure is described in the following steps (see Figures 15.4 and 15.5):

Step 1. The UE sends the SIP **REGISTER** message to the P-CSCF through UTRAN, SGSN, and GGSN. The request includes the *home domain name* of the UE. In SIP, **REGISTER** is issued by a client (UE in our example) to the server (UMTS network) with an address at which the client can be reached for a SIP session.

Step 2. Based on the home domain name, the P-CSCF performs address translation (through a DNS-based mechanism) to find the I-CSCF

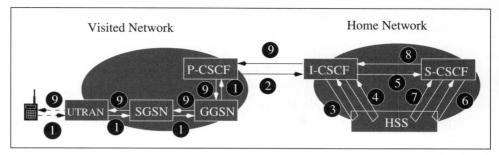

Figure 15.4 Application Level Registration

address. Then it proxies the **REGISTER** message to the I-CSCF at the home network. Note that there may be multiple I-CSCFs within an operator's network.

Step 3. Based on the subscriber identity received from the P-CSCF and the home domain name, the I-CSCF determines the HSS address. Note that if there are two or more HSSs in the home network, the I-CSCF needs to query the *subscription location function* to find the HSS address.

The I-CSCF sends the **Cx-Query** message to the HSS. The HSS checks if the subscriber has been registered. Then it returns the **Cx-Query**

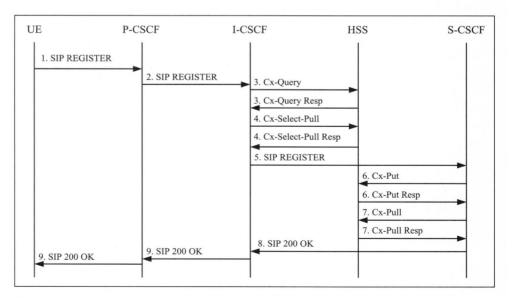

Figure 15.5 Message Flow for Application-Level Registration.

Resp message to the I-CSCF. By the end of this step, the user has been authenticated.

Step 4. The I-CSCF sends the **Cx-Select-Pull** message to the HSS to obtain the required S-CSCF capability information (supported service set and protocol version number). Based on the service network indication and subscriber identity provided by the I-CSCF, the location service of the HSS returns the information of the required S-CSCF capabilities through the **Cx-Select-Pull-Resp** message.

Based on the information provided by the HSS, the I-CSCF selects the name of an appropriate S-CSCF. This S-CSCF must be in the home network.

Step 5. The I-CSCF sends the **REGISTER** request to the S-CSCF. The request includes the HSS name as a parameter.

Step 6. Using **Cx-Put**, the S-CSCF sends its name and subscriber identity to the HSS, which will be used by the HSS to route mobile terminated calls to the S-CSCF. The HSS acknowledges with the **Cx-Put Resp** message.

Step 7. The S-CSCF obtains the subscriber data from the HSS through the **Cx-Pull** and **Cx-Pull-Resp** message exchange. The subscriber data is stored in the S-CSCF, and includes supplementary service parameters, the application server address, triggers, and so on.

Step 8. The S-CSCF determines if the home contact name is the S-CSCF name or the I-CSCF name. If the contact name is for the S-CSCF, then the P-CSCF can access the S-CSCF directly, and the internal configuration of the home network is known to the outside world. If the contact name is for the I-CSCF, then the P-CSCF can only access the S-CSCF indirectly through the I-CSCF. In this case, the home network configuration is hidden.

The S-CSCF sends its address and the home contact name to the I-CSCF through the SIP **200 OK** response message.

Step 9. With the **200 OK** message, the I-CSCF returns the home contact name (either the I-CSCF or the S-CSCF address) to the P-CSCF. The P-CSCF stores the home contact name, and forwards the **200 OK** message to the UE, indicating that registration is successful.

Note that if the registration information expires or the registration status changes, the UE initiates *re-registration*. The re-registration procedure is basically the same as the registration procedure described above except for the following:

- In Step 3, the HSS determines that the user is currently registered, and the S-CSCF name is sent back to the I-CSCF directly. Thus, Step 4 can be omitted.

- Steps 6 and 7 may be skipped if the S-CSCF detects that this procedure is for re-registration.

- In Step 8, the S-CSCF only sends the home contact name to the I-CSCF. The S-CSCF address need not be sent to the I-CSCF.

If the UE roams to a new network or is turned off, then application-level de-registration is performed. The details can be found in [3GP04i].

15.3.2 CS Mobile Call Origination

CS mobile call origination is similar to that in UMTS R99. This procedure does not involve the CSCFs and the HSS. Before CS call origination, the UE has already attached to the UMTS CS domain, and has registered to the VLR of an MSC server. CS call origination is described in the following steps (see Figures 15.6 and 15.7):

Step 1. The call request from the UE is forwarded to the MSC server through the UTRAN. The VLR function of the MSC server performs originating service control through the support of the CAMEL Service Environment (CSE). Assume that the request is accepted.

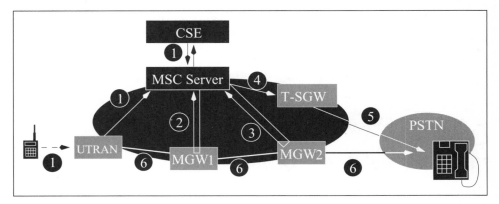

Figure 15.6 CS Call Origination

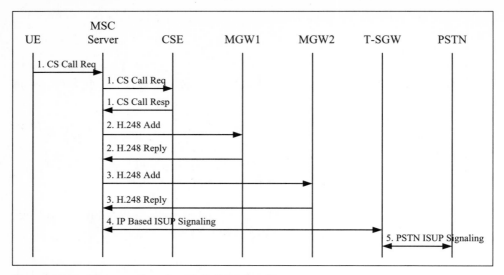

Figure 15.7   Message Flow for CS Call Origination

Step 2. Based on the location of the UE, the MSC server issues the H.248 **Add** message (similar to the MGCP **CreateConnection** message) to select the first media gateway (MGW1) that connects to the UE through the UTRAN. This MGW is responsible for QoS provisioning and conversion between the ATM and IP protocols. MGW1 returns the H.248 **Reply** message .

Step 3. Assume that the called party is in the PSTN. The MSC server selects the second media gateway (MGW2) that serves as the termination to the PSTN. Note that this second media gateway can be MGW1 itself. Signaling based on protocols such as H.248 is performed to reserve the MGW2 resources; that is, the ports to MGW1 and the ports to the PSTN.

Steps 4 and 5. The MSC server and the PSTN exchange the ISUP call setup messages through the T-SGW.

When the call setup procedure is complete, the voice path is UE ↔ UTRAN ↔ MGW1 ↔ MGW2 ↔ PSTN (see path (6) in Figure 15.6).

15.3.3   PS Mobile Call Origination

Before PS call origination, the UE has already attached to the UMTS PS domain, and application-level registration is performed so that an S-CSCF is assigned to the UE (i.e., the user profile has been fetched from the HSS and

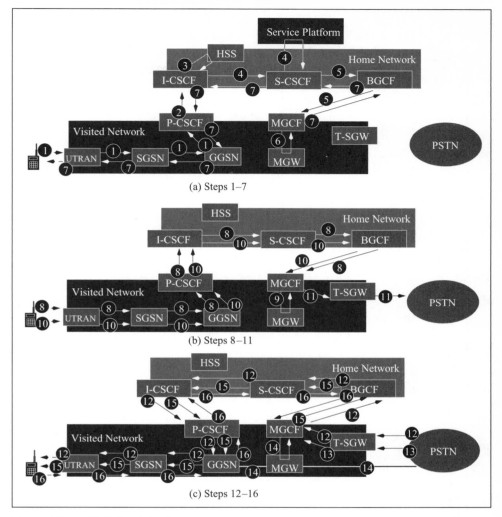

Figure 15.8 PS Call Origination

stored in the S-CSCF). Assume that the UE is in a visited network. The PS mobile call origination to the PSTN is described as follows (see Figures 15.8 and 15.9):

Step 1. The UE sends a SIP **INVITE** request to the P-CSCF. The P-CSCF and the UE must be located in the same network. The **INVITE** message is used to initiate a SIP media session with an initial Session Description Protocol (SDP). The SDP provides session information (for example, RTP payload type, addresses, and ports) to potential session participants.

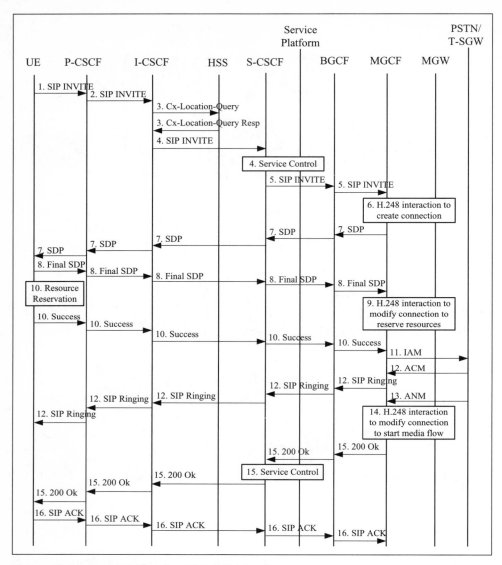

Figure 15.9 Message Flow for PS Call Origination

Step 2. The P-CSCF resolves the UE's home network address (suppose that it is the I-CSCF name stored in Step 9, Section 15.3.1), and forwards the **INVITE** message to the I-CSCF.

Step 3. The I-CSCF provides ICGW and AH functionalities, and interrogates the location service of the HSS through the **Cx-Location-Query**

and **Resp** message exchange to obtain S-CSCF signaling transport parameters.

Step 4. The I-CSCF relays the **INVITE** message to the S-CSCF through the *Mw interface*. The S-CSCF will act as the host to the call control logic. It validates the service profile of the subscriber, and may contact the service platform to perform origination service control.

Step 5. The S-CSCF translates the destination address and determines that the call will break out to the PSTN. It therefore forwards the **INVITE** message to the BGCF in the home network. If the MGW is located in the home network, the BGCF sends the **INVITE** message to the MGCF in the home network. For the case where the MGW is in the visited network, there are two possibilities. The BGCF may forward the **INVITE** message to the visited BGCF (which selects an MGCF in the visited network). Alternatively, the BGCF may directly forward the **INVITE** message to the MGCF in the visited network, as in this example.

Step 6. Through the H.248 **Add** and **Reply** message pair exchange, the MGCF determines the MGW capabilities and allocates the MGW ports for the call connection.

Step 7. The MGCF returns the **183 Session Progress** message to the P-CSCF. The message contains the SDP that indicates the media stream capabilities of the called party. The P-CSCF authorizes the required resources for this session and forwards the **183 Session Progress** message (with the SDP) to the UE through the signaling path established by the **INVITE** message.

Step 8. The UE determines the final set of media streams, and sends the final SDP to the MGCF through the **PRACK** (Provisional Ack) message.

Step 9. The MGCF issues the H.248 **Modify** message (similar to the MGCP **ModifyConnection** message; see Section 10.1.1), which instructs the MGW to reserve the required resources for the media streams.

Step 10. After Step 8, the UE reserves the resources for this session through the PDP context activation procedure. Then it sends the **Resource Reservation Successful** message (the **UPDATE** message) to the MGCF.

Step 11. The MGCF sends the **Initial Address Message** (IP IAM) to the T-SGW. The T-SGW translates the **IP IAM** message into the ISUP **IAM** message and forwards it to the PSTN. The **IAM** requests the PSTN to set up the PSTN call path toward the called party.

Step 12. The PSTN establishes the call path, alerts the called party, and returns the ISUP **Address Complete Message** (ACM) to the T-SGW. This message is translated into the **IP ACM** message and is forwarded to the MGCF. The MGCF sends the ring-back message (SIP **180 Ringing**) to the UE. The **ACM** message indicates that the path to the destination has been established.

Step 13. When the called party answers, the PSTN sends the ISUP **Answer Message** (ANM) to the T-SGW. The SGW translates the message into the **IP ANM** message and forwards it to the MGCF.

Step 14. The MGCF instructs the MGW to create the bidirectional connection through the H.248 **Modify** and **Reply** message pair exchange.

Step 15. After Step 13, the MGCF sends the **200 OK** response to the S-CSCF. The S-CSCF may perform service control for the call. Then it forwards the **200 OK** response to the UE through the P-CSCF (which approves the usage of the reserved resources). The UE starts the media flow for this session.

Step 16. The UE forwards the final SIP **ACK** message to the MGCF (through P-CSCF, I-CSCF, and S-CSCF).

Although several nodes are visited in call setup signaling, the shortest path is established for the user plane media stream (i.e., UE ↔ UTRAN ↔ SGSN ↔GGSN ↔ MGW ↔ PSTN).

15.3.4 PS Mobile Call Termination

Before PS mobile call termination, the UE has already attached to the UMTS PS domain, and has completed application-level registration. The PDP context has been established for SIP signaling message delivery. We assume that the UE is in the home network. The call termination procedure is given in the following steps (see Figures 15.10 and 15.11):

Step 1. The originating switch of the PSTN uses ISUP **IAM** to initiate a call. That is, the PSTN sends the **IAM** to the T-SGW. The T-SGW sends the **IP IAM** message to the MGCF in the home network.

Step 2. Through the H.248 **Add** and **Reply** message exchange, the MGCF communicates with the MGW to reserve the resources (MGW ports) for user plane media streams. Note that how to determine the MGW and how to ensure the most optimal routing toward this MGW is an open question.

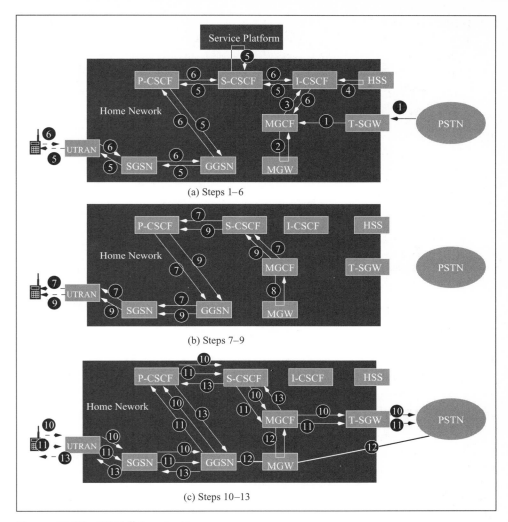

Figure 15.10 PS Call Termination

Step 3. The MGCF translates the destination address and determines that the called party is in the home network. The MGCF sends the SIP **INVITE** request (including the initial SDP) to the I-CSCF.

Step 4. The I-CSCF exchanges the **Cx-Location-Query** and **Cx-Location-Query-Resp** message pair with the HSS to obtain the location information of the called party.

Step 5. The I-CSCF sends the **INVITE** message to the UE through S-CSCF, P-CSCF, GGSN, SGSN, and UTRAN. In this signaling flow, the S-CSCF

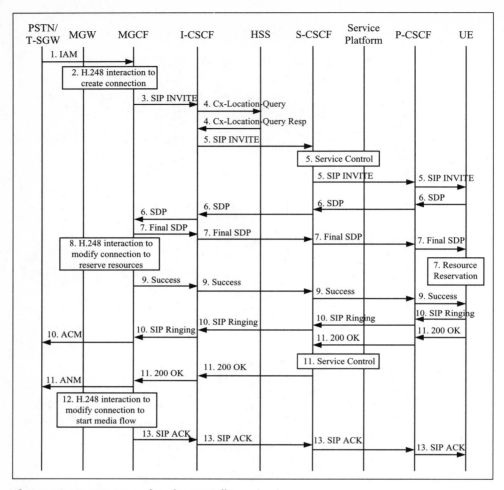

Figure 15.11   Message Flow for PS Call Termination

validates the service profile, and may contact the service platform to perform termination service control.

Step 6. The INVITE message received by the UE indicates the requested media flows. Based on this information, the UE determines the resources to be allocated for the media streams. The UE then sends the SIP **183 Session Progress** message with the SDP to the MGCF through UTRAN, SGSN, GGSN, P-CSCF, and S-CSCF. The SDP information indicates the UE's media stream capabilities. Since different types of UEs may support different media types, the UE's capabilities have an impact on the SDP description. In this signaling flow (which was established by the INVITE message in Steps 3 and 5), the P-CSCF authorizes the necessary resources for this session.

Step 7. Based on the UE's capabilities, the MGCF determines the allocated media streams for this session, and sends the final SDP response to the UE. The UE initiates resource reservation for this session through the PDP context activation procedure [3GP05q, Lin 01b].

Step 8. After the MGCF has sent the SDP in Step 7, it instructs the MGW to reserve the resources for the session.

Step 9. The MGCF may perform service control as appropriate, and sends the Resource Reservation Successful message (that is, the UPDATE message) to the UE. If the UE has completed resource reservation in Step 7, then it alerts the subscriber (called person).

Step 10. The UE sends the SIP 180 Ringing message to the MGCF. The MGCF sends the IP ACM message to the T-SGW, and the T-SGW sends the ISUP ACM to the PSTN. At this point, the PSTN knows that the call path toward the UE has been established.

Step 11. When the called person of the UE answers, the final 200 OK response is sent from the UE to the MGCF through P-CSCF and S-CSCF. The UE starts the media flow for this session. In this signaling flow, the P-CSCF indicates that the reserved resources should be committed, and the S-CSCF may perform service control as appropriate. When the MGCF receives the 200 OK, it sends the IP ANM to the T-SGW, and the T-SGW sends the ISUP ANM to the PSTN.

Step 12. The MGCF sends the H.248 Modify message to the MGW, which instructs the MGW to make a bidirectional connection between the GGSN and the PSTN node.

Step 13. The MGCF then returns the SIP ACK message to the UE.

The user plane media streams are connected between the UE and the PSTN through path UE $\leftrightarrow$ UTRAN $\leftrightarrow$ SGSN $\leftrightarrow$ GGSN $\leftrightarrow$ MGW $\leftrightarrow$ PSTN.

15.4   Open Service Access

Existing telecommunications services are considered a part of a network operation's domain, and the development of services are achieved by, for example, *Intelligent Network (IN)* technology [Bel95]. By introducing Internet and mobility into the telecommunications networks, more flexible and efficient approaches are required for mobile service deployment. Such approaches must allow network operators and enterprises to increase revenues via third-party applications and service providers. To achieve these goals, standardization bodies such as the 3GPP CN5 [3GP04k, 3GP05o, 3GP04a], ETSI SPAN12, ITU-T SG11, and the Parlay Group have been defining the *Open Service Access* (*OSA*) specifications [Moe03, Wal02]. OSA provides unified service creation and execution environments to speed up service deployment that is independent of the underlying mobile network technology.

In OSA, network functionality offered to applications is defined by a set of *Service Capability Features* (*SCFs*). Services can be implemented by applications accessing the *Service Capability* (SC) through the standardized OSA *Application Programming Interface* (*API*). As illustrated in Figure 15.12, the OSA consists of three parts:

- Applications are implemented in one or more *Application Servers* (*ASs*; Figure 15.12 (1)).

- *The framework* (*FW*; see Figure 15.12 (2)) authorizes applications to utilize the *Service Capabilities* (*SC*; Figure 15.12(5)) in the network. That is, an application can only access the OSA API via the FW for services.

- *Service Capability Servers* (*SCSs*; Figure 15.12 (3)) provide the applications with access to underlying network functionality through SCFs (Figure 15.12 (4)). These SCFs, specified in terms of interface classes and their methods, are offered by SCs within networks (and under network control). SCs are bearers needed to realize services.

The FW is considered one of the SCSs, and is always present, one per network. The FW provides access control functions to authorize the access to SCFs or service data for any API method invoked by an application, with specified security level, context, domain, et cetera. Before any application can interact with a network SCF, an off-line service agreement must be established. Once the service agreement exists, mutual authentication can be performed between the application and the FW. Then the application can be authorized by the FW to access a specific SCF (in OSA, authentication must precede authorization). Finally, the application can use the *discovery* function to obtain information on authorized network SCFs. The discovery function can be used at any time after successful authentication. SCFs offered by an SCS are typically registered at the FW. This information is retrieved when the application invokes the discovery function. The FW allows OSA to go beyond traditional IN technology through openness, discovery,

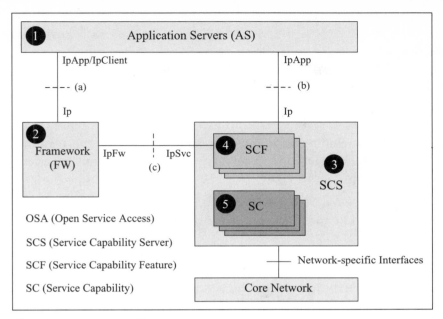

Figure 15.12 OSA Architecture

and integration of new features. Based on *Telecommunications Informa-tion Networking Architecture (TINA)* [Ber00], the FW provides controlled access to the API by supporting flexibility in application location and busi-ness scenarios. Furthermore, the FW allows multi-vendorship and even the inclusion of non-standardized APIs, which is crucial for innovation and ser-vice differentiation.

An SCS can be deployed as a stand-alone node in the network or directly on a node in the core network. In the distributed approach, the OSA gateway node contains the FW and zero or more SCS components. Other SCSs are im-plemented in different nodes. It is possible to add more SCSs and distribute the load from different applications over multiple SCSs. At the service selec-tion phase, the FW may divert one application to one SCS and another to a different SCS. With middleware such as CORBA, it is possible to distribute the load on a session basis without the application being aware that different sessions involve different SCSs. In some APIs, it is possible to add multiple application callbacks to the SCS so that the SCS can distribute the load of multiple sessions over different applications running on different servers. To allow applications from the visited networks to use the SCSs in the home network, all communications between the application server and the SCSs must be the secured through, for example, the Secure Socket Layer (SSL) or the IP Security Protocol (IPsec). Examples of OSA SCFs are given as follows:

- *Call and session control SCFs* provide capabilities for setting up basic calls or data sessions as well as manipulating multimedia conference calls.

- *User and terminal-related SCFs* allow obtaining information from the end-user (including user location and status) and the terminal capabilities—playing announcements, sending short text messages, accessing mailboxes, and so on.

- *Management-related SCFs* provision connectivity QoS, access to end-user accounts, and application/data usage charging.

Interaction between an application and an SCS is always initiated by the application. In some scenarios, it is required to initiate the interaction from the SCS. An example is the *call screening* service. Suppose that the network routes a call to a user who has subscribed to this OSA service. Before the call reaches the user, the call screening application needs to be invoked. This issue is resolved by the OSA *request of event notification* mechanism. Initially, the application issues an OSA interface class method (API call) to the SCS. This OSA method allows the SCS to invoke the application (for example, call screening) through a callback function when it receives events (for example, incoming calls) from the network related to the application.

Since functionality inside a telecommunications network can be accessible via the OSA APIs, applications can access different network capabilities using a uniform programming paradigm. To be accessible to a side developer community, the APIs should be deployed based on open *information technology* (*IT*). Three OSA API classes are defined among the applications, the FW, and the SCFs (in the SCSs):

1. Interface classes between the applications and the FW (Figure 15.12(a)) provide control of access to the network and integrity management; specifically, they provide applications with functions such as authentication, authorization, and discovery of network functionality. The FW-side interfaces to be invoked by the applications are prefixed with `Ip`. The application-side interfaces to be called back by the FW are prefixed with `IpApp` or `IpClient`.

2. Interface classes between the applications and the SCFs (Figure 15.12 (b)) allow the applications to invoke network functionality for services.

The SCF-side interfaces to be invoked by the applications are prefixed with `Ip`. The application-side interfaces to be called back by the SCFs are prefixed with `IpApp`.

3. Interface classes between the FW and the SCFs (Figure 15.12 (c)) provide the mechanisms for SCF registration and a multi-vendor environment. The SCF-side interfaces to be used by the FW are prefixed with `IpSvc`. The FW-side interfaces to be used by the SCFs are prefixed with `IpFw`.

The SCSs implement the OSA server side of the API, and the applications implement the OSA client side of the API. An application should communicate with an SCS through a standard IT middleware infrastructure such as *Common Object Request Broker Architecture (CORBA)* [Pya 98]. Details of OSA implementation can be found in [Cho05].

15.5   Efficiency of IP Packet Delivery

To support a UMTS all-IP network, the IP packets must be delivered efficiently in the radio access bearer. Specifically, the UTRAN should meet the objectives of spectral efficiency and error robustness. Radio spectrum efficiency is significantly affected by IP packet overhead. The sizes of the combined packet (IP/UDP/RTP) headers are at least 40 bytes for IPv4 and at least 60 bytes for IPv6. Furthermore, the header part will result in more error protection than the payload because a header loss will require to discard the corresponding packet, and no error concealment or mitigation can be applied to the header. For applications such as VoIP, the voice payload is typically shorter than the packet header. Packing more speech frames into one packet will reduce the relative overhead. However, the voice delay is increased and thus the voice quality is degraded. A more appropriate solution is to utilize the *header adaptation techniques*, which reduce the size of the header before radio transmission. Reduction of the header's size is achieved in two ways:

- *Header compression* removes redundancy in the originally coded header information. In a UMTS all-IP network, the header compression schemes must be developed based on the radio link reliability characteristics (that is, high error rates and long round-trip times). Such schemes may be able to compress the packet headers down to 2 bytes [3GP99]. To exercise header compression between the UE and the network, each of them maintains consistent compression and decompression. Initially, the uncompressed headers are transmitted. For the subsequent packets, the compressed headers (the "differences" from the previous headers) are delivered over the air. The decompressor uses the received compressed information and the knowledge of previous headers to reconstruct the next headers. Header compression should achieve efficiency (the average header size is minimized), robustness (no packet is lost due to header compression), and reliability (the decompressed header is identical to the header before compression). A major disadvantage of this technique is that the compressed headers have variable sizes, which introduce extra overhead for end-to-end security (IPsec) and bandwidth management. Note that the header compression mechanism is typically implemented in the UTRAN.

■ *Header stripping* removes header field information and thereby losing functionality. Packet headers are stripped before radio transmission and regenerated at the receiving end. Essentially, only the payload is transmitted, but some additional header-related information needs to be transmitted to enable the header regeneration. The degree of header transparency depends on the amount of transmitted header-related information. No header error protection is needed when the header information is completely removed. When the payload has constant size, bandwidth management issue can be simplified since the payloads can be carried on a constant bit rate channel. This approach also mitigates QoS issues such as delay and jitter. Note that the header stripping approach is typically adopted in the GERAN.

In selecting the header reduction solution, one should consider the impact on transparency and robustness to errors. Three approaches are considered for *User Plane Adaptation (UPA)*:

Full opacity (no adaptation): The original packet header is sent over the air to achieve full transparency. The UPA supports IPsec on an end-to-end basis. High overhead of the header results in very poor spectrum efficiency.

Payload opacity (header adaptation only): The header is compressed or stripped. The UPA understands the header structure but not the payload.

No opacity (full adaptation): The UPA knows the structure of the headers and the payload. Header can be compressed or stripped. Payload transmission is optimized by techniques such as unequal bit protection. Channel and error coding is optimized for the payload structure. To support VoIP, we may use header compression with equal bit protection in the payload.

15.6 Concluding Remarks

This chapter described a UMTS all-IP approach for the core network architecture. We introduced the core network nodes, and then elaborated on application-level registration, circuit-switched call origination, packet-switched call origination, and packet-switched call termination.

UMTS should guarantee end-to-end QoS and radio spectrum efficiency. To provide the expected QoS across domains, operators must agree on the deployment of common IP protocols. The common IP protocols impact roaming (that is, the interfaces between core networks) and the communications between terminals and networks in order to support end-to-end QoS and achieve maximum interoperability. QoS signaling and resource allocation schemes should be independent of the call control protocols. The details can be found in [Lin 01b, 3GP04i, 3GP04f]. To support radio spectrum efficiency, IP header compression or stripping is required, as discussed in Section 15.5. This chapter mentioned standard GPRS interfaces, including Gn, Gp, and Gi. Details of these interfaces can be found in Chapter 18 in [Lin 01b].

We also discussed the business issues for UMTS [UMT00a, UMT00b, UMT00c]. The wireless Internet services offered by UMTS (for example, location-based services) will influence people's lifestyles, which creates a new understanding of telecommunications, enables new technologies, and spurs future demand. It is clear that we cannot directly apply the fixed Internet model to wireless Internet. The model of "almost-free" fixed Internet access should be re-investigated. Wireless Internet needs a new model so that end-users will be willing to pay for secure, convenient wireless data services. Accurate customer segmentation, customer value, and availability of services at the right price will be key factors to success. Specifically, prices should be pitched at affordable levels for the target segments. Profitability will be strongly dependent on the packaging of the applications and the presentation of content based on the characteristics of people in different regions of the world and their particular needs.

Another driver for customers to use 3G services is transparency of charging that allows better cost control via easy-to-use interfaces. There must be a strong link between the users' perceived QoS and the rates. QoS support across multiple networks will require new forms of commercial agreements between operators, which will be radically different from the traditional peer-to-peer agreements from the Internet world or the roaming agreements known from the 2G mobile world.

15.7 Questions

1. Describe the differences between UMTS R00 options 1 and 2. Why do you need two alternatives for a UMTS all-IP network? Does it a temporary or a long-term arrangement?

2. In the all-IP network architecture, how is the charging gateway classified vertically and horizontally?

3. Describe the three alternatives that UMTS all-IP network can interact with application platforms.

4. Describe how QoS is handled in the all-IP network in the vertical and the horizontal layers.

5. Describe the CSCF functions. Can we implement the MGCF functions in CSCF so that we can simplify IMS by eliminating the MGCF component?

6. Does it make sense to have CSCF interact with the R99 HLR? If so, which component in CSCF is responsible for this function? What protocols should be added?

7. Section 15.2.1 pointed out that message screening can be implemented in the ICGW. The same function can be implemented in the GGSN, as described in Chapter 4. Show the tradeoffs for these two approaches.

8. Describe the information maintained in the HSS. What extra pieces of information in the HSS are not found in the HLR (see Section 9.4)?

9. In Chapter 4, the SGSN needs to make a DNS query. Can this DNS function be implemented in the address protocol termination function of HSS?

10. In IMS, which network node(s) use the SIP, the MAP, and the MGCP/H.248 protocols?

11. What is the subscription location function? Does this function exist in UMTS R99?

12. What is CSCF selection? Why and when is this feature needed?

13. What is P-CSCF discovery? Which network node conducts this function for a UE?

14. Draw the message flow for IMS re-registration flow.

15. Draw the message flow for call setup from the CS domain to the PS domain in UMTS R00 option 2 by assuming that the called party is in the home network.

16. At Step 2 of Section 15.3.4, how can the MGCF select the MGW for optimal routing?

17. Draw the message flow for PS mobile call termination where (a) the calling party is in the PSTN and the called party is in the visited IMS network, and (b) the calling party is a GSM-IP user (Section 10.1) and the called party is in the home IMS network.

18. Describe the three major components of OSA, and the interfaces among these components. Why are these components essential to a service network?

19. In OSA, one can build some basic service features (SCFs/SCs), and then use these basic features to develop new OSA service features. Give an example.

20. Show how iSMS (Chapter 1) can be integrated into the OSA architecture.

21. Describe two approaches for header-size reduction. Compare the tradeoffs of these two approaches.

22. Describe user plane adaption for header reduction.

Issues for the IP Multimedia Core Network Subsystem

Based on the architecture described in Chapter 15, this chapter investigates two issues for the IP Multimedia Core Network Subsystem (IMS). The first issue regards Interrogating Call Session Control Function (I-CSCF) access. In IMS, any incoming call will first arrive at the I-CSCF. The I-CSCF queries the Home Subscriber Server (HSS) to identify the Serving CSCF (S-CSCF) of the called mobile user. The S-CSCF then sets up the call to the called mobile user. In Section 16.1, we investigate the performance of the IMS incoming call setup. We also describe cache schemes with fault tolerance to speed up the incoming call setup process. Our study indicates that the I-CSCF cache can significantly reduce the incoming call setup delay, and checkpointing can effectively enhance the availability of I-CSCF.

The second issue relates to IMS authentication. The 3GPP defines two-pass authentication whereby both the GPRS and the IMS authenticate the Mobile Station (MS) or the User Equipment (UE). In Section 16.2, we integrate both the GPRS and the IMS authentications into a one-pass procedure. At the IMS level, authentication is implicitly performed in IMS registration. Our approach may save up to 50% of the IMS registration/authentication traffic, as compared with the 3GPP two-pass procedure. We formally prove that the one-pass procedure correctly authenticates IMS users.

Caching in I-CSCF

the IMS provides multimedia services by utilizing the *Session Protocol* (*SIP*; see Chapter 12). By redrawing Figure 15.1, Figure 16.1 illustrates a simplified UMTS network architecture that emphasizes the IMS. As shown in this figure, the IMS user data traffic is transported through the *Media Gateways* (*MGWs*). As described in Chapter 15, IMS signaling is carried out by the *Proxy-Call Session Control Function* (*P-CSCF*), the *Interrogating CSCF* (*I-CSCF*), and the *Serving CSCF* (*S-CSCF*). The I-CSCF determines how to route incoming calls to the S-CSCF and then to the destination UEs. When a UE attaches to the GPRS/IMS network and performs PDP context activation, a P-CSCF is assigned to the UE. The P-CSCF contains limited address translation functions to forward the requests to the I-CSCF. By exercising the IMS registration, an S-CSCF is assigned to serve the UE. This S-CSCF supports the signaling for call setup and supplementary services control. This section investigates the performance of the IMS incoming call setup. Specifically, we describe several cache schemes to speed up the incoming call setup process.

16.1.1 Standard IMS Registration and Call Setup

We first reiterate simplified versions of the registration and the incoming call setup procedures for UMTS IMS described in Section 15.3. Then we propose

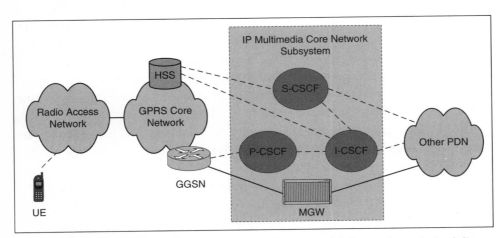

Figure 16.1 Simplified UMTS Network Architecture (Solid lines: data; dashed lines: signaling; dashed lines between UE and RAN: radio interface)

a cache scheme and two checkpoint schemes that speed up the incoming call setup process.

Suppose that a UE already obtained the IP connectivity through the PDP context activation, and has performed at least one IMS registration. The UE may issue re-registration due to, for example, movement among different service areas. In order to compare with the cache schemes described in this section, Figure 16.2 redraws Figure 15.5 to illustrate the registration message flow of the basic scheme (called B-RP). For simplicity, we remove P-CSCF from the registration path. We also omit Steps 3 and 6 in Figure 15.5 by assuming that the S-CSCF is known in advance. The **Cx-Query** and **Cx-Select-Pull** message pair at Step 4 of Figure 15.5 is the same as the **User Authorization Request (UAR)** and **User Authorization Answer (UAA)** message pair at Step 2 in Figure 16.2. The **Cx-Put** and **Cx-Select-Pull** message pair at Step 7 of Figure 15.5 is the same as the **Server Assignment Request (SAR)** and **Server Assignment Answer (SAA)** message pair at Step 4 in Figure 16.2.

The IMS incoming call (that is, call termination) setup described in Section 15.3.4 is referred to as the basic incoming call setup (B-ICS). The message flow in Figure 15.11 is simplified in Figure 16.3, and is reiterated below:

Step 1. The caller sends the **INVITE** message to the I-CSCF. The initial media description offered in the Session Description Protocol (SDP) is contained in this message.

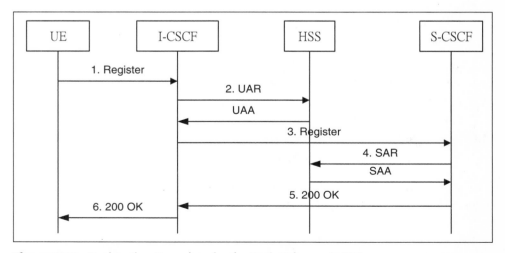

Figure 16.2 Registration Procedure for the Basic Scheme (B-RP)

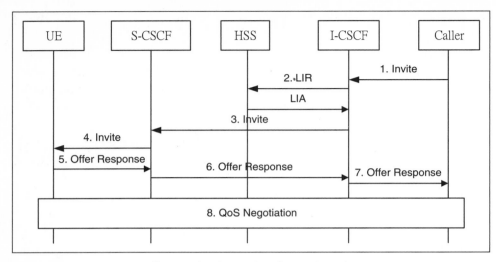

Figure 16.3 Incoming Call Setup for the Basic Scheme (B-ICS)

Step 2. The I-CSCF exchanges the Cx **Location Info Request (LIR)** and **Location Info Answer (LIA)** message pair with the HSS to obtain the S-CSCF name for the destination UE.

Steps 3 and 4. The I-CSCF forwards the **INVITE** message to the S-CSCF. Based on the *user profile* of the destination UE, the S-CSCF invokes whatever service logic that is appropriate for this session setup attempt. Then it sends the **INVITE** message to the UE (through the P-CSCF of the IMS network where the UE resides).

Steps 5–7. The UE responds with an answer to the offered SDP. This **Offer Response** message is passed along the established session path back to the caller.

Step 8. The QoS for this call is negotiated between the originating network (of the caller) and the terminating network (of the UE). The details are given in Section 15.3.

16.1.2 IMS Registration and Call Setup with Cache

This subsection shows how to utilize cache mechanisms at the I-CSCF to speed up the incoming call setup process. We first describe a basic cache scheme. To enhance availability and reliability, we then consider two checkpoint schemes that immediately recover the I-CSCF cache after an I-CSCF crash (failure).

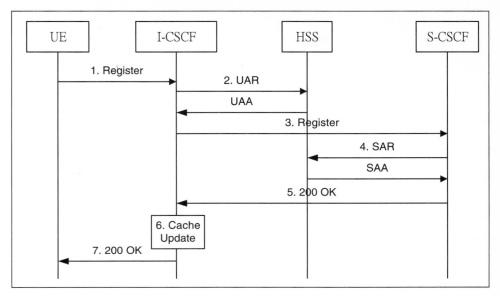

Figure 16.4 Registration with Cache Update for the C (C1 and C2) Schemes (C-RP)

Basic Cache Scheme (The C Scheme)

Figure 16.4 illustrates the registration message flow for the C scheme (called C-RP). C-RP is the same as B-RP except that when the **200 OK** Message is sent from the S-CSCF to the I-CSCF, the (UE, S-CSCF) mapping (called the S-CSCF record) is saved in the cache (Step 6 in Figure 16.4). The incoming call setup for the C scheme (C-ICS) is illustrated in Figure 16.5. In C-ICS, the **Location Info Request** and **Location Info Answer** message pair exchanged (Step 2 of B-ICS in Figure 16.3) is replaced by a cache retrieval (Step 2, in Figure 16.5) to obtain the S-CSCF address. If an I-CSCF failure occurs and the whole cache content is lost, then the S-CSCF records are gradually rebuilt through the IMS registration procedure. If an incoming call arrives earlier than the registration, then B-ICS must be executed to set up the call.

To immediately recover the I-CSCF cache after a failure, we may save the content of the cache (only for the modified records) into a backup storage. This operation is called *checkpoint*. The following subsections describe two checkpoint schemes.

Checkpoint Scheme 1 (The C1 Scheme)

This scheme periodically saves the cache into the backup. When an I-CSCF failure occurs, the cache content is restored from the backup. Therefore,

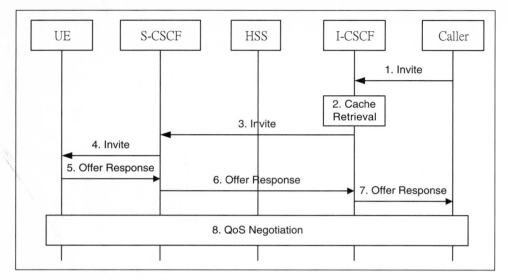

Figure 16.5 Incoming Call Setup with Cache Retrieval for C (C1 and C2) Schemes (C-ICS)

in normal operation, the registration procedure and the incoming call setup procedure for the C1 scheme are the same as those for the C scheme.

After a failure, the incoming call setup procedure is the same as C-ICS (see Figure 16.5) if the S-CSCF is up to date (called a *cache hit*). Note that between a failure and the previous checkpoint, the S-CSCF record of an MS may be modified. In this case, the S-CSCF may be obsolete when an incoming call arrives (called a *cache miss*), and the call setup message flow is as illustrated in Figure 16.6. The first three steps of this message flow are the same as C-ICS. Since the UE already moved from the old S-CSCF to the new S-CSCF, at the end of Step 3, the old S-CSCF replies with the **404 Not Found** message to the I-CSCF. The I-CSCF then retrieves the new S-CSCF information from the HSS and sets up the call, following Steps 2–8 of B-ICS in Figure 16.3.

It is clear that after an I-CSCF failure, the call setup cost for the C1 scheme is very expensive if a cache miss occurs. We resolve this issue by introducing the C2 scheme.

Checkpoint Scheme 2 (The C2 Scheme)

Like the C1 scheme, this scheme periodically checkpoints the cache content into the backup. Furthermore, an S-CSCF record in the backup is invalidated if the corresponding record in the cache is modified. The C2 registration procedure is the same as C-RP except for Step 6 in Figure 16.4. In this step,

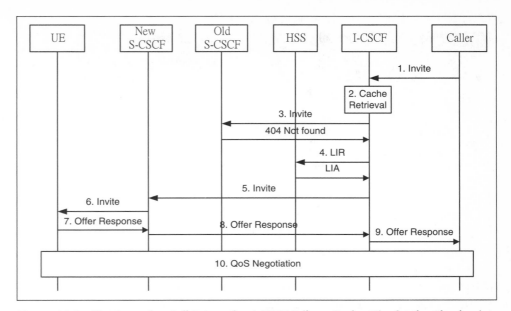

Figure 16.6 First Incoming Call Setup after I-CSCF-Failure: Cache Miss for the Checkpoint Scheme 1

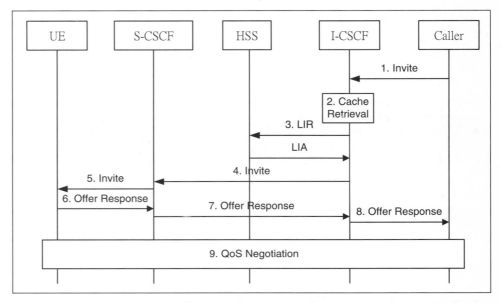

Figure 16.7 First Incoming Call Setup after I-CSCF-Failure: Cache Miss for Checkpoint Scheme 2

Table 16.1 Caching and checkpointing operations

SCHEME	PERIODIC CHECKPOINTING	BACKUP RECORD INVALIDATION	CACHE RESTORATION AFTER I-CSCF FAILURE
Basic	no	no	no
Cache	no	no	no
Checkpoint 1	yes	no	yes
Checkpoint 2	yes	yes	yes

we check if the S-CSCF record at the backup has been invalidated since the last checkpoint. If so, no extra action is taken. If not, the record in the backup is invalidated. Therefore, if multiple registrations for the same UE occur between two checkpoints, the S-CSCF record in the backup is only invalidated for the first registration. As shown in Figure 16.7, after an I-CSCF failure, the C2 scheme knows exactly which S-CSCF records are invalid. For the first incoming call after the failure, if the S-CSCF record is valid, then the call setup procedure follows C-ICS in Figure 16.5. Conversely, if the S-CSCF record is invalid, the procedure follows B-ICS in Figure 16.3.

Features of the B, C, C1, and C2 schemes are summarized in Tables 16.1 and 16.2 In [Lin05b], we analyzed the registration and incoming call setup by using both analytic and simulation models (see also questions 1 and 2 in Section 16.4). The performance study indicated that by utilizing the I-CSCF

Table 16.2 IMS registration and call setup

SCHEME	REGISTRATION	NORMAL INCOMING CALL SETUP	FIRST INCOMING CALL SETUP AFTER FAILURE
Basic	B-RP	B-ICS	B-ICS
Cache	C-RP (B-RP + cache update)	C-ICS (B-ICS without HSS query)	B-ICS
Checkpoint 1	C-RP	C-ICS	Cache hit: C-ICS; Cache miss: B-ICS plus extra access to S-CSCF
Checkpoint 2	C-RP + possible backup record invalidation	C-ICS	Cache hit: C-ICS; Cache miss: B-ICS

cache, the average incoming call setup time can be effectively reduced, and a smaller I-CSCF timeout threshold can be set to support early detection of incomplete call setups.

16.2 Integrated Authentication for GPRS and IMS

The UMTS all-IP network supports IP multimedia services through IMS. According to the Third Generation Partnership Project (3GPP) specifications, authentication is performed at both the GPRS and the IMS networks before an ME or UE can access the IMS services. Many steps in this 3GPP "two-pass" authentication procedure are identical.

In UMTS, when an MS sends an "Initial L3 message" (for example, location update request, connection management service request, routing area update request, attach request, paging response, et cetera) to the SGSN, the SGSN may be triggered to authenticate the user. The GPRS authentication procedures are described in Section 9.1. In addition to the GPRS authentication, it is necessary to authenticate the MS before it can access the IMS services. Without IMS authentication, a mobile user who passes the the GPRS authentication can easily fake being another IMS user. Details of the fake procedure will be elaborated in Section 16.2.1. IMS authentication is performed between the *IMS Subscriber Identity Module* (*ISIM*) in the MS and the AuC in the home network [3GP05p]. This procedure is basically the same as the GPRS authentication. In this procedure, the CSCF first sends a multimedia authentication request to the HSS/AuC with the *IP Multimedia Private Identity* (*IMPI*) of the MS, and receives a response with an array of Authentication Vectors (AVs).

This step is skipped if the CSCF already has the AV array. The CSCF then invokes the IMS authentication and key agreement procedure with an authentication vector. The MS authenticates the network through the received AUTN, and the CSCF authenticates the MS using the *Signed Result* (*SRES*) pair; that is, RES and XRES. In our description, the SRES generated by the MS is RES, and the SRES generated by the network is XRES. A detailed message flow of this procedure is given in Section 16.2.1

Although both GPRS and IMS authentications are necessary, most of the steps in these two authentication passes are duplicated. In other words, the two-pass authentication proposed in 3GPP TS 33.203 [3GP05p] is not efficient. In this section, we describe a one-pass authentication procedure that effectively combines both the GPRS and the IMS authentications. We prove that this simplified one-pass authentication procedure correctly authenticate IMS users.

16.2.1 3GPP Two-Pass Authentication

This subsection describes the 3GPP two-pass authentication procedure. We elaborate on IMS authentication and explain why authentication must be performed in both the GPRS and the IMS levels.

IMS Authentication

After performing GPRS authentication (Section 9.1) and PDP context activation (Sections 3.1.1 and 4.3), the MS can request the IMS services through the registration procedure illustrated in Figure 16.8 (see also Section 16.1.2). In this procedure, the MS interacts with the S-CSCF, possibly through the P-CSCF and the I-CSCF. To simplify our discussion, Figure 16.8 uses the term "CSCF" to represent the proxy, interrogating, and service functions of the CSCF. The registration procedures described in Sections 15.3 and 16.1.1 omit the authentication details. Here we address the authentication aspect of

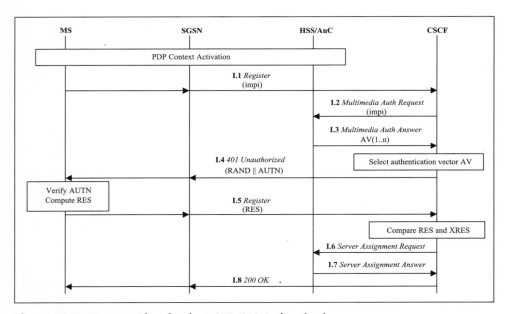

Figure 16.8 Message Flow for the 3GPP IMS Authentication

IMS. The complete IMS registration procedure can be obtained by merging both Figures 16.8 and 16.2. IMS authentication/registration is implemented by the SIP and Cx protocols [3GP05b, 3GP05a], and consists of the following steps:

Step I.1. The MS sends the SIP **REGISTER** message to the CSCF (with the parameter IMPI $= impi$) through the SGSN.

Step I.2. Assume that the CSCF does not have the AVs for the MS. The CSCF invokes the authentication vector distribution procedure by sending the Cx **Multimedia Authentication Request** message to the HSS/AuC (with the parameter IMPI $= impi$).

Step I.3. The HSS/AuC uses $impi$ to retrieve the record of the MS and generate an ordered array of AVs. The HSS/AuC sends the AV array to the CSCF through the Cx **Multimedia Authentication Answer** message.

Step I.4. The CSCF selects the next unused authentication vector from the ordered AV array and sends the parameters RAND and AUTN (from the selected authentication vector) to the MS through the SIP **401 Unauthorized** message.

Step I.5. The MS checks whether the received AUTN can be accepted. If so, it produces a response RES. The MS sends this response back to the CSCF through the SIP **REGISTER** message. The CSCF compares the received RES with the XRES. If they match, then the authentication and key agreement exchange is successfully completed.

Step I.6. The CSCF sends the Cx **Server Assignment Request** message to the HSS/AuC.

Step I.7. Upon receipt of the **Server Assignment Request**, the HSS/AuC stores the CSCF name and sends the Cx **Server Assignment Answer** message to the CSCF.

Step I.8. The CSCF sends the **200 OK** message to the MS through the SGSN, and the IMS registration procedure is completed.

In the above procedure, Steps I.1–I.5 exercise authentication, and Steps I.6–I.8 perform registration.

Fraudulent IMS Usage

Although the GPRS and the IMS authentications are implemented in different protocols (that is, SS7 MAP and SIP/Cx, respectively), many steps of these two authentication procedures are duplicated (see Table 16.3).

Table 16.3 Identical steps in GPRS and IMS authentication

GPRS AUTHENTICATION (SS7 MAP)	IMS AUTHENTICATION (SIP/Cx)
MAP Send Authentication Info Request; Parameter: IMSI	I.2: Multimedia Authentication Request Parameter: IMPI
MAP Send Authentication Info Response; Parameter: AV[1..n]	I.3: Multimedia Authentication Answer Parameter: AV[1..n]
User Authentication Request Parameter: RAND||AUTN	I.4: 401 Unauthorized Parameter: RAND||AUTN
User Authentication Response Parameter: RES	I.5: REGISTER Parameter: RES
GMM Attach Accept	I.8: 200 OK

Unfortunately, these redundant steps are required. That is, after GPRS authentication, it is necessary to authenticate the MSs again at the IMS level. Without IMS authentication, an IMS user may pretend to be another IMS user. Consider the example in Figure 16.9, where there are two MSs. MS-A has the IMSI value *imsi-A* and the IMPI value *impi-A*. MS-B has the IMSI value *imsi-B* and the IMPI value *impi-B*. Suppose that MS-B is a legal GPRS user and has passed the GPRS authentication (by using *imsi-B*) to obtain GPRS network access. If no IMS authentication is required, MS-B may perform IMS registration by sending the CSCF the **REGISTER** request that includes the MS-A's IMPI value *impi-A* as a parameter. The CSCF will consider this IMS registration as a legal action activated by MS-A. Therefore, MS-B can illegally access the IMS services of MS-A. This example shows that IMS-level authentication is required to prevent illegal access to the IMS services. In the next section, we describe a one-pass authentication procedure for both GPRS and IMS authentications. This approach significantly reduces the number of accesses to the HSS/AuC.

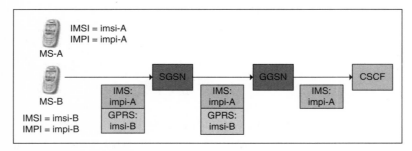

Figure 16.9 Illegal IMS Registration

16.2.2 One-Pass Authentication Procedure

This subsection describes a one-pass authentication procedure (performed at the GPRS level) to authenticate an IMS user that does not explicitly perform the IMS-level authentication. In this approach, the SGSN implements a *SIP Application Level Gateway (ALG)* [Che05b] that modifies the format of SIP messages. We first describe the SIP message flow of the one-pass procedure. Then we provide a brief cost comparison between the one-pass and the two-pass procedures.

SIP Message Flow

After GPRS authentication, the MS performs PDP context activation to obtain GPRS access. Then the MS registers to the IMS through Steps I*.1–I*.4 illustrated in Figure 16.10:

Step I*.1. The MS sends the SIP **REGISTER** message to the SGSN with the parameter IMPI = $impi$. Note that after PDP context activation, the SGSN can identify the IMSI of the MS that transmits the GPRS packets [3GP05q]. The SIP ALG in the SGSN adds the IMSI value (that is, $imsi$) of the MS in the **REGISTER** message and forwards it to the CSCF. Details of a possible SIP ALG implementation can be found in [Che05b].

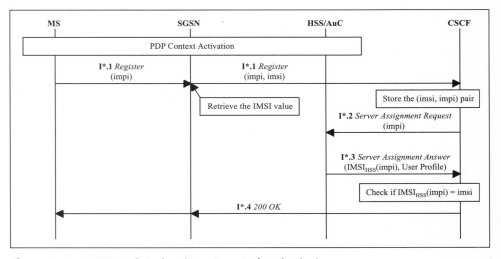

Figure 16.10 IMS Registration (One-Pass Authentication)

Step I*.2. The CSCF stores the $(imsi, impi)$ pair in the MS record, and sends the Cx **Server Assignment Request** message to the HSS/AuC with the parameter IMPI $= impi$. Note that if the CSCF has stored the $(imsi, impi)$ pair before, then Steps I*.2 and I*.3 are skipped.

Step I*.3. The HSS/AuC uses the received IMPI value $impi$ as an index to retrieve the IMSI and the user profile of the MS. We denote $IMSI_{HSS}(impi)$ as the IMSI value retrieved from the HSS/AuC. The HSS/AuC stores the CSCF name and sends the Cx **Server Assignment Answer** to the CSCF, with the parameters $IMSI_{HSS}(impi)$ and the user profile.

Step I*.4. The CSCF checks whether the value $imsi$ and $IMSI_{HSS}(impi)$ are the same. If so, the CSCF sends the SIP **200 OK** message to the SGSN and the authentication is considered successful. If $IMSI_{HSS}(impi) \neq imsi$, then it implies that the registration is illegal (that is, the scenario illustrated in Figure 16.9 occurs). Suppose that $IMSI_{HSS}(impi) = imsi$. The SGSN forwards the **200 OK** message to the MS, and the IMS registration procedure is successfully completed.

In Section 16.2.3, we prove that Step I*4 correctly determines whether a registration is legal or illegal.

Cost Analysis

Based on the simplified procedures described in the previous sections, Table 16.4 compares the steps executed in the one-pass and the two-pass authentication procedures. Suppose that the expected SIP message delivery cost between an MS and the CSCF is one unit, and the expected Cx message delivery cost between the CSCF and the HSS/AuC is δ units. It is anticipated that $\delta < 1$ for the following two reasons:

- The CSCF and the HSS/AuC exchange the Cx messages through the IP network. Conversely, besides the IP network overhead, SIP communications between the MS and the CSCF involves the GPRS core network and the UTRAN radio network.

- The CSCF and the HSS/AuC are typically located at the same location, whereas the MS is likely to reside at a remote location.

It is clear that the expected IMS registration C_1 for the one-pass procedure (see Figure 16.10) is

$$C_1 = 2 + 2\delta \tag{16.1}$$

Table 16.4 Comparing the one-pass and the two-pass authentication procedures in IMS Registration

ONE-PASS PROCEDURE	TWO-PASS PROCEDURE

I*.1: REGISTER
  Parameters: *impi* and *imsi*
	I.1: REGISTER
	  Parameter: *impi*

—	I.2: Multimedia Authentication Request Parameter: *impi*
—	I.3: Multimedia Authentication Answer Parameter: AV[1..n]
—	I.4: 401 Unauthorized Parameter: RAND||AUTN
—	I.5: REGISTER Parameter: RES
I*.2: Server Assignment Request	I.6: Server Assignment Request
I*.3: Server Assignment Response	I.7: Server Assignment Response
I*.4: 200 OK	I.8: 200 OK

Note that Step I*.1 needs to trigger SIP ALG for SIP message analysis. Since this action is executed in the micro-kernel of the SGSN, the overhead can be ignored as compared with SIP message exchange. Similarly, the extra cost of the $IMSI_{HSS}(impi)$ and *imsi* comparison at Step I*.4 can be ignored. Our analysis assumes that the (*imsi*, *impi*) pair does not exist at Step I*.1. Therefore, Steps I*.2 and I*.3 are always executed. This assumption favors the two-pass procedure.

In the two-pass procedure, if the distribution of authentication vectors from the HSS/AuC to the SGSN (Steps I.1–I.4 in Figure 16.8) is performed, then the expected IMS registration cost $C_{2,1}$ is expressed as

$$C_{2,1} = 4 + 4\delta \tag{16.2}$$

If the authentication vector distribution is not executed in the two-pass procedure, then the expected IMS registration cost $C_{2,2}$ is expressed as

$$C_{2,2} = 4 + 2\delta \tag{16.3}$$

Like the UMTS periodic location update described in Section 2.6, IMS registration is periodically performed. In Steps I.2 and I.3 of the two-pass procedure, an AV array of size n (where $n \geq 1$) is sent from the HSS/AuC to the CSCF. Therefore, one out of the n IMS registrations incurs execution of Steps I.2 and I.3. From (16.2) and (16.3), the expected IMS registration cost

C_2 for the two-pass procedure is

$$C_2 = \left(\frac{1}{n}\right) C_{2,1} + \left(\frac{n-1}{n}\right) C_{2,2} = 4 + \left(\frac{n+1}{n}\right) 2\delta \qquad (16.4)$$

From (16.1) and (16.4) the improvement S of the one-pass procedure over the two-pass procedure is

$$S = \frac{C_2 - C_1}{C_2} = \frac{n + \delta}{2n + (n+1)\delta} \qquad (16.5)$$

Figure 16.11 plots S as a function of n and δ. The figure indicates that the one-pass procedure can save up to 50 percent of the SIP/Cx traffic for IMS registration/authentication, as compared with the two-pass procedure. Another significant advantage of the one-pass procedure is that it consumes much less AVs (about 50 percent less) than the two-pass procedure.

One may argue that implementation of a SIP ALG is required in the one-pass procedure. Since IMS is based on SIP, a SIP ALG is required for other purposes (see an example in Section 14.1.2). Therefore, the one-pass procedure only incurs a little extra cost for implementing SIP ALG.

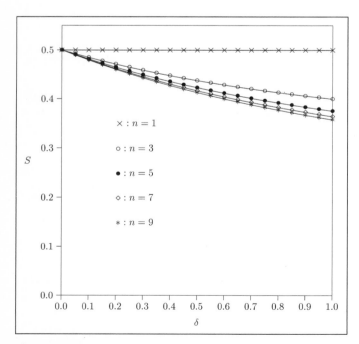

Figure 16.11   Improvement S of the One-Pass Procedure over the Two-Pass Procedure

16.2.3 Correctness of the One-Pass Procedure

In this subsection, we prove that the one-pass authentication procedure correctly authenticates IMS users. In UMTS, every MS maintains the attributes IMSI, IMPI, and the pre-shared secret key (K) in its SIM card. Consider an MS with IMSI $= imsi$, IMPI $= impi$, and K $= k$. To simplify our discussion, we assume that these parameters are grouped into a set $R_{MS} = \{imsi, impi, k\}$ in the SIM card of the MS. Define functions $IMSI_{MS}$, $IMPI_{MS}$, and K_{MS} such that for any $x \in R_{MS}$,

$$IMSI_{MS}(x) = imsi, \text{ where } imsi \text{ is the IMSI value in } R_{MS} \qquad (16.6)$$
$$IMPI_{MS}(x) = impi, \text{ where } impi \text{ is the IMPI value in } R_{MS} \qquad (16.7)$$
$$K_{MS}(x) = k, \text{ where } k \text{ is the K value in } R_{MS} \qquad (16.8)$$

Based on the above definitions, it is clear that, for example,

$$IMSI_{MS}(impi) = IMSI_{MS}(k) = imsi$$

Similarly, for every MS, the HSS/AuC maintains a record R_{HSS} that consists of attributes IMSI, IMPI, and K. That is, for an MS who has legal GPRS and IMS accesses,

$$R_{HSS} = \{imsi, impi, k\} = R_{MS}$$

Like (16.5)–(16.8), we define functions $IMSI_{HSS}$, $IMPI_{HSS}$, and K_{HSS} such that for $x \in R_{HSS}$,

$$IMSI_{HSS}(x) = imsi, \text{ where } imsi \text{ is the IMSI value in } R_{HSS} \qquad (16.9)$$
$$IMPI_{HSS}(x) = impi, \text{ where } impi \text{ is the IMPI value in } R_{HSS} \quad (16.10)$$
$$K_{HSS}(x) = k, \text{ where } k \text{ is the K value in } R_{HSS} \qquad (16.11)$$

In 3GPP TS 23.060 [3GP05q] (also Section 9.1) and 3G 33.203 [3GP05p], MS authentication at the GPRS and the IMS levels are based on the following theorem.

Theorem 16.2.1

Suppose that an MS claims that it has the IMSI value $imsi$ and the IMPI value $impi$. Then

(a) The MS is a legal GPRS user if $K_{MS}(imsi) = K_{HSS}(imsi)$.
(b) The MS is a legal IMS user if $K_{MS}(impi) = K_{HSS}(impi)$.

Note that Theorem 16.2.1 does not hold if an illegal user already possesses the SIM information of a legal user (for example, by duplicating the SIM card through the SIM card reader [Fen04]). This issue is addressed in Section 9.2. Here, we assume that such fraudulent usage does not occur. 3GPP GPRS authentication procedure checks if both a GPRS user and the HSS/AuC have the same pre-shared secret key K using Theorem 16.2.1 (a) and Fact 1 (a) below. Similarly, the 3GPP IMS authentication procedure (i.e., Steps I.1–I.8) checks whether both an IMS user and the HSS/AuC have the same pre-shared secret key using Theorem 16.2.1 (b) and Fact 1 (b).

Fact 1. (a) For an MS claiming IMSI $=$ $imsi$, if XRES $=$ RES, then $K_{MS}(imsi) = K_{HSS}(imsi)$.

Fact 1. (b) For an MS claiming IMPI $=$ $impi$, if XRES $=$ RES, then $K_{MS}(impi) = K_{HSS}(impi)$.

Now we prove that the one-pass authentication correctly authenticates the IMS users, that is, the one-pass procedure checks if $K_{MS}(impi) = K_{HSS}(impi)$. From the definitions of the $IMSI_{HSS}$ and K_{HSS} functions, i.e., (16.9) and (16.11), it is trivial to have the following fact:

Fact 2. For any IMPI value $impi$, if $IMSI_{HSS}(impi) = imsi$, then $K_{HSS}(impi) = K_{HSS}(imsi)$.

With Fact 2, correctness of the one-pass authentication procedure is guaranteed according to the following two theorems.

Theorem 16.2.2

Suppose that

(a) an MS with the IMSI value $imsi$ has passed the GPRS authentication; that is,

$$K_{MS}(imsi) = K_{HSS}(imsi) \qquad (16.12)$$

(b) the MS claims that its IMPI value is $impi$, and

(c) the network maps $impi$ to the IMSI value $imsi$; that is,

$$IMSI_{HSS}(impi) = imsi \qquad (16.13)$$

Then the MS is a legal IMS user. In other words,

$$K_{MS}(impi) = K_{HSS}(impi) \qquad (16.14)$$

Proof: From hypothesis (a), $imsi \in R_{MS}$. In hypothesis (b), the MS claims that it has the IMPI value $impi$, which implies that $impi \in R_{MS}$. From (16.8),

$$K_{MS}(imsi) = K_{MS}(impi) \qquad (16.15)$$

From Fact 2 and (16.13) in hypothesis (c), we have

$$K_{HSS}(impi) = K_{HSS}(imsi) \qquad (16.16)$$

From (16.12) in hypothesis (a) and (16.16) we have

$$K_{MS}(imsi) = K_{HSS}(impi) \qquad (16.17)$$

From (16.17) and (16.15) we have

$$K_{MS}(impi) = K_{HSS}(impi)$$

In other words, if hypotheses (a)–(c) hold, an MS is a legal IMS user with IMPI $= impi$.

Q.E.D.

Theorem 16.2.3

The one-pass authentication procedure correctly authenticates the IMS users; that is, for an MS claiming the IMPI value $impi$, the one-pass procedure recognizes the MS as a legal IMS user if $K_{MS}(impi) = K_{HSS}(impi)$.

Proof: After the GPRS authentication has been executed, the network verifies that $K_{MS}(imsi) = K_{HSS}(imsi)$; i.e., (16.12) in Theorem 16.2.2. is satisfied.

At Step I*.1, the MS claims that its IMPI value is $impi$, and therefore the network assumes that $K_{MS}(imsi) = K_{MS}(impi)$; i.e., (16.15) in Theorem 16.2.2 is satisfied.

At Step I*.4, the one-pass authentication checks if $IMSI_{HSS}(impi) = imsi$, i.e., (16.13) in Theorem 16.2.2 is checked. If so, $K_{MS}(impi) = K_{HSS}(impi)$ as a direct consequence of Theorem 16.2.2, and the authentication procedure recognizes the MS as a legal user (according to Theorem 16.2.1). Otherwise, the authentication fails.

In other words, the one-pass procedure follows Theorem 16.2.1 to authenticate an MS.

Q.E.D.

16.3   Concluding Remarks

Based on [Lin05b], this chapter investigated the performance of the IMS incoming call setup, and described the cache schemes with fault tolerance to speed up the incoming call setup process. Our study in [Lin05b] indicated that by utilizing the I-CSCF cache, the average incoming call setup time can be effectively reduced, and a smaller a I-CSCF timeout threshold can be set to support early detection of incomplete call setups. To enhance fault tolerance, the I-CSCF cache is periodically checkpointed into a backup storage. When an I-CSCF failure occurs, the I-CSCF cache content can be restored from the backup storage. Since the checkpointing process is conducted in the background, this activity does not affect the incoming call setup delays. As a final remark, if both the I-CSCF and the HSS fail, the S-CSCF records can only be recovered from the backup. In this case, our checkpoint schemes can significantly enhance the availability and fault tolerance of the IMS network.

This chapter also investigated an efficient IMS registration procedure without explicitly performing tedious authentication steps [Lin05c]. As specified by the 3GPP, after a UMTS mobile user has obtained GPRS network access through GPRS authentication, the "same" authentication procedure must be executed again at the IMS level (during IMS registration) before it can receive the IP multimedia services. This chapter described a one-pass authentication procedure that only needs to perform GPRS authentication. At the IMS registration, the one-pass procedure performs several simple operations to verify whether a user is legal. We proved that the one-pass procedure correctly authenticates IMS users. Compared with two-pass authentication, one-pass authentication saves two to four SIP/Cx message exchanges among the MS, the SGSN, the CSCF, and the HSS/AuC. Our study in [Lin05c] indicates that this new approach can save up to 50 percent of the network traffic generated by the IMS registration. This approach also saves 50 percent of the storage for buffering the authentication vectors.

16.4   Questions

1. The I-CSCF cache mechanism in Section 16.1 is modeled in this and the next questions. This question investigates the costs for the C1 and the C2 schemes. Figure 16.12 illustrates the timing diagram for the registration and checkpointing activities of a UE . At t_0, t_2, and t_4, the UE

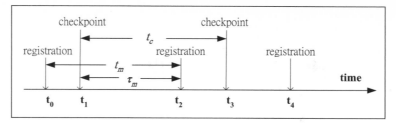

Figure 16.12 Timing Diagram for Registration and Checkpointing

issues registration requests either because it attaches to the network or it moves from one service area to another service area. The inter-registration intervals $t_2 - t_0$, $t_4 - t_2$, etc., are represented by a random variable t_m. In this figure, periodic checkpoints are performed at t_1 and t_3, where the checkpointing interval is represented by a random variable t_c. The interval τ_m between a checkpoint and the next registration (e.g., $t_2 - t_1$) is called the *excess life* of the inter-registration interval. At a checkpoint, only the modified S-CSCF records are saved into the backup. Let p_u be the probability that the S-CSCF record for the UE is saved at a checkpoint. Then

$$p_u = \Pr[t_c > \tau_m]$$

It is clear that the checkpoint cost increases as p_u increases. Two types of checkpoint intervals are often considered. *Fixed* checkpointing performs checkpoints with a fixed period $1/\mu$. In *exponential* checkpointing, the inter-checkpointing interval has the exponential distribution with the mean $1/\mu$. Assume that the inter-registration intervals t_m have the exponential distribution with the mean $1/\lambda$ (that is, the registration stream forms a Poisson process). For an arbitrary time interval T, the number X of registrations occurring in this period has a Poisson distribution. That is

$$\Pr[X = x, T = t] = \left[\frac{(\lambda t)^x}{x!} \right] e^{-\lambda t} \qquad (16.18)$$

and

$$\Pr[t_c > \tau_m] = 1 - \Pr[X = 0, T = t_c] \qquad (16.19)$$

From (16.19), the probability p_u for fixed checkpointing is expressed as

$$p_u = 1 - e^{-\frac{\lambda}{\mu}} \qquad (16.20)$$

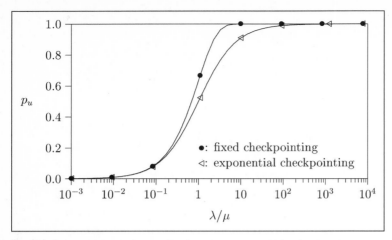

Figure 16.13 Comparing Fixed and Exponential Checkpointing (Poisson Registration Stream with Rate λ)

Similarly, the probability p_u for exponential checkpointing is expressed as

$$p_u = 1 - \int_{t_c=0}^{\infty} e^{-\lambda t_c} \mu e^{-\mu t_c} dt_c = \frac{\lambda}{\lambda + \mu} \qquad (16.21)$$

Figure 16.13 plots p_u for fixed and exponential checkpointing approaches based on (16.20) and (16.21). The figure indicates that p_u for fixed checkpointing is larger than that for exponential checkpointing (that is, exponential checkpointing yields better performance than fixed checkpointing). From now on, we only consider exponential checkpointing. Consider the inter-registration interval random variable t_m with the mean $1/\lambda$, the density function $f(\cdot)$, and the Laplace Transform $f^*(s)$. The excess life τ_m has the distribution function $R(\cdot)$, density function $r(\cdot)$, and the Laplace Transform $r^*(s)$.

Since exponential checkpointing is a Poisson process, t_1 in Figure 16.12 is a random observer of the t_m intervals. From the excess life theorem (see questions 1 and 2 in Section 9.6),

$$r^*(s) = \left(\frac{\lambda}{s}\right)[1 - f^*(s)] \qquad (16.22)$$

Assume that t_m is a Gamma random variable with the mean $1/\lambda$, the variance V, and the Laplace Transform

$$f^*(s) = \left(\frac{1}{V\lambda s + 1}\right)^{\frac{1}{V\lambda^2}} \qquad (16.23)$$

Based on (16.22) and (16.23), derive p_u. Based on the derived equation, plot p_u against V. You should observe that p_u decreases as the variance V increase. This phenomenon is explained as follows. When the registration behavior becomes more irregular, you will observe more checkpoint intervals with many registrations, and more checkpoint intervals without any registration. In other words, smaller p_u is observed. Therefore, the checkpointing performance is better when the registration activity becomes more irregular (that is, when V is large).

2. This question investigates the costs for the incoming call setup described in Section 16.1. We analyze the first incoming call setup after an I-CSCF failure. Note that the checkpoint action is a background process, and the cost for retrieving the cache (for example, Step 6 in Figure 16.4) and the cost for saving an S-CSCF record into the backup is negligible as compared with the communications cost between the I-CSCF and the HSS (the I-CSCF cache operation is typically 1,000 times faster than the inter I-CSCF and HSS communications). Therefore, we will ignore the I-CSCF cache operation costs in the incoming call setup study.

Let t_H be the round-trip transmission delay between the I-CSCF and the HSS, let t_S be the round-trip delay between the I-CSCF and the S-CSCF, and let t_M be the round-trip delay between the S-CSCF and the UE. Let T_x be the incoming call setup delay from the I-CSCF to the UE (without QoS negotiation) for the "x" scheme (where $x \in \{B, C, C1, C2\}$).

Consider the B scheme. In Figure 16.3, t_H is the delay for Step 2; t_S is the delay for Steps 3 and 6; and t_M is the delay for Steps 4 and 5. We have

$$T_B = t_H + t_S + t_M$$

In Figure 16.5, t_S is the delay for Steps 3 and 6; and t_M is the delay for Steps 4 and 5. We have

$$T_C = T_{C1} = T_{C2} = t_S + t_M$$

Suppose that t_H, t_S, t_M have the same expected values.

For the "x" scheme, let T_x^* be the round-trip transmission delay of the first incoming call setup after an I-CSCF failure. The delay T_x^* is derived as follows. For the B scheme, the first incoming call setup is not affected by the I-CSCF failure. That is,

$$T_B^* = T_B = t_H + t_S + t_M \tag{16.24}$$

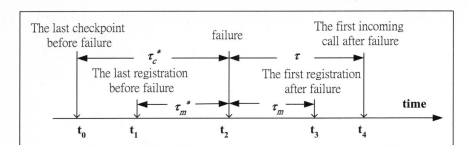

Figure 16.14 Timing Diagram Before and After an I-CSCF Failure

For the C scheme, the cache in the I-CSCF is cleared after the failure.

- If the first event after the failure is an incoming call, then $T_C^* = T_B^*$.
- If the first event is a registration event, then the S-CSCF record is restored in the cache before the first incoming call arrives. When the incoming call arrives, $T_C^* = T_C$.

Let α be the probability that the first event after I-CSCF failure is a registration. Then

$$T_C^* = \alpha T_C + (1 - \alpha)T_B = T_C + (1 - \alpha)t_H \qquad (16.25)$$

The probability α is derived as follows. Consider the timing diagram in Figure 16.14. Suppose that an I-CSCF failure occurs at time t_2. For a UE, the first registration after the failure occurs at t_3 and the first incoming call after the failure occurs at t_4. Let $\tau = t_4 - t_2$ and $\tau_m = t_3 - t_2$. Since the failure occurring time t_2 is a random observer of the inter-call arrival times and the inter-registration times, τ is the excess life of an inter-call arrival time. Also, τ_m is the excess life of an inter-registration time. Suppose that the call arrivals to the UE are a Poisson process with the rate γ. Then from the excess life theorem, τ has the exponential distribution with the mean $1/\gamma$. Similarly, τ_m has the density function $r(\cdot)$ and the Laplace Transform $r^*(s)$ as expressed in (16.22). Therefore,

$$\alpha = \Pr[\tau > \tau_m] = \left(\frac{\lambda}{\gamma}\right)\left[1 - f^*(\gamma)\right] \qquad (16.26)$$

For the C1 scheme, two cases are considered when the first incoming call after the failure occurs:

- **Case I:** The S-CSCF record restored from the backup is invalid (with probability β) and the first event after the failure is an incoming call (with probability $1 - \alpha$). In this case (see Figure 16.6), $T_{C1}^* = T_B + t_S$.

- **Case II:** The S-CSCF record restored from the backup is valid (with probability $1 - \beta$), or the restored record is invalid (with probability β) and the first event after the failure is an IMS registration (with probability α). In this case, $T_{C1}^* = T_C$.

Based on the above two cases, we have

$$T_{C1}^* = T_C + (1 - \alpha)\beta(t_H + t_S) \tag{16.27}$$

For the C2 scheme (see the message flow in Figure 16.7), we have

$$T_{C2}^* = T_C + (1 - \alpha)\beta t_H \tag{16.28}$$

Derive β and then compute T_C^*, T_{C1}^*, and T_{C2}^*.

3. Compare the message flows in Figures 15.11 and 16.3. For example, describe which message pair in Figure 15.11 is the same as the **LIR/LIA** pair in Figure 16.3.

4. Does the C2 scheme work if the I-CSCF for registration and the I-CSCF for an incoming call are different? (Note that several I-CSCFs may be deployed in an IMS network.)

5. Why is the IMS authentication model needed after a subscriber is authenticated by GPRS? How can a legal GPRS user conduct fraudulent IMS usage if the IMS authentication is not enforced?

6. In Section 16.2, the one-pass procedure is used by the network to authenticate the mobile users. Can this procedure be extended so that the user is also able to authenticate the IMS? Why or why not?

7. Redraw the two-pass and the one-pass algorithms by considering the extra details provided in Section 15.3. Based on your drawing, conduct the cost analysis as in Section 16.2.2.

8. Draw the IMS registration message flow by merging Figures 16.8 and 16.2.

## CHAPTER 17

A Proxy-based Mobile
Service Platform

Rapid advances in mobile devices, wireless networking, and messaging technologies have provided mobile users with a plethora of alternatives to access the Internet. Examples of these devices and protocols include the following:

- Palm PDA with Web Clipping

- Cellular phones with *Wireless Application Protocol* (*WAP*) [OMA04]

- Short Message Service (SMS) email devices that support Post Office Protocol Version 3 (POP3) [Mye 96] or Internet Message Access Protocol (IMAP) [Cri96] (e.g., Blackberry [Bla05] and AT&T PocketNet phone)

- Pocket PCs that support AOL Instant Messenger (AIM) [Ame05]

Unfortunately, these approaches do not interwork with each other easily. Mobile users face a dilemma: They want the convenience of various mobile accesses to critical services, but suffer from managing the complexity of incompatible devices and user interfaces. Wireless Internet is much more complicated than simply accessing the Internet wirelessly. Wireless users, being mostly mobile, have different needs, motivations, and capabilities than wired users. For example, a mobile user is usually in a multi-tasking mode (accessing the Internet while attending a meeting or shopping in the mall). On the

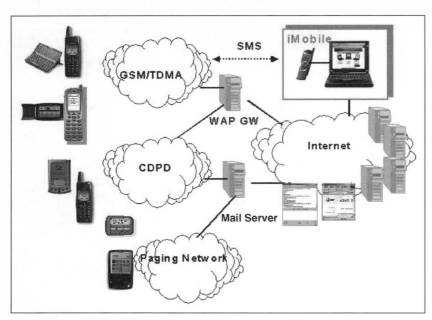

Figure 17.1   Personal Mobile Service Network

other hand, a mobile user may not always access the Internet wirelessly, and a wireless user may not be mobile at all. The mobile accesses (e.g., checking stock quotes or weather, or finding a nearby restaurant) are usually bursty in nature and very task-oriented. To access diverse services, different identities are utilized, e.g., cellular phone numbers and instant messaging screen names are more meaningful to mobile users than office phone numbers and static IP addresses. This chapter describes *iMobile* [Rao 01b, Che05a], a user-friendly environment for mobile Internet. The iMobile system was developed by AT&T and FarEasTone. As shown in Figure 17.1, iMobile runs on a computer with connections to the Internet and a wireless modem with two-way SMS, which is a generalization of iSMS, described in Chapter 1. Devices can communicate with iMobile through various protocols and access networks. Several examples are given below:

- GSM/TDMA phones with two-way SMS can communicate with iMobile through an SMS driver hosted on iMobile.
- *Cellular Digital Packet Data* (*CDPD*) devices (such as AT&T Pocket-Net phone and Palm V with the Omnisky modem) can use WAP to access iMobile through the Internet.

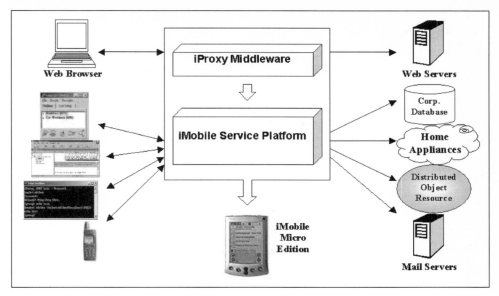

Figure 17.2 iMobile, iProxy and iMobile Micro Edition

- Email devices such as Blackberry [Bla05] can use the standard email protocols on the CDPD network or a two-way paging network to communicate with iMobile.

- PC and PDA devices can use AOL instant messenger or web browsers to interact with iMobile.

iMobile receives messages and commands from these devices, accesses Internet services and information on behalf of the mobile users, and then relays messages or Internet content back to the destination devices (which can be different from the sending devices).

Figure 17.2 illustrates the iMobile architecture that hides the complexity of multiple devices and content sources from mobile users. This goal is achieved by utilizing a programmable proxy server called iProxy [Rao99]. iProxy provides an environment for hosting agents and personalized services, which are implemented as reusable building blocks in Java. Since iProxy provides a built-in web server, an iProxy agent can be invoked as a regular *Common Gateway Interface (CGI)* program. It also allows scripts embedded inside web pages, which invoke agents to perform specific tasks. iProxy was originally designed as middleware between user browsers and web servers. It maintains *user profiles* and enhances intelligence of a traditional proxy server to provide personalized value-added services such as filtering, tracking, and archiving [Che97, Rao00]. iProxy provides customization and personalization features, which are essential to supporting iMobile services.

To support interactions among mobile devices in heterogeneous networks, [Rao 01b] proposed a lightweight mobile service platform called *iMobile Micro Edition (ME)*, which minimizes the requirements of system resources and is suitable for execution on mobile devices. iMobile ME communicates through a message center called *iMobile Router*, which stores and delivers messages for mobile devices. This architecture provides a platform for iMobile-based peer-to-peer computing.

This chapter elaborates on the design guidelines and implementation of iMobile. We first describe the iProxy middleware. Then we introduce the iMobile architecture and show the user and device profiles used in personalization and transcoding services. Finally, we investigate iMobile-based peer-to-peer computing.

17.1   iProxy Middleware

Being a middleware for web applications, iProxy [Rao99] can be installed as a personal proxy server running on an end-user's machine. iProxy provides standard web proxy functions for accessing, caching, and processing web data. It allows users to select the routes to access web servers (for example, choosing different proxies for different hosts), archive web pages, record and analyze web access history, and manage the proxy's cache. iProxy also provides hooks to plug in new functions for processing data received from web servers. For example, the web data can be condensed, compressed, encrypted, or patched. The iProxy filters convert pages back to their original forms, or synthesize new pages from web data and personal information stored on the local disk. With a built-in web server, iProxy can accept Hypertext Transfer Protocol (HTTP) calls from the Internet and trigger local applications to perform tasks such as event notifications, alerts, and data pushing.

17.1.1   iProxy System Architecture

As shown in Figure 17.3, iProxy consists of four components:

- **The proxy component** receives a *Uniform Resource Locator (URL)* request from a web browser and forwards the request to the corresponding web server. When it receives the result from the web

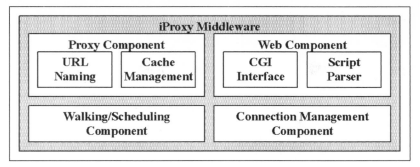

Figure 17.3   iProxy Architecture

server, it stores the returned page in the cache, and displays it to the user through the browser.

- **The web component** enables a user to access the iProxy configuration, invoke CGI programs, or execute a script embedded in a web page.

- **The walking/scheduling component** provides functions to trace the HTML tree structure asynchronously (for example, through background tracing or periodic tracing).

- **The connection management component** handles the socket connections for web accessing, inter-iProxy communication, and TCP/IP socket forwarding.

These components are described in detail in the following sections.

Proxy Component

iProxy maintains a cache to store frequently accessed web pages. When the proxy component receives a URL request from a browser, it returns the cached page for a cache hit. For cache-miss access, the proxy component forwards the request to the corresponding remote web server. When the requested page is returned, the proxy component stores the page in the cache, and then forwards it to the user. The proxy component consists of *URL naming* and *cache management* subcomponents. To provide new add-on services, the URL naming subcomponent extends the protocol exercised between the browser and the web server. Two new specifications, *action* and *view* parts, are added to the URL format, which result in the following new format:

```
http://view@hostname/filepath/filename.html?iProxy&action=K
```

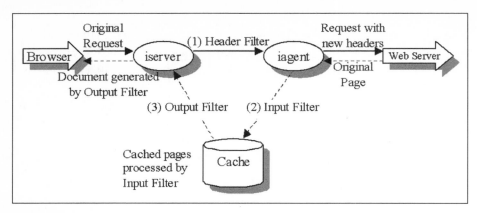

Figure 17.4   Proxy Component Filters

The action specification, starting with `?iProxy`, notifies iProxy to perform specific actions on the web page. An action may archive a page with a time stamp, store the page in a specific package, or invoke a CGI program to modify the page. The view specification is used to access web pages cached at a particular time or in a particular package. Examples of action/view are given in Section 17.1.2.

The cache management subcomponent maintains the cache repositories for web pages obtained from remote web servers. It may only keep the newest pages, as a normal proxy server does, or keep multiple versions according to the time stamps. Figure 17.4 shows the caching processes. For a web request, the *iserver* process receives the request and creates an *iagent* process to acquire and cache the web page. After the web page is cached, the iserver process forwards the cached page to the browser. The cache management subcomponent may invoke three different filters in the caching processes:

- **Header Filter** modifies the URL requests received from the web browser. This filter may add new cookies or header fields, or modify the URL to change the access behavior. For example, changing the URL of a web image to a local image eliminates transmission overhead due to remote graphical access. This filter is useful for mobile devices with low bandwidths.

- **Input Filter** modifies a page before storing it in the cache. This filter provides the flexibility to customize, compress, encrypt, or translate a cached page.

■ **Output Filter** processes a cached web page before it is returned to the browser. This filter may attach additional HTML components into the cached pages, including system statistics or personal information. The output filter is also used to decompress, decrypt, or translate a cached page.

Web Component

The web component consists of two subcomponents: *CGI interface* and *script parser*. A URL request to iProxy is forwarded to the web component. If the request is a CGI program, the CGI interface subcomponent invokes the corresponding Java program. Conversely, if the request is a script file starting with `#!/iProxy/script` (see the example in Figure 17.17), the script parser subcomponent interprets the script and generates the page from the output of the script. The web component also exports the web interface to a user for accessing iProxy system resources indicated by a URL starting with `http://localhost/`. This root page (`http://localhost/`) is the entry point for accessing the system parameters and configurations. The proxy component defines rules for filter programs to indicate which filter should be invoked on a specific URL request and when. These rules reference the filters as CGI programs defined in the web component.

Walking/Scheduling Component

The walking/scheduling component traces the HTML tree structure and stores the pages in the cache or archive repository. An HTML walking component is invoked by the URL action specification or through the administration web pages—that is,

```
http://localhost/
```

Starting from a root page, the walking component parses the HTML structure, and tracks the hyperlinks and/or images in the page. The walking parameters determine the depth of tracing, the images for caching, and the pages to be traced locally (that is, within the same web site) or globally.

A walking result may be stored in three different repositories: *cache*, *archive*, or *package*. The cache repository keeps the newest cached pages; the archive repository maintains multiple versions with time stamps; and the package repository stores all walked pages in a single file. Caching the walking pages speeds up subsequent accesses. This feature also effectively supports off-line browsing because it keeps not only the visited pages but

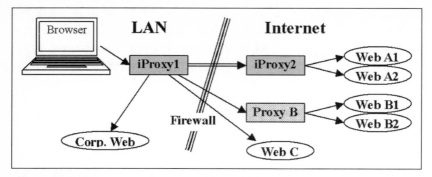

Figure 17.5   Web Access Routing and inter-iProxy Communications

also the subsequent hyperlinks in the cache. The archive repository memorizes the historical web pages for tracking, searching, and comparison. The package repository maintains the pages in different packages. A package can be moved around various servers to support user mobility. These repositories are handled by the cache management component and can be accessed using the view specification.

Connection Management Component

The connection management component handles the socket connections for web accessing, inter-iProxy communication, and TCP/IP socket forwarding. Figure 17.5 shows examples of routing paths for web access. In this figure, an iProxy server, iProxy1, and a corporation web server, Corp Web, are located in the same LAN. In the Internet, there is another iProxy server, iProxy2, a standard proxy server, Proxy B, and several web servers, A1, A2, B1, B2, and C. iProxy1 specifies *routing rules* to handle the requests for different web servers. For example, the routing rules can be as follows:

- **Rule 1:** Forward the requests for A1 and A2 to iProxy2.
- **Rule 2:** Forward the requests for B1 and B2 to Proxy B.

Based on these rules, iProxy2 handles the forwarded requests for A1 and A2, Proxy B handles the forwarded requests for B1 and B2, and iProxy1 handles the requests for Corp Web and Web C.

These routing rules are specified in the administration pages, which provide the flexibility to forward requests for particular web servers through other proxy servers. They are especially useful when the connections between proxy servers support the *keep-alive* feature.

17.1.2   Personal Services

Novel services can be provided by a client-side iProxy that has access to a user's private information, for example, the web access history and personal finance information. This section describes a few personal services implemented in iMobile.

Personal Web Archive

Although existing search engines allow users to find current pages, they may not allow locating and viewing the pages that were accessed in the past, except for those that are still kept in the browser's cache. Thank to inexpensive mass storage, a client-side iProxy can afford to archive all web pages that have been viewed before. These pages can be retrieved later without even bookmarking them. Tools like AltaVista Discovery [Ove05] can be used to index and search these web pages. Our experience showed that an active web browser user could create a web archive of roughly 80–100 MB per month, including images and all downloaded documents. This amounts to about 1 GB of storage per year for all pages a user has seen, which is now affordable to many customers.

Because iProxy intercepts HTTP requests, it can effectively extend a URL namespace by adding a time stamp in front of the regular HTTP address. For example, the URL of the 2005 February 11th version of AT&T's website is `http://+20050211@www.att.com/`. Even if the AT&T website has gone through major redesigns, the persistent URL (indexed by the time stamps) will always provide the same content.

Personal Web Page Reminders and Hot Sites

A client-side iProxy can analyze the logged web pages to provide convenient browsing services. Figure 17.6 shows two services implemented in iMobile:

- **TO-READ Homepages:** A user specifies the list of websites and corresponding frequencies they should be visited. iProxy checks the last visited dates and schedules a list of pages that will be visited by the user today.

- **HOT Sites:** iProxy computes the number of visits to each website and lists the top 10 websites with their last visiting dates whenever the user accesses the personal portal. The hot-site list allows a user to retrieve the last visit time of a favorite site (the timestamp is displayed

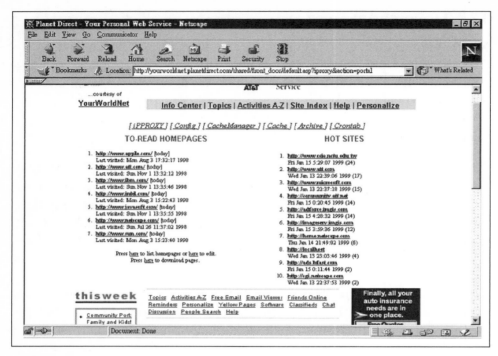

Figure 17.6 TO-READ Pages and Hot Sites

along the website link), access the latest version by clicking on the link, and compare the new version with the old one by using tools such as WebCiao or AIDE [Dou98].

My Stock Portfolio

Most portals allow users to specify stocks of interest and will display the latest prices when the user accesses the personalized page. However, these portals cannot compute personal current balances or net gains/losses unless the user provides confidential information such as the number of owned shares and the purchased dates. Such a practice is certainly not convenient to many users. Figure 17.7 shows a typical portfolio view on *Your WorldNet* [ATT05]. In this example, the user provides the following fake information in the portal server (see the upper table in Figure 17.7): one share for AOL Time Warner, Lucent, Microsoft, and AT&T stocks. No commission fees were paid and each stock was bought at $30.00.

Suppose that the real purchase price, commission fees, and number of shares each stock are stored on the client machine. By constructing an output

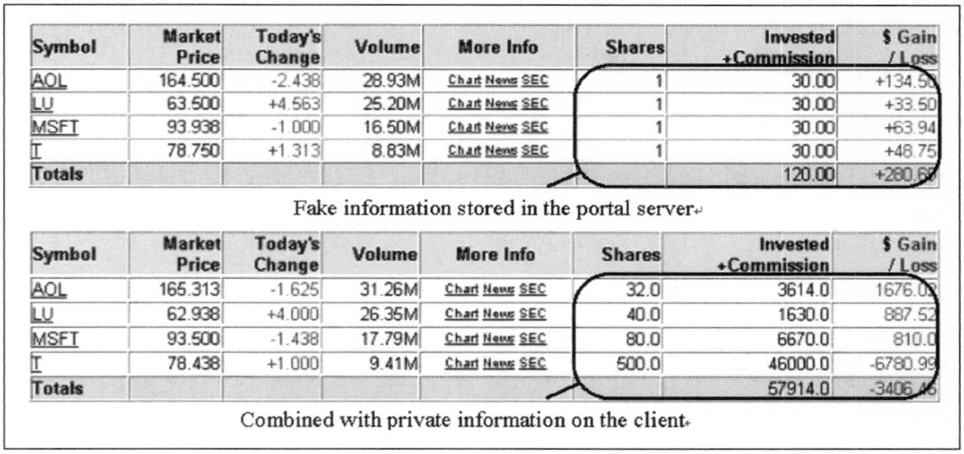

Symbol	Market Price	Today's Change	Volume	More Info	Shares	Invested +Commission	$ Gain / Loss
AOL	164.500	-2.438	28.93M	Chart News SEC	1	30.00	+134.50
LU	63.500	+4.563	25.20M	Chart News SEC	1	30.00	+33.50
MSFT	93.938	-1.000	16.50M	Chart News SEC	1	30.00	+63.94
T	78.750	+1.313	8.83M	Chart News SEC	1	30.00	+48.75
Totals						120.00	+280.69

Fake information stored in the portal server.

Symbol	Market Price	Today's Change	Volume	More Info	Shares	Invested +Commission	$ Gain / Loss
AOL	165.313	-1.625	31.26M	Chart News SEC	32.0	3614.0	1676.0
LU	62.938	+4.000	26.35M	Chart News SEC	40.0	1630.0	887.52
MSFT	93.500	-1.438	17.79M	Chart News SEC	80.0	6670.0	810.0
T	78.438	+1.000	9.41M	Chart News SEC	500.0	46000.0	-6780.99
Totals						57914.0	-3406.46

Combined with private information on the client.

Figure 17.7 My Portfolio with/without Private Information

filter for the stock page, iProxy can retrieve the private information and combine it with stock quotes provided by the remote portal site to compute the balance and net gain/loss. The output filter instructs iProxy to apply the Java class portfolio on the local server whenever the browser issues the corresponding HTTP request. The real numbers are visible to the client only (see the lower table in Figure 17.7), which are not revealed to the external portal server.

Portal Script

A proxy-based portal integrates the contents stored in a proxy server with those provided by a regular portal server such as *Your WorldNet*. Consider the following scenario in which a portal server works in concert with a client-side iProxy. Through the browser, a user sends an URL with a proxy directive:

```
http://www.att.net/?iProxy&action=portal
```

iProxy first retrieves the home page from www.att.net. This homepage has encoded iProxy directives that are used to process the local data and merge them with server content. iProxy then presents the personal portal page to the user. In order to provide a non-intrusive environment to other users who do not use iProxy, we embed the iProxy directives in HTML comments to generate the portal page. Consider the directives listed in Figure 17.8. iProxy intercepts these directives and performs necessary

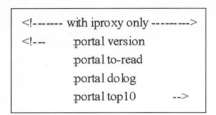

```
<!------- with iproxy only --------->
<!--      .portal version
          .portal to-read
          .portal dolog
          .portal top10        -->
```

Figure 17.8   A Portal Script Example

actions before returning the portal page. The directive `version` prints out the version number of iProxy. The directive `to-read` constructs the list of web pages scheduled to be read. The directive `dolog` analyzes the current web access log to produce the statistics. Finally, the directive `top10` presents the results on the personal portal. Browsers without iProxy support will ignore all directives embedded in the HTML comment.

17.2   iMobile Service Platform

Based on the iProxy middleware, iMobile provides a platform to support personalized mobile services. This platform aims to hide the complexity of multiple devices and content sources from mobile users. As shown in Figure 17.9, iMobile adds three agent abstractions to iProxy: *dev-let*, *info-let*, and *app-let*. These components communicate with each other through the *let engine*. This section describes the detailed interactions among these let agents in the iProxy environment.

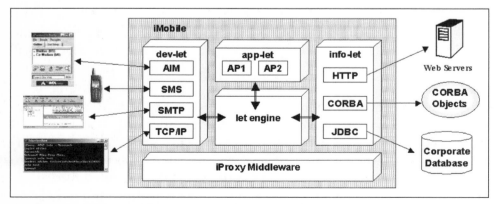

Figure 17.9   The iMobile Architecture

17.2.1   Dev-Let

A dev-let is a device controller that provides various protocol interfaces to user devices. Each dev-let interacts with the let engine through a well-defined interface. It receives user requests (character streams), and returns results in a *Multipurpose Internet Mail Extension (MIME)* type acceptable by the receiving device. Figure 17.9 shows four dev-let examples:

- The AIM dev-let is an AOL Instant Messenger client, which receives personal messages as requests and returns the result messages to the sender.

- The SMS dev-let uses short message services in GSM networks for message delivery.

- The mail dev-let is a mailer that receives requests from the mail server using POP3 and/or IMAP, and sends results to the email address of the sender using the *Simple Mail Transfer Protocol (SMTP)*.

- The TCP/IP dev-let accepts socket connections from the Internet, which interacts with mobile users as a character console.

The let engine manages these dev-lets and communicates with various mobile devices through these dev-lets.

When iMobile is started, a dev-let for each communication protocol is created, which listens to incoming requests delivered through that particular protocol. For example, the AIM dev-let starts an AIM client and listens to instant messages sent from other AIM clients.

The device driver for a particular protocol may co-locate with the dev-let or it may communicate with the dev-let through a TCP-based protocol. This approach allows a device driver to run on a remote machine other than the iMobile server. Figure 17.10 shows an example in which an SMS dev-let communicates with the GSM cellular phone attached to a remote PC through an SMS driver [Rao01a]. The protocol for the SMS driver is AT command described in Section 1.3.1. Mobile users send short messages to this cellular phone. The cellular phone then forwards the messages to iMobile for processing.

iMobile supports dev-lets that understand protocols including SMS, IMAP, AIM, ICQ, and Telnet. iMobile also supports WAP and HTTP through the iProxy HTTP interface. To allow email access to iMobile, the email dev-let periodically monitors messages arriving at a particular email account. A Telnet user can enter iMobile commands through a typical Unix or Windows terminal.

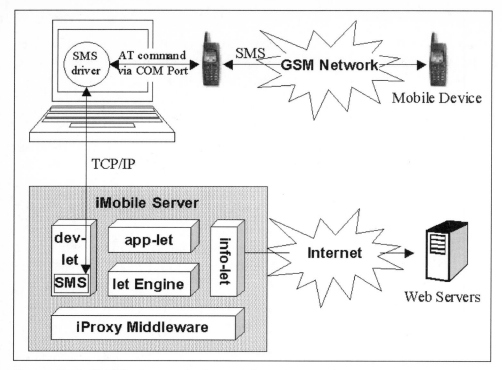

Figure 17.10 iMobile Communication Path for an SMS Device

While all iMobile devices and supported protocols have different user interfaces, every dev-let interacts with the let engine in a standard way. The naming of each device or destination address follows the URL naming format; that is, a protocol name followed by an account name or address. Examples for destination addresses include the following:

- `sms:+19737086242` (GSM phone)
- `aim:sunshine4` (AIM buddy name) and
- `mail:iProxy@research.att.com` (email identification)

Suppose that an iMobile user queries the AT&T (T) stock price. The user invokes the `quote` app-let by issuing the following message to iMobile:

quote T

If the request is sent using SMS on the GSM network, then the result will be returned as plain text to the receiving GSM phone. Conversely, if the mobile user wants to forward the result to the email account `herman@research.att.com`, then the GSM phone issues the following command:

forward mail:herman@research.att.com quote T

Since the email account understands the MIME type TEXT/HTML (according to the device profile to be elaborated later), the result will be delivered as an HTML file (which may include graphics) to `herman@research.att.com`.

The dev-let abstraction allows users in different networks to easily communicate with each other. For example, if a GSM subscriber wants to send a message to an AT&T PocketNet mail account `chen@mobile.att.net` on the CDPD network, and an AT&T TDMA phone number `908-500-0000` (Chen's cellular phone) through SMS, then the sender can use the message relay service supported by the `echo` app-let:

forward mail:chen@mobile.att.net,sms:+19085000000 echo call your boss

In this example, the sender really wants to reach a person, not a device. Since iMobile can map the user to devices (see Section 17.3.1), and it keeps track of the user's last access from a particular communication channel, we can use an alias to reach either all devices or the last device used by Chen.

17.2.2 Info-Let

The info-let abstraction extends the HTTP protocol and URL namespace to provide abstract views of various information spaces. An info-let may retrieve or modify an information space. Retrieved information may be passed to an app-let for further processing. Examples of information spaces include the following:

Stock quotes, weather, flight schedules, et cetera, are available on many web sites, but it would be better to retrieve such information from XML files or databases directly. Figure 17.11 shows an AIM client, Chen, that talks to an iMobile AIM agent. Chen issues the command

flight 001

to query the flight information on NorthWest airlines. The output includes time and gate information for each leg of the flight. The mapping from the flight command to NorthWest airlines is controlled by an app-let that consults the user profile of Chen. Also, the let engine invokes necessary transcoding to tailor the elaborate content on the web site to a format appropriate for the receiving AIM device. Figure 17.12 shows

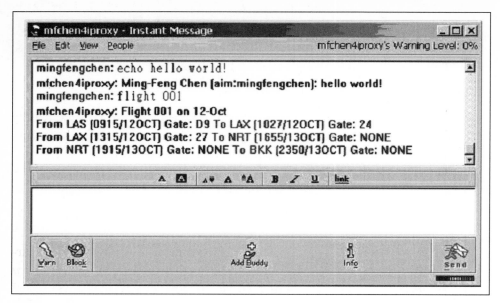

Figure 17.11 Flight Schedule Service

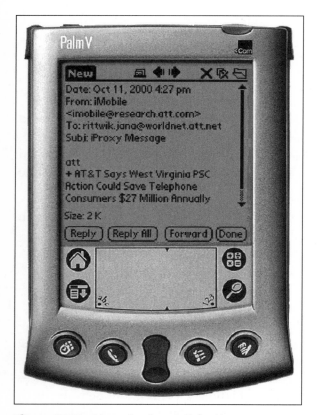

Figure 17.12 News Service on Palm V

a Palm V (with a wireless modem) that just sent an email to the iMobile email dev-let (`imobile@research.att.com`) with the command

   sitenews att

This command instructs iMobile to access the service provided by AT&T's Website News, which reports today's new hyperlinks on AT&T's website (`http://www.att.com`). The result is sent back as an email formatted for the Palm V PDA.

Corporate Database is typically accessed through the *Java Database Connectivity* (*JDBC*) and *Open Database Connectivity* (*ODBC*) interfaces. iMobile hosts a JDBC info-let that allows mobile users to access or update enterprise database information (employee data, marketing/sales data, system interface data, etc.) through *Structured Query Language* (SQL)-like queries. For example, Figure 17.13 shows how a user accesses an enterprise database through an AIM client to find a particular service and the date for that service. One can also access the same information from a PDA that supports AIM. In any case, it

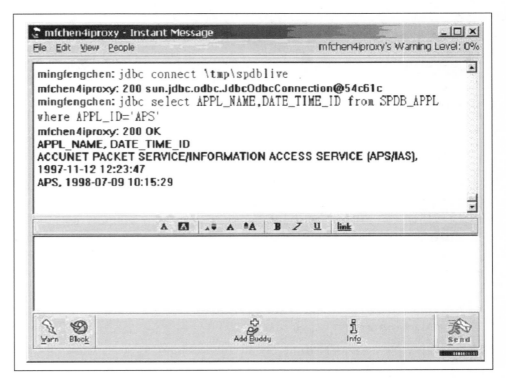

Figure 17.13   Corporate Database Access

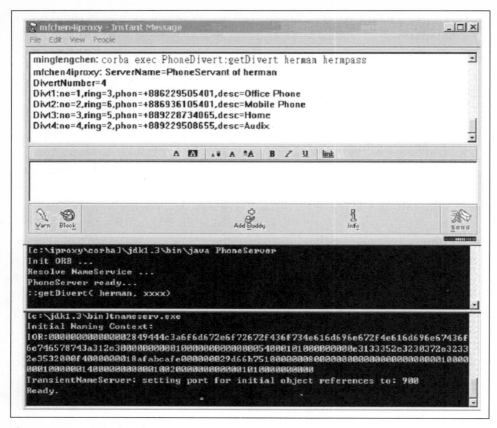

Figure 17.14   CORBA Object Access

is critical that only end-to-end solutions are used for mobile access to corporate databases.

Network/Infrastructure Resources are typically accessed through the *Common Object Request Broker Architecture (CORBA)* interface. CORBA is an architecture and specification for managing distributed program objects in a network. It allows programs at different locations to communicate through an *interface broker*. iMobile hosts a CORBA info-let that allows mobile users to request services from CORBA objects. Figure 17.14 shows how an AIM user obtains phone diversion information for the user Herman through the CORBA info-let that accesses a phone server.

X10 Home Network Devices for home appliances (such as lamps and the garage door opener) are connected on the same power line. The X10 device control signals are issued by a computer, and are delivered through the power line. iMobile hosts an X10 info-let that determines

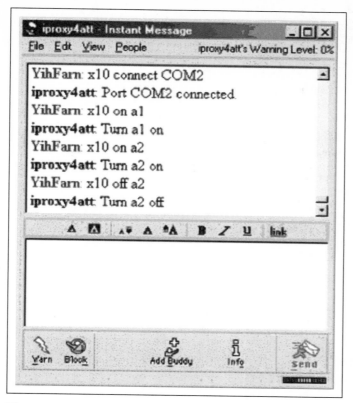

Figure 17.15   X10 Home Network Access

when and how to activate certain X10 devices for home environment control. Figure 17.15 shows how an iMobile user controls X10 devices remotely. The user instructs iMobile to locate a *firecracker* (a device that is capable of sending a radio signal to a transceiver device on the X10 network) through the serial port on the iMobile server. After the connection is established, the user sends the command

  x10 on a1

to turn on the fan (which is named device a1 on that particular X10 network) and

  x10 on a2

to turn on the coffee pot. The X10 interface on iMobile allows a mobile user to remotely control the home appliances from anywhere in the world with a cellular phone, an Instant Messaging client, or any mobile device supported by iMobile. This example demonstrates how an

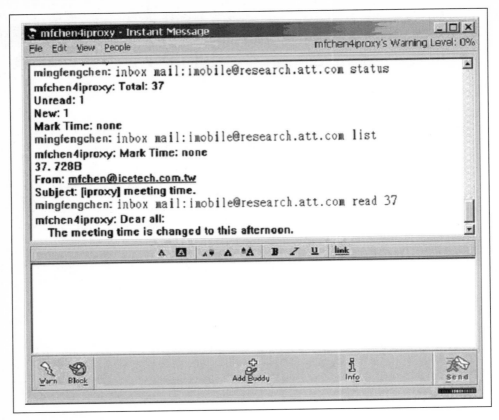

Figure 17.16 Email Service

info-let can be used to both retrieve and change the state of an information space.

Mail Servers are managed by an IMAP info-let called *inbox* that can access a user's email account. In this scenario, an encrypted email password is required for user authentication. In Figure 17.16, a mobile user checks the unread messages in his/her inbox. He/she then looks at the size (e.g., message 37 has 728 bytes), subject, and sender of every message before actually viewing it. Such an interaction style is typical for a mobile user using a communication device with limited bandwidth and screen space.

17.2.3 App-Let

An app-let implements service or application logic that processes contents from different sources and relays the results to various destination devices through dev-lets. An app-let may have complex interactions with info-lets.

```
#!/iproxy/script
# get the localtion information (zip)
:javabin infoLet zip getlocation
# get top10 movies (mlist)
:javabin infoLet mlist top10 movie
:foreach mtitle ${mlist}
# List the movie title
-- Movie: ${mtitle} --
:javabin infoLet thlist findTheater ${mtitle} ${zip}
:foreach theater ${thlist}
# List the theater
${theater}
:endfor
:endfor
```

Figure 17.17   Find-Me-A-Movie App-Let

Figure 17.17 shows a `FindMeAMovie` app-let implemented as an iProxy script. The app-let finds local theaters showing the top 10 movies by executing the following steps:

1. Obtain the location (zip code) of the cellular phone through the mobile location service.

2. Identify the top 10 movies from a movie database or web site.

3. For each of these movies, check if any local theater is showing that movie.

4. List the theaters.

17.3   User and Device Management

When the let engine receives commands from user devices, it translates the commands according to user aliases and profiles. This section describes device and user profiles.

17.3.1   Device Profile and Device-to-User Mapping

Every abstract device must register its profile information with the let engine. The format of a device name is `protocol:acct_id`. For example, an AIM device for a user webciao is `aim:webciao`. A device profile is a list of (case-insensitive) attribute-value pairs. The most important attribute is `dev.format.accept`, which determines the MIME type to be accepted by the device. iMobile uses this information to transcode original content to an appropriate format for this device. iMobile maintains a default profile for each device type. A device instance can overwrite the default profile with device-specific information. Consider the default profile for an email device. The profile has the following content:

```
dev.format.accept=text/html,*/*
dev.page.size=0
```

This profile indicates that the default MIME type is TEXT/HTML, but all MIME types (`*/*`) are acceptable. Also, the page size 0 means that there is no size limit for each message transmission. These values are inherited by all mail devices unless they are overwritten. For example, the above default values are not appropriate for emails used in cellular phones (say, AT&T's PocketNet phone). Instead, the device profile for that cellular phone uses the following rules:

```
dev.format.accept=text/plain
dev.page.size=230
```

which indicates that only the MIME type text/plain is appropriate and the phone does not accept messages longer than 230 characters. The device profile may also specify how and when to access this device. For example, a profile may include the following entries:

```
mail.store.checktime=10000
mail.store.url=imap://imobile:password@bigmail
mail.transport.url=smtp://bigmail
```

which specifies the frequency (every 10 seconds) for checking the email account (`store.checktime`), the account password (`store.url`), the transport protocol for sending email (`transport.url`), et cetera. Each device is mapped to a registered iMobile user. There are two reasons for this mapping:

- To ensure access for legitimate iMobile users
- To personalize a service based on the user profile

Examples of device-to-user mappings are shown below:

```
sms:+886936731826=herman
sms:+19087376842=chen
mail:dchang@research.att.com=difa
aim:webciao=chen
```

17.3.2 User Profile

iMobile authenticates a mobile user based on the device or protocol used by that user. An example of the iMobile user profile is given in Figure 17.18.

In this example, the user profile stores the user name, the password, and a list of devices that the user registers with iMobile. It also stores command and address aliases. When a user accesses iMobile through AIM using the identification (Id) webciao, iMobile determines from the user-device mapping that the user is "Chen". Therefore, iMobile will use the user profile of Chen to handle all subsequent service requests from this device. For example, a user may send the following short message using a GSM phone:

forward $mail.1 Q T

In this short message, the special character $ requests iMobile to map the named device (mail.1) to the corresponding entry in the profile. According to the user profile in Figure 17.18, iMobile interprets the short message as

forward mail:chen@research.att.com quote T

```
name=Chen
password=xf2gbH3
default=$mail.1
# my addresses
sms.1=sms:+19087376842
mail.1=mail:chen@research.att.com
mail.2=mail:imobile@mobile.att.net
mail.all=$mail.1,$mail.2
aim.1=aim:webciao
# command aliases
sms.cmd.q=quote
sms.cmd.sn=sitenews
# address aliases
sms.addr.cc=aim:chrischen
```

Figure 17.18 A User Profile Example

This user profile also stores the user's last access device in the default parameter. Other mobile users may send the following message to reach the user:

forward $chen echo call your boss

The alias ($ + username) requests iMobile to look up the last access device (mail.1) of Chen and interpret the message as

forward mail:chen@research.att.com echo call your boss

As a final remark, iMobile assumes that wireless networks (such as FarEasTone GSM or AT&T TDMA networks in North America) are reliable, which provide legal cellular phone Ids through short messages. This assumption is generally acceptable unless a cellular phone is stolen and the user did not lock the phone with a secure password (see Section 9.2). iMobile also trusts AOL authentication for non-critical services. Extra user authentication through iMobile is required if the user accesses iMobile through Telnet, WAP, or HTTP. The authentication information should be stored in the user profiles.

17.4 iMobile-based Peer-to-Peer Mobile Computing

Many resources available to a mobile device may not be readily available on any networked servers. Examples of the resources include location information, locally captured media files, and the device's exposure to surrounding resources, such as thermometers or X10 cameras that are wirelessly connected. With these resources, the *Peer-to-Peer* (*P2P*) computing paradigm will also enable mobile devices to directly exchange information with each other.

Figure 17.19 shows an example of a P2P mobile computing environment among four mobile users residing in different locations, with various access networks and devices. The mobile user Chen is in Paris with an iPAQ connected to the Internet through wireless LAN, possibly provided by a coffee shop. He may want to access a specific image captured by his friend Wei in New York with a Palm device connected to a CDPD/TDMA network. Chen is also interested in the location information of another user in San Antonio, and the address information of a friend in Paris, stored in the mobile device owned by his friend in San Diego. These contents, stored in individual mobile devices, are not available on any network-based servers. To access the contents, Chen must send the requests to other mobile devices. Because

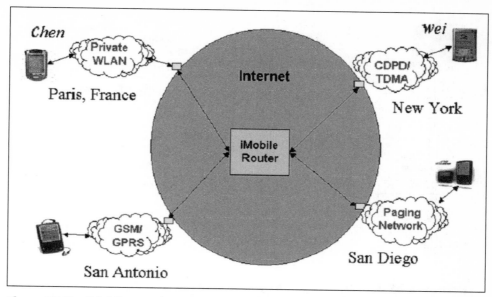

Figure 17.19 iMobile-Based Peer-to-Peer Mobile Computing

these mobile devices may not always connect to the Internet, we introduce *iMobile Router*, a network-based server that locates the mobile devices and routes the messages among them.

The iMobile-based P2P computing proposes a lightweight service platform called iMobile Micro Edition (ME), which provides personalized services on the mobile devices. iMobile ME is a simplified version of the iMobile platform described in the previous sections (which is referred to as the standard iMobile in this section). iMobile ME minimizes the requirements of system resources and is suitable for execution on mobile devices.

As shown in Figure 17.20, iMobile ME consists of two agent abstractions: dev-let and info-let. These components communicate with each other through the let engine. The let engine, dev-let, and info-let perform the same functions as those in the standard iMobile platform. The major differences between the iMobile ME and the standard iMobile platform are described below:

Data Encoding: To support a flexible output format in standard iMobile, the info-lets need to generate the results in MIME format and provide a transcoding mechanism if the result type is not acceptable to the destination device. In iMobile ME, the dev-lets and info-lets only support plain text. This simplification reduces the required communication bandwidth and the effort of data processing on mobile devices.

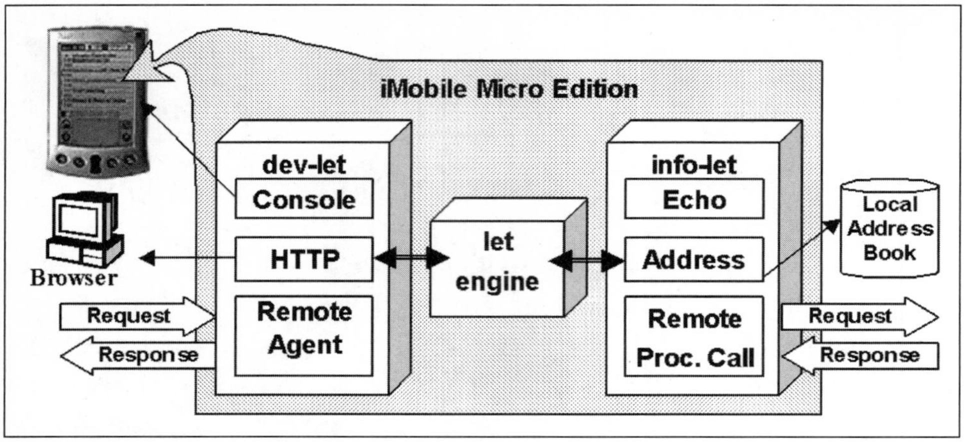

Figure 17.20   iMobile ME System Architecture

Removal of App-let: In standard iMobile, an app-let implements application logic by processing information obtained from one or more info-lets. This powerful abstraction allows the creation of complicated services by using the iProxy scripts that invoke functions defined in info-lets. Therefore, the info-lets must provide many functions to support the app-lets. Conversely, to reduce the overhead of processing the requests, iMobile ME removes the app-let component and the script parser. This simplification allows more straightforward execution.

Remote Access: iMobile ME introduces a `Remote Agent` dev-let and a `Remote Procedure Call` (RPC) info-let to support remote access among mobile devices. The `Remote Agent` dev-let accepts the requests from other mobile devices and returns the results to the senders. The RPC info-let forwards the local requests to the remote mobile devices and returns the results obtained from these remote devices. `Remote Agent` and `RPC` are not found in standard iMobile.

Message Queuing: A mobile device may be disconnected or connected with limited bandwidth, and it may be difficult to retain a long communication session between two interacting mobile devices. Therefore, each remotely accessible dev-let/info-let is extended with an *inbox queue*, which accumulates incoming messages, and an *outbox queue*, which buffers outgoing messages.

Message Routing: iMobile ME communicates with the iMobile Router to exchange queued messages. The iMobile Router is a network-based server, which routes requests and responses among mobile devices. It stores the messages in the queues for every iMobile ME, and synchronizes with the queues of all iMobile MEs.

Based on iMobile ME, examples of P2P services and the details of queue synchronization are elaborated as follows.

17.4.1 iMobile ME Services

Figure 17.20 shows examples for iMobile ME dev-lets and info-lets implemented in Java 2 Micro Edition (J2ME). An iMobile ME provides two basic dev-lets: `Console` and `Remote Agent`. The `Console` dev-let provides a pure-text console to send requests and display responses. The `Remote Agent` dev-let receives requests from other mobile devices and returns the results to these devices. If the mobile device is powerful enough to run a simple web server (for example, iPAQ), iMobile ME may also provide a HTTP dev-let to receive requests directly from web browsers.

An info-let exports the local resource of a mobile device to other devices. Some examples are listed below:

- The `Echo` info-let simply echoes the received string. This info-let is useful for checking the system and connections among mobile devices. The `Echo` info-let also provides round-trip delay statistics from one mobile device to another.

- The `Address` info-let provides a lookup interface for the address book database, which can be found in most mobile devices. Figure 17.21 illustrates the response shown on a Palm device for address lookup of a user named Huang.

- The `Remote Procedure Call` (RPC) info-let parses a request to obtain the `destination` and `command` parameters. Based on the parameters, the RPC forwards the command to the corresponding destination.

- The `Sensor` info-let exposes the location or environment information of a mobile device that has a built-in location determination system or sensor to obtain its surrounding environment information (such as temperature and moisture).

17.4.2 Queue Synchronization

iMobile ME stores the incoming and outgoing messages in queues and synchronizes with the queues of other iMobile MEs defined in the iMobile

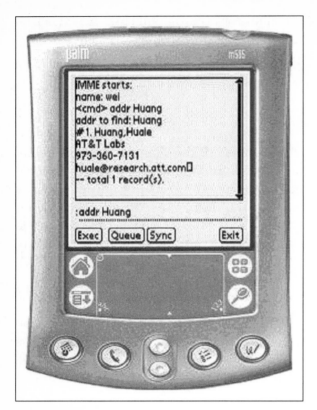

Figure 17.21 An Address Info-Let Example

Router. Queue synchronization is issued by an iMobile ME when it connects to the Internet. Every iMobile ME registers a unique Id to the iMobile Router and uses this Id to communicate with others. Figure 17.22 shows an example of a remote procedure call from iMobile ME1 to ME2, which includes four synchronization actions:

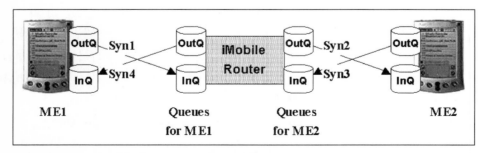

Figure 17.22 Queue Synchronization Example

1. A RPC request is issued from the `Console` dev-let in ME1. The `RPC` info-let receives the request and stores it in the outbox queue. Then ME1 issues the first synchronization that forwards the request to the iMobile Router (see Syn 1 in Figure 17.22). The iMobile Router buffers the request in the outbox queue for ME2 and waits for ME2 to retrieve the request.

2. When ME2 issues the second synchronization, it obtains the request from the iMobile Router (Syn 2 in Figure 17.22). ME2 executes the request by invoking the corresponding info-let, and stores the response in its outbox queue.

3. ME2 issues the third synchronization that sends the response to the iMobile Router (Syn 3 in Figure 17.22). The iMobile Router stores the response in the outbox queue for ME1.

4. ME1 issues the fourth synchronization to receive the response (Syn 4 in Figure 17.22). Finally, the `Console` dev-let shows the response on the screen of ME1.

Figure 17.23 shows an RPC request example issued by Chen to look up the address for Huang in Wei's device. This figure shows two screen shots that capture the interactions between two ME devices, Chen and Wei.

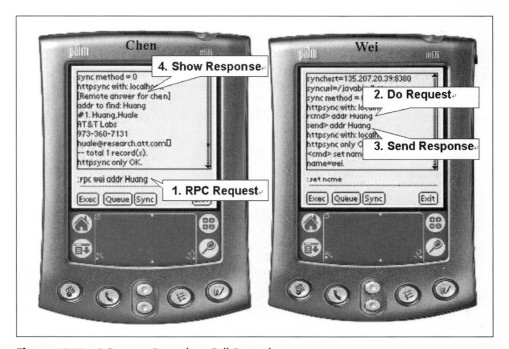

Figure 17.23 A Remote Procedure Call Example

17.5 Concluding Remarks

This chapter described iMobile, a proxy-based platform for developing mobile services for various mobile devices and wireless access technologies. iMobile introduces three abstractions on top of an agent-based proxy: dev-let, which interacts with various access devices and protocols; info-let which accesses multiple information spaces; and app-let, which implements application and service logic. The let engine arbitrates the communications among dev-lets, app-lets, and info-lets. The let engine also maintains user and device profiles for personalized services. The iMobile vision allows a mobile user to access vast amounts of information available on various wired and wireless networks regardless of where the user is and what device or communication protocol is available. This modular architecture allows developers to write device drivers, information access methods, and application logic independently of each other.

We also elaborated on a simplified iMobile platform called iMobile ME. The iMobile ME architecture provides a uniform architecture on mobile devices, and allows these devices to both communicate with and access resources from each other. As mobile devices become more powerful, iMobile ME will provide an ideal infrastructure to facilitate P2P mobile computing. An early version of this chapter appeared in the Wireless Communications and Mobile Computing Journal. The major contributors are Herman Rao and M.-F. Chen.

17.6 Questions

1. Describe the iMobile architecture. What is the role of iProxy in this architecture?

2. Does iMobile follow the OSA design guidelines described in Section 15.4? Explain why or why not.

3. Describe the four components of iProxy. Can you add/remove components to/from iProxy? Which component in iProxy is responsible for routing?

4. Describe the action and the view specifications. How can they integrated into the URL format? What happens if the node receiving the modified URL cannot recognize action/view?

5. Describe the filters defined in the proxy component of iProxy. Why does iProxy need these filters?

6. Describe the following personal services provided by iProxy: personal web archive, my stock portfolio, and portal script. Can you enhance these services by adding new features to them?

7. Describe dev-let, info-let, and app-let, and the relationship among them and the let engine.

8. Show how iSMS in Chapter 1 can be integrated into iMobile ME.

9. Describe the following info-lets: a corporate database and X10 home network devices. Can you extend these info-lets by adding more features to them?

10. Describe the iMobile user profile. Why is the user profile so important in provisioning services?

11. Describe how peer-to-peer mobile computing can be achieved in the iMobile platform. Why is iMobile ME more appropriate for peer-to-peer computing than standard iMobile?

12. Describe the major differences between standard iMobile and iMobile ME (for example, in terms of data encoding, remote access, removal of app-lets, and so on).

13. Describe the info-lets supported in iMobile ME.

Bibliography

[3CO00] 3COM Inc. A SIP Application Level Gateway for Network Address Translation. IETF draft-biggs-sip-nat-00, March 2000.

[3GP99] 3GPP. 3rd Generation Partnership Project; Technical Specification Group Services and Systems Aspects; Architectural for an All IP Network. Technical Report 3G TR 23.922 Version 1.0.0 (1999-10), 1999.

[3GP00a] 3GPP. 3rd Generation Partnership Project; Technical Specification Group Core Network; Feasibility Technical Reprot – CAMEL Control of VoIP Services. Technical Report 3G TR 21.978 Version 3.0.0 (2000-06), 3GPP, 2000.

[3GP00b] 3GPP. 3rd Generation Partnership Project; Technical Specification Group Services and System Aspects; Combined GSM and Mobile IP Mobility Handling in UMTS IP CN. Technical Report 3G TR 23.923 Version 3.0.0 (2000-05), 2000.

[3GP01] 3GPP2. 3rd Generation Partnership Project 2; 3GPP2 Access Network Interfaces Interoperability Specification. 3GPP2 A.S0001-A Version 2.0 (2001-06), 2001.

[3GP02a] 3GPP. 3rd Generation Partnership Project; Technical Specification Group Core Network; General Packet Radio Service (GPRS); GPRS Tunneling Protocol (GTP) across the Gn and Gp Interface. Technical Specification 3G TS 09.060 Version 7.10.0 (2002-12), 2002.

[3GP02b] 3GPP. 3rd Generation Partnership Project; Technical Specification Group RAN 3; Handovers for Real Time Services from PS-domain . Technical Report 3G TR 25.936, 3GPP, 2002.

[3GP02c] 3GPP. 3rd Generation Partnership Project; Technical Specification Group Services and Systems Aspects; Architectural Requirements for Release 1999. Technical Report 3G TS 23.121 Version 3.6.0 (2002-06), 3GPP, 2002.

[3GP02d] 3GPP2. 3rd Generation Partnership Project 2; Wireless IP Network Standard. 3GPP2 P.S0001-B Version 1.0 (2002-10), 2002.

[3GP03a] 3GPP. 3rd Generation Partnership Project; Feasibility Study on 3GPP System to Wireless Local Area Network (WLAN) Interworking. Technical Specification 3G TS 22.934 Version 6.2.0 (2003-09), 2003.

[3GP03b] 3GPP. 3rd Generation Partnership Project; Technical Specification Group Terminals; Technical Realization of Cell Broadcast Service (CBS). Technical Report 3G TS 23.041 Version 4.4.0 (2003-03), 3GPP, 2003.

[3GP03c] 3GPP2. 3rd Generation Partnership Project 2; IP Network Architecture Model for cdma2000 Spread Spectrum Systems. 3GPP2 S.R0037-0 Version 3.0 (2003-09), 2003.

[3GP04a] 3GPP. 3rd Generation Partnership Project; Technical Specification Core Network; Open Service Access (OSA); Application Programming Interface (API); Part 1: Overview; (Release 4). Technical Specification 3G TS 29.198-1 Version 6.3.1 (2004-12), 2004.

[3GP04b] 3GPP. 3rd Generation Partnership Project; Technical Specification Group Core Network; Mobile Station (MS) - Serving GPRS Support Node (SGSN); Subnetwork Dependent Convergence Protocol (SNDCP) (Release 6). Technical Specification 3G TS 44.065 V6.3.0 (2004-09), 2004.

[3GP04c] 3GPP. 3rd Generation Partnership Project; Technical Specification Group Core Network; Multimedia Messaging Service (MMS); Functional description; Stage 2. 3G TS 23.140, 2004.

[3GP04d] 3GPP. 3rd Generation Partnership Project; Technical Specification Group Core Network; Support of Mobile Number Portability (MNP); Technical Realisation; Stage 2. Technical Specification 3G TS 23.066 Version 6.0.0 (2004-12), 2004.

[3GP04e] 3GPP. 3rd Generation Partnership Project; Technical Specification Group Radio Access Network; UTRAN Overall Description. Technical Specification 3G TS 25.401 Version 6.5.0 (2004-12), 2004.

[3GP04f] 3GPP. 3rd Generation Partnership Project; Technical Specification Group Services and System Aspects; End-to-End Quality of

Service (QoS) concept and architecture (Release 5). Technical Specification 3G TS 23.207 Version 6.4.0 (2004-09), 2004.

[3GP04g] 3GPP. 3rd Generation Partnership Project; Technical Specification Group Services and Systems Aspects; 3G Security; Security Architecture. Technical Specification 3G TS 33.102 Version 6.3.0 (2004-12), 2004.

[3GP04h] 3GPP. 3rd Generation Partnership Project; Technical Specification Group Services and Systems Aspects; Functional Stage 2 Description of Location Services in UMTS for Release 1999. 3G TS 23.171 Version 3.11.0 (2004-03), 2004.

[3GP04i] 3GPP. 3rd Generation Partnership Project; Technical Specification Group Services and Systems Aspects; IP Multimedia Subsystem Stage 2. Technical Specification 3G TS 23.228 Version 6.8.0 (2004-12), 2004.

[3GP04j] 3GPP. 3rd Generation Partnership Project; Technical Specification Group Services and Systems Aspects; Telecommunication Management; Charging Management; Charging Data Description for the Packet Switched (PS) Domain (Release 5). Technical Report 3G TS 32.215 Version 5.7.0 (2004-09), 3GPP, 2004.

[3GP04k] 3GPP. 3rd Generation Partnership Project; Technical Specification Group Services and Systems Aspects; Virtual Home Environment/Open Service Access. Technical Report 3G TS 23.127 Version 6.1.0 (2004-06), 3GPP, 2004.

[3GP04l] 3GPP. 3rd Generation Partnership Project; Technical Specification Group Terminals; Technical realization of Short Message Service (SMS). Technical Report 3G TS 23.040 Version 6.5.0 (2004-09), 3GPP, 2004.

[3GP05a] 3GPP. 3rd Generation Partnership Project; Technical Specification Core Network; Cx and Dx Interfaces Based on the Diameter Protocol; Protocol Details. Technical Specification 3G TS 29.229 V6.4.0 (2005-03), 2005.

[3GP05b] 3GPP. 3rd Generation Partnership Project; Technical Specification Core Network; IP Multimedia Subsystem Cx and Dx Interfaces; Signaling Flows and Message Contents (Release 5). Technical Specification 3G TS 29.228 V6.6.1 (2005-04), 2005.

[3GP05c] 3GPP. 3rd Generation Partnership Project; Technical Specification Core Network; Mobile Application Part (MAP) specification (Release 1999). Technical Specification 3G TS 29.002 Version 6.9.0 (2005-03), 2005.

[3GP05d] 3GPP. 3rd Generation Partnership Project; Technical Specification Group Core Network; General Packet Radio Service (GPRS); GPRS Tunneling Protocol (GTP) across the Gn and Gp Interface (Release 6). Technical Specification 3G TS 29.060 Version 6.8.0 (2005-04), 2005.

[3GP05e] 3GPP. 3rd Generation Partnership Project; Technical Specification Group Core Network; Mobile Application Part (MAP) Specification. 3rd Generation Partnership Project. Technical Specification 3G TS 29.002 Version 6.9.0 (2005-03), 2005.

[3GP05f] 3GPP. 3rd Generation Partnership Project; Technical Specification Group Core Network; Mobile Radio Interface Layer 3 Specification; Core Network Protocols - Stage 3 (Release 6). 3GPP TS 24.008 Version 6.8.0 (2005-03), 2005.

[3GP05g] 3GPP. 3rd Generation Partnership Project; Technical Specification Group Core Network; Mobile Radio Interface Signalling Layer 3; General Aspects. 3GPP TS 24.007 Version 6.4.0 (2005-03), 2005.

[3GP05h] 3GPP. 3rd Generation Partnership Project; Technical Specification Group Core Network; Mobile Station - Serving GPRS Support Node (MS-SGSN); Logical Link Control (LLC) Layer Specification; (Release 6). Technical Specification 3G TS 44.064, Version 6.0.1 (2005-01), 2005.

[3GP05i] 3GPP. 3rd Generation Partnership Project; Technical Specification Group Core Network; Voice Group Call Service (VGCS) - Stage 2 (Release 4). 3G TS 43.068 Version 4.5.0 (2005-03), 2005.

[3GP05j] 3GPP. 3rd Generation Partnership Project; Technical Specification Group Radio Access Network; RRC Protocol Specification for Release 6. Technical Report 3G TS 25.331 Version 6.5.0 (2005-03), 3GPP, 2005.

[3GP05k] 3GPP. 3rd Generation Partnership Project; Technical Specification Group Radio Access Network; UTRAN Iu Interface Radio Access Network Applicaiton Part (RANAP) Signaling for Release 6. Technical Specification 3G TS 25.413 Version 6.5.0 (2005-03), 2005.

[3GP05l] 3GPP. 3rd Generation Partnership Project; Technical Specification Group Radio Access Network; UTRAN Iu Interface User Plane Protocols. Technical Specification 3G TS 25.415 Version 6.2.0 (2005-03), 2005.

[3GP05m] 3GPP. 3rd Generation Partnership Project; Technical Specification Group Services and System Aspects; Customized

Applications for Mobile Network Enhanced Logic (CAMEL); Service Description (Stage 1). Technical Specification 3G TS 22.078, Version 7.2.0 (2005-01), 2005.

[3GP05n] 3GPP. 3rd Generation Partnership Project; Technical Specification Group Services and System Aspects; Multimedia Messaging Service (MMS); Stage 1. 3rd Generation Partnership Project. Technical Specification 3G TS 22.140 Version 6.7.0 (2005-03), 2005.

[3GP05o] 3GPP. 3rd Generation Partnership Project; Technical Specification Group Services and System Aspects; Service Requirement for the Open Services Access (OSA); Stage 1. Technical Specification 3G TS 22.127, Version 6.8.0 (2005-03), 2005.

[3GP05p] 3GPP. 3rd Generation Partnership Project; Technical Specification Group Services and Systems Aspects; 3G Security; Access Security for IP-based Services. Technical Specification 3G TS 33.203 V6.6.0 (2005-03), 2005.

[3GP05q] 3GPP. 3rd Generation Partnership Project; Technical Specification Group Services and Systems Aspects; General Packet Radio Service (GPRS); Service Descripton; Stage 2. Technical Specification 3G TS 23.060 Version 6.8.0 (2005-03), 2005.

[3GP05r] 3GPP. 3rd Generation Partnership Project; Technical Specification Group Services and Systems Aspects; Network Architecture. Technical Specification 3G TS 23.002 Version 6.7.0 (2005-03), 2005.

[Abo99] Aboba, B., and Beadles, M. The Network Access Identifier. IETF RFC 2486, January 1999.

[ACA99] ACA. Mobile Number Portability in Australia. *Australian Communications Authority*, pages 1–45, 1999.

[ADC00] ADC. NewNet Connect7TM User Manual. Technical Report Part No. D-0116-US-214-000, ADC Telecommunication, 2000.

[Ahm03] Ahmavaara, K., Haverinen, H., Pichna, R. Interworking Architecture between 3GPP and WLAN Systems. *IEEE Wireless Communications Magazine*, 4, 2003.

[Ame97] American's Network. We don't want any. *American's Networks*, October, 1 1997.

[Ame05] American Online, Inc. AOL Instant Messenger. www.aim.com/, 2005.

[And03] Andreasen, F., Foster, B. Media Gateway Control Protocol (MGCP). Technical Report RFC 3435, Internet Engineering Task Force, 2003.

[ANS96] ANSI. American National Standard for Telecommunications—Signaling System Number 7: Message Transfer Part (MTP). Technical Report ANSI T1.111, ANSI, 1996.

[ANS00a] ANSI. American National Standard for Telecommunications—Signaling System Number 7: Integrated Services Digital Network (ISDN) User Part. Technical Report ANSI T1.113, ANSI, 2000.

[ANS00b] ANSI. American National Standard for Telecommunications—Signaling System Number 7: Transaction Capabilities Application Part (TCAP). Technical Report ANSI T1.114, ANSI, 2000.

[ANS01] ANSI. American National Standard for Telecommunications—Signaling System Number 7: Signaling Connection Control Part (SCCP). Technical Report ANSI T1.112, ANSI, 2001.

[Att98] Attenborough, N., Sandbach, J., Sadat, U., Siolis, G., Cartwright, M., and Dunkley, S. Feasibility Study and Cost Benefit Analysis of Number Portbility for Mobile Services in Hong Kong. Technical report, National Economic Research Associates, February 1998.

[ATT05] ATT. Your WorldNet. http://yourworldnet.planetdirect.com, 2005.

[Bel95] Bellcore. Advanced Intelligent Network 0.2 switching Systems generic requirement. Technical Report GR-1298-CORE, Bellcore, 1995.

[Ber98] Berners-Lee, T., et al. Uniform Resource Identifiers (URI): Generic Syntax. IETF RFC 2396, August 1998.

[Ber00] Berndt, H., et al. TINA: Its Achievements and Its Future Directions. *IEEE Communications Surveys and Tutorials*, (1), 2000.

[Bla05] BlackBerry. BlackBerry 7750 - Getting Started Guide. 2005.

[Blu 04] Bluetooth Forum. The Offical Bluetooth Website. http://www.bluetooth.com, 2004.

[Bos01] Bos, L., and Leroy, S. Toward an All-IP-based UMTS System Architecture. *IEEE Network*, 15(1):36–45, 2001.

[Cao02] Cao, G. Proactive Power-Aware Cache Management for Mobile Computing Systems. *IEEE Transactions on Computers*, 51(6):608–621, 2002.

[Cha98] Chang, M.-F., Lin, Y.-B., and Su, S.-C. Improving Fault Tolerance of GSM Network. *IEEE Network*, 1(12):58–63, 1998.

[Che97] Chen, Y.-F. and Koutsofios, E. WebCiao: A Website Visualization and Tracking System. *Proceedings of WebNet97*, 1997.

[Che99] Chetham, A. The Numbers Games – GSM in Hong Kong. *mobile comms. asia*, pages 6–8, February 1999.

[Che 05] Chen, Y.-K. and Lin, Y.-B. IP Connectivity for Gateway GPRS Support Node. *IEEE Wireless Communications*, 12(1):37–46, 2005.

[Che05a] Chen, M.F., Lin, Y.-B., Rao, R. C.-H., Wu, Q.C. A Mobile Service Platform Using Proxy Technology. *Wireless Communications and Mobile Computing Journal*, 2005.

[Che05b] Chen, W.E., Wu, Q.C., Pang, A.-C., and Lin, Y.-B. Design of SIP Application Level Gateway for UMTS. *Design and Analysis of Wireless Networks*, edited by Pan, Y., and Xiao, Y. Nova Science Publishers, 2005.

[Che05c] Cherry, S. Seven Myths about Voice over IP. *IEEE Sepectrum*, 42(3):46–51, 2005.

[Cho97] Cho, Y.-J., Lin, Y.-B., and Rao, C.-H. Reducing the Network Cost of Call Delivery to GSM Roamers. *IEEE Network*, 11(5):19–25, September/October 1997.

[Cho05] Chou, C.-M., Hsu, S.-F., Lee, H.-Y., Lin, Y.-C., Lin, Y.-B. Lin, and R.S. Yang. CCL OSA: A CORBA-based Open Service Access System. *International Journal of Wireless and Mobile Computing*, 2005.

[Cis04] Cisco. *Cisco SIP Proxy Server Administrator Guide, Version 2.2*. 2004.

[Col01] Collins, D. *Carrier Grade Voice Over IP*. McGraw-Hill, Singapore, 2001.

[Cri96] Crispin, M. Internet Message Access Protocol. IETF RFC 2060, December 1996.

[Daw 98] Dawson, F. vCard MIME Directory Profile. Technical Report RFC 2426, Internet Engineering Task Force, 1998.

[Don00] Donovan, S. The SIP INFO Method. IETF RFC 2976, October 2000.

[Dou98] Douglis, F., Ball, T., Chen, Y.-F., and Koutsofios, E. The AT&T Internet Differencing Engine: Tracking and Viewing Changes on the Web. *World Wide Web Journel*, 1(1), 1998.

[EIA97] EIA/TIA. Cellular Radio-telecommunications Intersystem Operations (Rev. D). Technical Report IS-41, EIA/TIA, 1997.

[ETS93a] ETSI/TC. Location Registration Procedures. Technical Report Recommendation GSM 03.12, ETSI, 1993.

[ETS93b] ETSI/TC. Restoration Procedures, Version 4.2.0. Technical Report Recommendation GSM 03.07, ETSI, 1993.

[ETS97a] ETSI/TC. Alphabets and Language-Specific Information. Technical Specification. Technical Report Recommendation GSM 03.38 Version 5.6.0 (Phase 2+), ETSI, 1997.

[ETS97b] ETSI/TC. European Digital Cellular Telecommunications System (Phase 2+); GPRS Base Station System (BSS) - Serving GPRS Support Node (SGSN) Interface; Gb Interface Layer 1. Technical Report Recommendation GSM 08.14 Version 6.0.0, ETSI, 1997.

[ETS98a] ETSI. Telecommunications and Internet Protocol Harmonization over Networks (TIPHON): Description of Technical Issues. Technical Report TR 101 300 V1.1.5, ETSI, 1998.

[ETS98b] ETSI/TC. AT Command Set for GSM Mobile Equipment (ME). Recommendation GSM 07.07 Version 7.8.0, 1998.

[ETS 98c] ETSI/TC. Use of Data Terminal Equipment-Data Circuit Terminating; Equipment (DTE-DCE) Interface for Short Message Service (SMS) and Cell Broadcast Service (CBS). Recommendation GSM 07.05 Version 7.0.1, 1998.

[ETS 99] ETSI/TC. Mobile Radio Interface Signaling Layer 3 (Phase 2), Version: 7.1.0. Technical Report Recommendation GSM 04.07, ETSI, 1999.

[ETS 00a] ETSI/TC. European Digital Cellular Telecommunications System (Phase 2+); GPRS Base Station System (BSS) - Serving GPRS Support Node (SGSN) BSS GPRS Protocol (BSSGP). Technical Report Recommendation GSM 08.18 Version 8.3.0, ETSI, 2000.

[ETS 00b] ETSI/TC. GPRS Service description Stage 2. Technical Report Recommendation GSM GSM 03.60 Version 7.4.1 (Phase 2+), ETSI, 2000.

[ETS 00c] ETSI/TC. GPRS Tunnelling Protocol (GTP) across the Gn and Gp Interface. Technical Report Recommendation GSM GSM 09.60 Version 7.3.0 (Phase 2+), ETSI, 2000.

[ETS02] ETSI. ISDN; SS7; ISUP; Enhancements for Support of Number Portability. Technical Report Recommendation EN 302 097 v1.2, ETSI, 2002.

[ETS 04] ETSI/TC. Mobile Radio Interface Layer 3 Specification. Recommendation GSM 04.08, 2004.

[Fan00a] Fang, Y., Chlamtac, I., and Fei, H. Analytical Results for Optimal Choice of Location Update Interval for Mobility Database Failure Restoration in PCS Networks. *IEEE Transactions on Parallel and Distributed Systems*, 2000.

[Fan00b] Fang, Y., Chlamtac, I., and Lin, Y.-B. Portable Movement Modeling for PCS Networks. *IEEE Transactions on Vehicular Technology*, 49(4), 2000.

[Fen04] Feng, V. W.-S., Wu., L.-Y., Lin, Y.-B., and Chen, W. E. WGSN:

WLAN-based GPRS Environment Support Node with Push Mechanism. *The Computer Journal*, 47(4):405–417, 2004.

[Fos03] Foster, M., McGarry, T., and Yu, J. Number Portability in the GSTN: An Overview. RFC 3482, 2003.

[Gra98] Granberg, O. GSM on the Net. *Ericsson Review*, (4):184–191, 1998.

[Gro03] Groves C., et. al. Gateway Control Protocol (Version 1). Technical Report RFC 3525, Internet Engineering Task Force, 2003.

[Haa98] Haas, Z., and Lin, Y.-B. On Optimizing the Location Update Costs in the Presence of Database Failures. *ACM Wireless Networks Journal*, 4(5):419–426, 1998.

[Ham99] Hamzeh, K., Pall, G., Verthein, W., Taarud, J., Little, W., and Zorn, G. Point-to-Point Tunneling Protocol (PPTP). IETF RFC 2637, July 1999.

[Han94] Hanks, S., Li, T., Farinacci, D., and Traina, P. Generic Routing Encapsulation over IPv4 networks. IETF RFC 1702, October 1994.

[Han98] Handley, M., and Jacobson, V. SDP: Session Description Protocol. IETF RFC 2327, April 1998.

[Hei99] Heinanen, J., Baker, F., Weiss, W., and Wroclawski, J. Assured Forwarding PHB Group. IETF RFC 2597, June 1999.

[Hol 04] Holma, H., and Toskala, A., eds. *WCDMA for UMTS*. 3rd Edition. John Wiley & Sons, 2004.

[Hop 74] Hopcroft, J. E. and Ullman, J. D. *Introduction to Automata Theory, Language, and Computation*. Addison-Wesley, 1974.

[Hor98] Horrocks, J., and Rogerson, D. *Implementing Number Portability*. Ovum, Ltd., 1998.

[How 88] Howard, J., Kazar, M., Menees, S., Nichols, D, Satyanarayanan, M., Sidebotham, R., and West, M. Scale and Performance in a Distributed File System. *ACM Transactions on Computer Systems*, 6(1):51–58, February 1988.

[Hun 04] Hung, H.-N., Lin, Y.-B., Lu, M.-K., and Peng, N.-F. A Statistic Approach for Deriving the Short Message Transmission Delay Distributions. *IEEE Transactions on Wireless Communications*, 3(6), 2004.

[Hun05] Hung, H.-N., Lin, Y.-B., Peng, N.-F., and Sou, S.-I. Connection Failure Detection Mechanism of UMTS Charging Protocol. *IEEE Transactions on Wireless Communications*, 2006.

[IDA00a] IDA. Methodology for Determining Fixed and Mobile Inter-Operator Number Portability Charges. *Info-communications Development Authority of Singapore (IDA)*, pages 1–45, 2000.

[IDA00b] IDA. Number Portability In Singapore: the Technical Approach. *Info-communications Development Authority of Singapore (IDA)*, pages 1–6, 2000.

[IEE01] IEEE. IEEE Standard for Local and Metropolitan Area Networks-Port-Based Network Access Control . Technical Report IEEE Std 802.1X, Institute of Electrical and Electronics Engineers, 2001.

[IET94] IETF. Address Allocation for Private Internets. IETF RFC 1597, 1994.

[IET99a] IETF. Dynamic Host Configuration Protocol. IETF RFC 2131, June 1999.

[IET99b] IETF. Layer Two Tunneling Protocol "L2TP". IETF RFC 2661, 1999.

[IET00] IETF. Remote Authentication Dial In User Service (RADIUS). IETF RFC 2865, 2000.

[IET02] IETF. M3UA: MTP3 User Adaptation Layer. IETF RFC 3332, 2002.

[IET03] IETF. PPP Extensible Authentication Protocol (EAP). IETF RFC 2284, March 2003.

[IET04a] IETF. Extensible Authentication Protocol (EAP). IETF RFC3748, 2004.

[IET04b] IETF. Extensible Authentication Protocol Method for GSM Subscriber Identity Modules (EAP-SIM). Technical Report Internet Draft draft-haverinen-pppext-eap-sim-16, Internet Engineering Task Force, December 2004.

[Ipt05] Iptel. SIP Express Router. www.iptel.org/ser/, 2005.

[ISO99] ISO/IEC. Information Technology-Security Techniques - Entity Authentication - Part 4: Mechanisms Using a Cryptographic Check Function. Technical Report ISO/IEC 9798-4, ISO/IEC, 1999.

[ITU97] ITU-T. Transaction Capabilities Formats and Encoding. Technical Report Recommendation Q.773, ITU-T, 1997.

[ITU98] ITU. Call Signalling Protocols and Media Stream Packetization for Packet Based Multimedia Communication Systems. Technical Report ITU-T H.225 (Version 3), International Telecommunication Union, 1998.

[ITU00] ITU. Gateway Control Protocol. Technical Report ITU-T H.248, International Telecommunication Union, 2000.

[ITU03] ITU. Packet-based Multimedia Communications Systems. Technical Report ITU-T H.323, Version 5, International Telecommunication Union, 2003.

[Jac99]   Jacobson, V., Nichols, K., and Poduri, K. An Expedited Forwarding PHB. IETF RFC 2598, June 1999.

[Joh70a]   Johnson, N.L. *Continuous Univariate Distributions-1*. John Wiley& Sons, 1970.

[Joh70b]   Johnson, N.L. *Continuous Univariate Distributions-2*. John Wiley& Sons, 1970.

[Kel79]   Kelly, F. P. *Reversibility and Stochastic Networks*. John Wiley & Sons, 1979.

[Kim 03]   Kim, S., et al. Interoperability between UMTS and cdma2000 Networks. *IEEE Transactions on Wireless Communication*, pages 22–28, February 2003.

[Lin 94]   Lin, Y.-B. Determining the User Locations for Personal Communications Networks. *IEEE Transactions Vehicular Technology*, 43(3):466–473, 1994.

[Lin95]   Lin, Y.-B. Failure Restoration of Mobility Databases for Personal Communication Networks. *ACM-Baltzer Journal of Wireless Networks*, 1:365–372, 1995.

[Lin96a]   Lin, Y.-B. A Cache Approach for Supporting Life-Time Universal Personal Telecommunication Number. *ACM-Baltzer Wireless Networks*, 2:155–160, 1996.

[Lin96b]   Lin, Y.-B. Mobility Management for Cellular Telephony Networks. *IEEE Parallel & Distributed Technology*, 4(4):65–73, November 1996.

[Lin97a]   Lin, Y.-B. Modeling Techniques for Large-Scale PCS Networks. *IEEE Communications Magazine*, 35(2), February 1997.

[Lin97b]   Lin, Y.-B., Mohan, S., Sollenberg, N., and Sherry, H. An Improved Adaptive Algorithm for Reducing PCS Network Authentication Traffic. *IEEE Transactions Vehicular Technology*, 46(3):588–596, 1997.

[Lin01a]   Lin, Y.-B. Eliminating Overflow for Large-Scale Mobility Databases in Cellular Telephone Networks. *IEEE Transactions on Computers*, 50(4):356–370, April 2001.

[Lin 01b]   Lin, Y.-B., and Chlamtac, I. *Wireless and Mobile Network Architectures*. John Wiley & Sons, 2001.

[Lin 01c]   Lin, Y.-B., Haung, Y.-R., Chen, Y.-K., and Chlamtac, I. Mobility Management: From GPRS to UMTS. *Wireless Communications and Mobile Computing*, 1(4), 2001.

[Lin 02a]   Lin, Y.-B. A Multicast Mechanism for Mobile Networks. *IEEE Communications Letters*, 5(11):450–452, 2002.

[Lin 02b] Lin, Y.-B., M.-F. Chen, and Rao, H. C.-H. Potential Fraudulent Usage in Mobile Telecommunications Networks. *IEEE Transactions on Mobile Computing*, 1(2), 2002.

[Lin03a] Lin, Y.-B., and Chen, Y.-K. Reducing Authentication Signaling Traffic in Third Generation Mobile Network. *IEEE Transactions on Wireless Communication*, 2(3), 2003.

[Lin03b] Lin, Y.-B., Chlamtac, I., and Yu, H.-C. Mobile Number Portability. *IEEE Network*, 17(5):8–17, 2003.

[Lin03c] Lin, Y.-B., Lai, W.-R., and Chen, J.-J. Effects of Cache Mechanism on Wireless Data Access. *IEEE Transactions on Wireless Communications*, 2(6), 2003.

[Lin05a] Lin, Y.-B. Per-user Checkpointing for Mobility Database Failure Restoration. *IEEE Transactions on Mobile Computing*, 4(2):189–194, 2005.

[Lin05b] Lin, Y.-B. and Tsai, M.-H. Caching in I-CSCF of UMTS IP Multimedia Subsystem. *IEEE Transactions on Wireless Communications*, 2006.

[Lin05c] Lin, Y.-B., Chang, M.-F., Hsu, M.-T., and Wu, L.-Y. One-Pass GPRS and IMS Authentication Procedure for UMTS. *IEEE Journal on Selected Areas in Communications*, 23(6):1233–1239, 2005.

[Lou04] Loughney, J., Sidebottom, G., Goene, L., et. al. Signaling Connection Control Part User Aaptation Layer (SUA). Technical Report RFC 3868, Internet Engineering Task Force, 2004.

[Lyn02] Lynch, G. Will LNP Rule Make Carriers Smarter? *America's Network Weekly*, August 2002.

[Mit87] Mitrani, I. *Modeling of Computer and Communication Systems*. Cambridge University Press, 1987.

[Moe03] Moerdijk, A.-J., and Klostermann, L. Opening the Networks with Parlay/OSA: Standards and Aspects Behind the APIs. *IEEE Network*, May/June 2003.

[MRT] MRTG. MRTG: Multi Router Traffic Grapher. http://people.ee.ethz.ch/oetiker/webtools/mrtg/.

[Mye 96] Myers, J. and Rose, M. Post Office Protocol - Version 3. IETF RFC 1939, May 1996.

[NAN98] NANC. Local Number Portability Administration Working Group Report on Wireless Wineline Integration. Technical Report, North American Numbering Council, May 1998.

[Nel 88] Nelson, M., Welch, B., and Ousterhout, J. Caching in the Sprite Network File System. *ACM Transactions on Computer Systems*, 6(1), February 1988.

[Nic98] Nichols, K., et al. Definition of the Differentiated Services Field (DS Field) in the IPv4 and IPv6 Headers. IETF RFC 2474, December 1998.

[Nok 97] Nokia. Smart Messaging Specification Version 1.0.0. Technical Report, Nokia, 1997.

[Nov01] Novak, L. and Svensson, M. MMS-Building on the success of SMS. *Ericsson Review*, (3):102–109, 2001.

[Oft97a] Oftel. Number Portability Costs and Charges. Technical Report, The Office of Telecommunications, January 1997.

[Oft97b] Oftel. Number Portability in the Mobile Telephony Market. Technical Report, The Office of Telecommunications, July 1997.

[OMA04] OMA. The Wireless Application Protocol 2.0 Conformance Release. Technical report, Open Mobile Alliance, 2004.

[Ong99] Ong, L., Rytina, I., Garcia, M., Schwarzbauer, H., Coene, L., Lin, H., Juhasz, I., Holdrege, M., Sharp, C. Framework Architecture for Signaling Transport. Technical Report RFC 2719, Internet Engineering Task Force, 1999.

[Ove05] Overture Services, Inc. AltaVista Discovery. www.altavista.com/, 2005.

[Pan04] Pang, A.-C., Lin, Y.-B., Tsai, H.-M., and Agrawal, P. Serving Radio Network Controller Relocation for UMTS All-IP Network. *IEEE Journal on Selected Areas in Communications*, 22(4):617–629, 2004.

[Pan04a] Pang, A.-C., Chen, J.-C., Chen, Y.-K., and Agrawal, P. Moility and Session Management: UMTS vs. cdma2000. *IEEE Wireless Communications*, 1(4):30–44, 2004.

[Par 02] Park, J.-H. Wireless Internet Access for Mobile Subscribers Based on the GPRS/UMTS Network. *IEEE Communications Magazine*, pages 38–49, April 2002.

[Per 98] Perkins, C. E. *Mobile IP: Design Principles and Practices*. Addison-Wesley, 1998.

[Per 00] Perkins, C. E. and Calhoun, P. Mobile IPv4 Challenge/Response Extensions. Internet Engineering Task Force RFC 3012, November 2000.

[Per 02] Perkins, C. E. PPP Authentication Protocols. Internet Engineering Task Force RFC 3344, August 2002.

[Pya 98] Pyarali, I., and Schmidt, D. C. An Overview of the CORBA Portable Object Adapter. *ACM StandardView Magazin*, 6(1), March 1998.

[Rao99]   Rao, H., Chen, Y.-F., Chen, M.-F., and Chang, J. iProxy: A Programmable Proxy Server. *Proceedings of the WebNet99 Conference*, 1999.

[Rao00]   Rao, H., Chen, Y.-F., and Chen, M.-F. A Proxy-Based Web Archiving Service. *Proceedings of the Middleware Symposium*, 2000.

[Rao01a] Rao, C.H., Chang, D.-F., and Lin, Y.-B. iSMS: An Integration Platform for Short Message Service and IP Networks. *IEEE Network*, 15(2):48–55, 2001.

[Rao 01b] Rao, H., Chen, Y.-F., Chang, D.-F, and Chen, M.-F. iMobile: A Proxy-based Platform for Mobile Services. *The First ACM Workshop on Wireless Mobile Internet*, 2001.

[Rao 03] Rao, H., Cheng, Y.-H., Chang, K.-H., and Lin, Y.-B. iMail: A WAP Mail Retrieving System. *Information Sciences*, (151):71–91, 2003.

[Ros96]   Ross, S. M. *Stochastic Processes*. John Wiley & Sons, 1996.

[Ros02]   Rosenberg, J. et al. SIP: Session Initiation Protocol. IETF RFC 3261, June 2002.

[Rus02]   Russell, T. *Signaling System No. 7*. McGraw-Hill, 2002.

[Sal04]   Salkintzis, A.K. Interworking Techniques and Architectures for WLAN/3G Integration Toward 4G Mobile Data Networks. *IEEE Wireless Communications*, 11:50–61, 2004.

[Sat 90] Satyanarayanan, M., et al. Coda: A Highly Available File System for a Distributed Workstation Environment. *IEEE Transactions on Computers*, 39(4):447–459, 1990.

[Sch96a] Schulzrinne. H. RTP Profile for Audio and Video Conferences with Minimal Control. IETF RFC 1890, 1996.

[Sch96b] Schulzrinne, H., et al. RTP: A Transport Protocol for Real-Time Applications. IETF RFC 1889, January 1996.

[Sch99]   Schulzrinne, H., and Rosenberg, J. The IETF Internet Telephony Architecture. *IEEE Communications*, pages 18–23, May 1999.

[Sil 01] Silberschatz, A., Galvin, P., and Gagne, G. *Operating System Concepts*. 6th edition. Reading, Massachusetts: Addison-Wesley, 2001.

[Sim 94] Simpson, W. The Point-to-Point Protocol (PPP). IETF RFC 1661, July 1994.

[Sim 95] Simpson, W. IP in IP Tunneling. IETF RFC 1853, October 1995.

[Sim 96] Simpson, W. PPP Challenge Handshake Authentication Protocol (CHAP). Internet Engineering Task Force RFC 1994, August 1996.

[Sto 93] Stone, H. *High-Performance Computer Architecture*. Addison-Wesley, Reading, Massachusetts, 1993.

[Str00] Strewart, R., et al. Stream Control Transmission Protocol. IETF RFC 2960, October 2000.

[Tho98] Thomson, S., and Narten, T. IPv6 Stateless Address Autoconfiguration. IETF RFC 2462, December 1998.

[Tri] Trillium. www.trillium.com/.

[UMT00a] UMTS Forum. Enabling UMTS/Third Generation Services and Applications. Technical Report 11, UMTS, 2000. See www.umts-forum.org.

[UMT00b] UMTS Forum. Shaping the Mobile Multimedia Future - An Extended Vision from the UMTS Forum. Technical Report 10, UMTS, 2000. See www.umts-forum.org.

[UMT00c] UMTS Forum. The UMTS Third Generation Market – Structuring the Service Revenues Opportunities. Technical Report 9, UMTS, 2000. See www.umts-forum.org.

[Wal02] Walkden, M., Edwards, N., Foster, D., Jankovic, M., Odadzic, B., Nygreen, G., Moiso, C., Tognon, S.M., de Bruijn, G. Open Service Access: Advantages and Opportunities in Service Provisioning on 3G Mobile Networks Definition and Solution of Proposed Parlay/OSA Specification Issues. Technical Report Project P1110 Technical Information EDIN 0266-1110, ETSI, 2002.

[Wat81] Watson, E.J. *Laplace Transforms and Applications*. Birkhauserk, 1981.

[Woo04] Wook, H., Han, J., Huh, M., Park, S., and Kang, S. Study on Robust Billing Mechanism for SIP-based Internet Telephony Services. *The 6th International Conference on Advanced Communication Technology*, 2004.

[Yin 99] Yin, J., Alvisi, L., Dahlin, M., and Lin, C. Volume Leases for Consistency in Large-Scale Systems. *IEEE Transactions on Knowledge and Data Engineering*, 11(4), 1999.

[Yu 99] Yu, H.-C., Lin, Y.-B., Chen, K.C., Liu, C.J., Liu, K.C., and Hsu, J.S. A Study of Number Portability in Taiwan. Technical Report, National Chiao Tung University, April 1999.

Index